# Natural Pigments

## Sources, Properties, and Health Benefits

nova
science publishers
www.novapublishers.com

# Biochemistry Research Trends

**More information about this series can be found at**
https://novapublishers.com/product-category/series/biochemistry-research-trends/

# Food Science and Technology

**More information about this series can be found at**
https://novapublishers.com/product-category/series/food-science-and-technology-series/

**Rashidi Othman**
**Qurratu Aini Mat Ali**
**Razanah Ramya**
**Farah Ayuni Mohd Hatta**
**Wan Syibrah Hanisah Wan Sulaiman**
**and Nur Hanie Mohd Latiff**
Editors

# Natural Pigments

## Sources, Properties, and Health Benefits

nova
science publishers
www.novapublishers.com

DOI: https://doi.org/10.52305/GARO1418



## NOTICE TO THE READER



Additional color graphics may be available in the e-book version of this book.

**Library of Congress Cataloging-in-Publication Data**

ISBN: 979-8-90134-055-4 (Hardcover)
ISBN: 979-8-90134-119-3 (eBook)

*Published by Nova Science Publishers, Inc. † New York*

# Contents

# Preface

Natural pigments have long been an integral part of human culture and industry, from their use in ancient art to their role in modern sustainability practices. *Natural Pigments: Sources, Properties, and Health Benefits* aims to highlight the multifaceted potential of these pigments, which go beyond their aesthetic value. As we continue to seek sustainable solutions in various fields—from food preservation to eco-friendly architecture—the unique properties of natural pigments offer transformative opportunities. With contributions from renowned experts in diverse disciplines, this volume explores how these pigments are sourced, their chemical properties, and their profound health benefits. Through this collection, we hope to inspire greater awareness and further innovation, fostering a deeper appreciation for the natural world and its gifts. By exploring the intersection of traditional knowledge and modern science, we aim to uncover new applications that align with both sustainability goals and human health, driving positive change in industries and everyday life.

# Chapter 1

# Natural Black Pigment: Tannins


## Rashidi Othman*, PhD

Department of Landscape Architecture, Kulliyyah of Architecture and Environmental Design, International Islamic University Malaysia, Kuala Lumpur, Malaysia


## Abstract


Tannins, a group of naturally occurring polyphenolic compounds, have garnered significant attention in recent years due to their potential as sustainable alternatives to synthetic dyes. Derived from various plant sources, tannins offer a biodegradable and non-toxic solution for a wide range of applications in industries including textiles, cosmetics, pharmaceuticals, and food packaging. This chapter explores the extraction methods, characterization, and applications of tannin-based pigments, with a focus on their environmental sustainability and performance. The extraction of tannins is explored through several traditional and advanced techniques. Aqueous extraction remains a cost-effective and eco-friendly method, while more advanced methods like microwave-assisted extraction (MAE), ultrasound-assisted extraction (UAE), and enzyme-assisted extraction (EAE) are being developed to improve extraction efficiency, reduce environmental impact, and increase yields. These modern techniques offer faster and more sustainable alternatives for large-scale production. The chapter also delves into the measurement of tannins and black pigments, including the Folin-Ciocalteu assay for tannin quantification, colorimetry for pigment analysis, and lightfastness and wash-fastness testing to assess stability. Additionally, the blackness (My) and jetness (Mc) of black pigments are calculated to evaluate their intensity and undertone, important factors in applications requiring specific visual qualities, such as automotive



* Corresponding Author's Email: rashidi@iium.edu.my



In: Natural Pigments: Sources, Properties, and Health Benefits
Editors: Rashidi Othman, Qurratu Aini Mat Ali, Razanah Ramya et al.
ISBN: 979-8-90134-055-4

coatings. Finally, the chapter addresses the sustainability of tannin-based pigments, including their low toxicity, biodegradability, and their role in reducing water consumption and waste. The use of waste materials like agricultural residues contributes to a circular economy, while ongoing research aims to enhance color stability and extend the applications of tannin-based pigments in various industries. Through these advancements, tannin-based pigments are positioned to play a significant role in the growing demand for sustainable and eco-friendly alternatives to traditional synthetic colorants.

**Keywords**: tannins, natural colorants, sustainable dyeing, black pigments, sustainability

## Introduction

Black pigments, as one of the oldest and most essential colors used by humans, have a long and fascinating history. The pursuit of black pigment spans across ancient cultures, where it was derived from various natural sources, including minerals, plants, and animal-based products. Early civilizations used charcoal, soot, and naturally occurring mineral oxides such as magnetite and manganese to create black pigments for pottery, artwork, and textiles. These early black pigments were not only valued for their deep, lasting color but also for their practical application in preserving artworks and cultural relics over time (Smith & Brown, 2017; Dean, 2016).

The use of plant-based black pigments, however, emerged as an important alternative with the discovery of certain plant compounds like tannins. Tannins, derived from various plant materials such as oak galls, sumac, and bark, are unique for their ability to produce rich, dark hues, especially when combined with iron salts or other mordants. Their natural occurrence in plants like oak trees, chestnuts, and even some berries has made them a valuable source for sustainable black pigments, particularly in industries focused on eco-friendly and biodegradable materials (Jones, 2019; Lee & Kim, 2020; Shi et al., 2021). Tannin-based pigments, though overshadowed by synthetic dyes in modern times, have a historical legacy of being used in traditional dyeing and tanning processes, providing a sustainable and environmentally friendly alternative (Tan et al., 2021; Williams, 2023).

Tannins, a diverse group of naturally occurring polyphenolic compounds, are ubiquitous in the plant kingdom. Known for their astringency and bitter taste, tannins play a critical role in plant defense, acting as deterrents to herbivores and microorganisms (Jackson & Morgan, 2022). While often overlooked, these compounds hold significant promise as natural pigments, particularly in the form of black dyes. The resurgence of interest in natural dyes in recent years can be attributed to the growing concerns over the environmental and health impacts of synthetic chemicals, which have spurred a renewed interest in plant-based dyes like tannins (Gomez et al., 2021). Tannins have been used for centuries, especially in traditional dyeing and tanning processes, where they impart rich, deep hues to textiles, leather, and inks (Tan et al., 2021).

As environmental concerns have grown and the need for sustainable alternatives to synthetic dyes has become more urgent, natural pigments such as tannins are regaining popularity. Unlike synthetic black pigments, which often rely on petroleum-based processes and can contribute to environmental pollution, tannins offer a renewable and eco-friendly alternative. Additionally, their biodegradability and non-toxicity position them as an ideal choice for industries focused on sustainability, from textile and leather production to art and cosmetics (Pereira et al., 2020; Gomez et al., 2021). Furthermore, tannins are abundant in many plant species and are easily accessible, making them a practical resource for large-scale pigment production (Lee & Kim, 2020).

Tannin-based black pigments also align with the growing trend of using materials with minimal environmental impact in the production of consumer goods. Their use extends beyond textiles and into applications in food, cosmetics, and pharmaceuticals, where their natural properties offer unique colorants and functional benefits (Fearon, 2022; Talu, 2023). The demand for sustainable pigments is likely to increase as consumers and industries alike prioritize environmentally conscious practices in product development (Nguyen et al., 2022).

This chapter will explore the potential of tannins as a natural black pigment, examining their chemical properties, the sources from which they are derived, and their applications across various industries. By understanding the role of tannins in pigment production, we can gain deeper insight into their environmental, economic, and cultural significance in the context of sustainable material design.

## Chemical Composition of Tannins

Tannins are a group of polyphenolic compounds widely distributed in the plant kingdom. They are characterized by their ability to bind with proteins, which leads to their astringency and their role in plant defense mechanisms. Chemically, tannins are complex molecules, primarily classified into two main types: hydrolyzable tannins and condensed tannins. The distinction between these two types plays a crucial role in determining their applications as natural dyes, including black pigments (Bharathi & Ponnusamy, 2017; Tan et al., 2021).

## Types of Tannins

1. *Hydrolyzable Tannins:*
   - These tannins are esters of a polyol (such as glucose) and phenolic acids, primarily gallic acid or ellagic acid. When hydrolyzed, they release these phenolic acids into the solution, making them more reactive and susceptible to degradation (Ali et al., 2020).
   - Gallic Acid: Gallic acid is the most common phenolic acid found in hydrolyzable tannins and contributes to their strong affinity for metal ions, making them ideal for dyeing when combined with iron salts (Sun et al., 2019; Shi et al., 2021).
   - Ellagic Acid: Found in the galls of oak trees, sumac, and pomegranate, ellagic acid is known for its stability and antioxidant properties, which contribute to the stability of tannin-based pigments (Pereira et al., 2020).
2. *Condensed Tannins:*
   - Also known as proanthocyanidins, these tannins are polymers of flavonoid units (e.g., catechins and epicatechins), linked by carbon-carbon bonds. Unlike hydrolyzable tannins, they are not easily broken down by hydrolysis but can form complex structures that are relatively stable (Wang et al., 2022).
   - Flavonoid Units: These are the building blocks of condensed tannins, and their structure includes benzene rings with hydroxyl groups, which enhance the pigmenting ability and reactivity towards metal ions (Wang et al., 2022; Williams, 2023).
   - Polymeric Nature: The polymeric structure of condensed tannins allows them to form complex networks that contribute to their deep color intensity when used in dyeing, particularly in the production of dark hues like black (Tan et al., 2021).

## Chemical Structure and Functionality

Tannins' chemical structure plays a pivotal role in their interaction with metal ions, which is essential for their function as pigments. The key features of tannins that make them effective pigments include:

1. *Phenolic Hydroxyl Groups:*
   - Tannins contain multiple hydroxyl groups (-OH) attached to aromatic rings, which are responsible for their ability to form hydrogen bonds with other molecules. This property allows tannins to react with metal ions (such as iron, copper, and aluminum) to form stable-colored complexes, a key feature in their role as pigments (Pereira et al., 2020; Lee & Kim, 2020).
   - *Iron Binding*: When tannins are combined with iron salts, they form deep black complexes that are commonly used in dyeing and tanning. The intensity of the color depends on the concentration of tannins and the specific metal ion used (Fearon, 2022).
2. *Astringency and Precipitation of Proteins:*
   - The astringent nature of tannins is due to their ability to precipitate proteins. This property is particularly useful in the tanning industry, where tannins are used to convert animal skins into leather. The same property contributes to their ability to form stable, long-lasting color when applied to fabrics and other materials (Gomez et al., 2021; Sun et al., 2019).
3. *Stability and Durability:*
   - Tannins are known for their relative stability when exposed to light, oxygen, and other environmental factors, especially compared to synthetic dyes. This makes tannin-based pigments ideal for use in applications where durability is essential, such as textiles, art, and leather goods (Nguyen et al., 2022).
   - *UV Resistance*: Some tannins, particularly those derived from gallnuts, exhibit strong resistance to UV degradation, making them suitable for long-lasting black pigments (Tan et al., 2021).

## Molecular Interactions in Pigment Formation

The process of pigment formation using tannins often involves the interaction between tannin molecules and metal ions, which leads to the development of a stable, dark color. The most common metal ion involved in the creation of black pigments is *iron* ($Fe^{2+}$). When tannins are exposed to iron salts (e.g., ferric chloride or ferrous sulfate), a chemical reaction occurs, resulting in the formation of a *tannin-iron complex.* This complex is responsible for the deep, rich black color that characterizes tannin-based pigments (Williams, 2023; Pereira et al., 2020).

1. *Tannin-Iron Complex:*
    - The binding of tannin molecules to iron ions through their phenolic hydroxyl groups forms a complex that results in a dark-colored precipitate. The stability of this complex is influenced by factors such as pH, temperature, and the concentration of both the tannin and iron salts (Fearon, 2022).
    - The intensity of the black color depends on the concentration of the tannins and the interaction with the metal ions, which can be fine-tuned to produce various shades of black (Bharathi & Ponnusamy, 2017).
2. *Other Metal Ions:*
    - In addition to iron, tannins can also form complexes with other metal ions such as aluminum, copper, and chromium, leading to variations in color intensity and hue. For instance, tannin-aluminum complexes may produce browner or darker hues, while tannin-copper complexes could yield greenish tones, although iron remains the most common metal used for black pigments (Wang et al., 2022; Williams, 2023).

## Applications of Tannin-Based Pigments

The chemical properties of tannins make them highly suitable for use in various pigmenting applications:

- *Textile Dyeing*: Tannins have been historically used to dye fabrics, especially for producing black and brown hues. The color intensity and stability depend on the tannin source and the mordanting process, with iron being the most commonly used mordant (Gomez et al., 2021; Tan et al., 2021).

- *Leather Tanning:* Due to their ability to precipitate proteins, tannins have been used for centuries in the tanning process to convert raw hides into durable leather (Tan et al., 2021; Williams, 2023).
- *Art and Craft*: Tannins are widely used in the production of inks and paints, offering rich, deep hues that are both eco-friendly and long-lasting (Fearon, 2022).

## Source of Tannins

Tannins are naturally occurring compounds found in various plant parts, including the bark, leaves, fruits, and galls of certain species. They are produced by plants primarily as a defense mechanism against herbivores, fungi, and pathogens. The sources of tannins are abundant and diverse, and different species produce tannins with varying chemical compositions, which in turn affect their potential applications as pigments.

### *Common Plant Sources of Tannins*

Tannin-rich plants are distributed globally, with certain species being more abundant in specific regions. Europe and North America are home to a variety of oak species, which are a major source of tannins. Asia and the Middle East have a long tradition of using sumac and pomegranate in dyeing. Australia has numerous *Acacia* species, whose tannins are integral to local industries such as leather tanning and textile dyeing (Nguyen et al., 2022). In Malaysia, tannin-rich plants are abundant and have been integral to traditional practices in dyeing, tanning, and even medicine. Malaysia's tropical climate fosters a rich diversity of plant species, many of which contain high levels of tannins. The traditional use of tannins in local industries, particularly in batik dyeing and leather tanning, showcases the region's long history of utilizing natural resources. The following sections highlight common plants rich in tannins that have contributed to the development of natural black pigments. The extraction of tannins from plant materials generally involves drying, grinding, and then using water or alcohol-based solvents to separate the tannins from the plant material.

**Table 1.** Tannin-Rich plant species (2000–2025) from Mangrove, heath forest, tropical, and aquatic ecosystems

***Mangrove Ecosystem***

| Plant Source | Tannin Type | Tannin Content | Applications | References |
|---|---|---|---|---|
| *Avicennia marina* | Hydrolyzable (Gallic Acid) | High tannin content in leaves and bark | Dyeing (brown hues), antimicrobial properties, traditional medicine | Nor et al., 2020; Aziz et al., 2021 |
| *Bruguiera gymnorrhiza* | Condensed (Proanthocyanidins) | High in bark and leaves | Dyeing (dark hues), ink production, medicinal applications | Noor et al., 2018; Siti et al., 2021 |
| *Ceriops tagal* | Hydrolyzable (Ellagic Acid) | High tannin content in bark and fruit | Dyeing (black hues), medicinal uses | Tan et al., 2021 |
| *Sonneratia alba* | Condensed (Proanthocyanidins) | High in bark and leaves | Dyeing (brown to reddish hues), sustainable uses | Tan et al., 2021 |
| *Lumnitzera racemosa* | Hydrolyzable (Tannic Acid) | Tannins in bark and leaves | Dyeing (brown hues), ink production | Noor et al., 2018 |

***Heath Forest Ecosystem***

| Plant Source | Tannin Type | Tannin Content | Applications | References |
|---|---|---|---|---|
| Vaccinium vitisidaea | Hydrolyzable (Ellagic Acid) | High tannin content in berries, leaves, stems | Dyeing (brown to red hues), antimicrobial, health supplements | Tan et al., 2020 |
| Rhododendron spp. | Condensed (Proanthocyanidins) | High in leaves, bark, and flowers | Dyeing (dark colors), anti-inflammatory, antimicrobial | Neran et al., 2025; Mu et al., 2023 |
| Leucobryum glaucum (Moss) | Hydrolyzable (Gallic Acid) | Moderate tannin content in moss and leaves | Dyeing, antimicrobial, traditional medicine | Tan et al., 2020 |
| Pieris spp. | Hydrolyzable (Tannic Acid) | Moderate in leaves and stems | Dyeing (brown hues), anti-inflammatory, antimicrobial | Zhang et al., 2019 |

***Tropical Species***

| Plant Source | Tannin Type | Tannin Content | Applications | References |
|---|---|---|---|---|
| Areca catechu | Condensed (Proanthocyanidins) | High in fruit and bark | Dyeing (black to dark brown hues), medicinal uses | Fearon (2022); Tan et al., 2021 |
| Acanthophonax trifoliatum | Hydrolyzable (Tannic Acid) | Moderate in leaves and roots | Dyeing (brown to reddish hues), local medicinal uses | Meena et al., 2018 |
| Briedelia stipularis | Hydrolyzable (Tannic Acid) | High in bark | Traditional dyeing (brown pigments), ink production | Shah et al., 2020; |
| Citrus medica | Hydrolyzable (Gallic Acid) | Moderate in fruit peel and leaves | Dyeing (brown and yellow hues), antioxidant properties | Yadav et al., 2016; Tan et al., 2021 |
| Cyanometra ramiflora | Hydrolyzable (Tannic Acid) | Moderate in bark and leaves | Dyeing (brown to black hues), medicinal applications | Kumar et al., 2020 |
| Garcinia mangostana | Hydrolyzable (Ellagic Acid) | Tannins in fruit rind and bark | Dyeing (purple to brown hues), medicinal uses | Sun et al., 2019; Wijaya et al., 2021 |
| Heliotropium trigosum | Condensed (Proanthocyanidins) | High in leaves and stems | Dyeing (brown hues), traditional uses | Ramana et al., 2017; Tan et al., 2021 |
| Peltophorum pterocarpum | Condensed (Proanthocyanidins) | High in bark and leaves | Dyeing (brown hues), medicinal uses, tannin extraction | Vasudevan et al., 2020 |

***Aquatic Ecosystem***

| Plant Source | Tannin Type | Tannin Content | Applications | References |
|---|---|---|---|---|
| Nymphaea spp. | Hydrolyzable (Gallic Acid) | Tannins found in leaves and flowers | Dyeing (brown to reddish hues), antimicrobial, traditional uses | Sun et al., 2020; Bhowmik et al., 2019 |
| Pontederia cordata | Condensed (Proanthocyanidins) | High tannin content in leaves and stems | Dyeing (brown hues), antimicrobial, antioxidant properties | Alviano et al., 2017; Thilakarathna et al., 2021 |
| Lemna minor | Hydrolyzable (Gallic Acid) | Tannins found in leaves | Dyeing (green to yellowish hues), bioremediation, antioxidant properties | Singh et al., 2020; Wang et al., 2021; Mohammad et al., 2020 |
| Salvinia molesta | Hydrolyzable (Ellagic Acid) | High in tannins in leaves and roots | Dyeing (green to brown hues), bioremediation, antimicrobial | Mohan et al., 2021; Gao et al., 2020 |

The extraction method and the plant source significantly influence the quality and type of tannins obtained, which in turn affects the final pigment produced (Tan et al., 2021; Fearon, 2022). The use of mordants, such as iron salts, is crucial in enhancing the color intensity and stability of the tannin-based pigments (Pereira et al., 2020; Shi et al., 2021). Table 1 summarizes plant species known for their high tannin content, including those found in mangrove, heath forest, tropical, and aquatic ecosystems. The species are categorized by their tannin types (hydrolyzable or condensed), with detailed information on their tannin content and applications in dyeing, medicinal uses, and other industries. References from 2000 to 2025 are provided for each species, offering insights into their ecological roles and practical applications.

## Methods of Tannin Extraction

The extraction of tannins is a critical process for harnessing their potential as pigments, particularly in the production of eco-friendly dyes. Tannins can be extracted from various plant sources, including bark, leaves, galls, and fruits, each of which requires tailored extraction methods to achieve optimal yields. Here, we explore the traditional and modern extraction methods used in the tannin industry.

### *Aqueous Extraction*

Aqueous extraction, often regarded as the most traditional method, utilizes water as a solvent to extract tannins from plant materials. This method works well with plant parts that are rich in tannins, such as bark, galls, and leaves, and is often used for hydrolyzable tannins (e.g., gallotannins, ellagitannins).

Process: Plant material is dried and ground into small particles, which increases the surface area and facilitates easier extraction of tannins. The ground material is then immersed in warm or hot water (usually ranging between 60°C to 90°C) for several hours. Boiling can also be employed, but this risks tannin degradation if not carefully controlled. After extraction, the tannin-rich solution is filtered to remove solid plant matter, and the tannins are concentrated through evaporation or freeze-drying. Table 2 summarizes the advantages, limitations, recent advances and applications of tannins aqueous extraction.

**Table 2.** Aqueous extraction of tannins

| Category | Details |
|---|---|
| Advantages | Cost-effective and environmentally friendly: Water is a cheap and abundant solvent, and no toxic chemicals are used. Simple and widely used for large-scale extraction, especially in regions with abundant plant resources. |
| Limitations | The yield from aqueous extraction is often lower compared to solvent-based methods, and the tannins may not be as pure. Longer extraction times and higher temperatures can lead to degradation of sensitive tannins, especially hydrolyzable tannins like gallotannins. |
| Recent Advances | Temperature control and ultrasound-assisted heating have been explored to enhance the efficiency of aqueous extraction without degrading the tannins |
| Applications | Commonly used for textile dyeing and eco-friendly art supplies where a moderate yield is acceptable. |
| References | Tan et al., 2021; Shi et al., 2020 |

***Solvent-Based Extraction***

Solvent-based extraction uses organic solvents such as ethanol, methanol, acetone, or ethyl acetate to dissolve and extract tannins from plant material. This method is particularly useful for extracting condensed tannins (e.g., proanthocyanidins) and hydrolyzable tannins, which may not be as soluble in water.

*Process*: Similar to aqueous extraction, plant material is first dried and ground. The material is then soaked in a solvent, typically at room temperature or slightly elevated temperatures. The solution is filtered to remove residual plant matter, and the solvent is evaporated under reduced pressure to leave behind concentrated tannins.

Table 3 summarizes the advantages, limitations, recent advances and applications related to solvent-based tannin extraction.

***Supercritical Fluid Extraction (SFE)***

Supercritical fluid extraction (SFE) is an advanced, environmentally friendly technique that uses supercritical $CO_2$ (carbon dioxide) to extract tannins from plant material. The solvent properties of $CO_2$ at high pressure and temperature enable it to dissolve non-polar and polar compounds, including tannins.

*Process:* Plant material is placed inside a chamber, and $CO_2$ is pressurized to its supercritical state (above 31.1°C and 73.8 bar). The supercritical $CO_2$ acts as a solvent, extracting tannins as it passes through the plant material. After extraction, the $CO_2$ is depressurized, separating the tannins from the gas, and the solvent is recycled.

**Table 3.** Solvent-Based extraction of tannins

| Category | Details |
|---|---|
| Advantages | Higher yield of tannins compared to aqueous extraction, especially for condensed tannins. Faster than aqueous extraction, requiring less time. Selective extraction: Different solvents can selectively extract different types of tannins based on solubility. |
| Limitations | The use of organic solvents requires careful handling, as many solvents are flammable and pose environmental risks. Solvent recovery systems are necessary to reduce waste and make the process more eco-friendly. |
| Recent Advances | Ethanol-based extraction is gaining popularity because it is biodegradable and less toxic compared to other solvents like acetone or methanol. Solvent mixtures are being explored for higher yield and specificity in extracting different types of tannins. |
| Applications | Used in the pharmaceutical and cosmetic industries for creating natural colorants and bioactive compounds. Used in textile dyeing, where higher tannin concentrations are desired for deep colors. |
| References | Pereira et al., 2020; Zhang et al., 2019 |

**Table 4.** Supercritical Fluid Extraction (SFE) of Tannins

| Category | Details |
|---|---|
| Advantages | Eco-friendly: $CO_2$ is non-toxic, recyclable, and does not leave harmful residues.<br>High selectivity and efficiency in extracting both hydrolyzable and condensed tannins.<br>No solvent waste: $CO_2$ is recycled in the process, reducing environmental impact. |
| Limitations | High initial setup costs for specialized equipment. Requires expertise in handling supercritical $CO_2$ and setting optimal conditions for extraction. |
| Recent Advances | Research is focused on improving the efficiency of $CO_2$ extraction by combining it with co-solvents (e.g., ethanol) to enhance tannin solubility.<br>Ongoing efforts are making SFE more affordable and commercially viable for large-scale operations. |
| Applications | Increasingly used in the food, cosmetic, and pharmaceutical industries for natural colorants and bioactive compounds. High-purity tannin extracts are produced for use in artistic paints, textile dyeing, and sustainable packaging. |
| References | Van Der Walt et al., 2021; Tan et al., 2020 |

Table 4 summarizes the advantages, limitations, recent advances and applications, related to Supercritical Fluid Extracion (SFE).

## Factors Affecting Tannin Extraction

The efficiency of tannin extraction depends on various factors, including the type of plant material, extraction method, temperature, pH, and extraction time (Tan et al., 2021). These factors influence the yield, quality, and stability of the extracted tannins. Table 5 summarises the factors affecting tannin extraction.

**Table 5.** Factors affecting tannin extraction

| Factor | Details | Examples / Notes |
|---|---|---|
| Species & Plant Part | Different species and plant parts contain varying tannin levels. | Oak galls & sumac leaves → rich in hydrolyzable tannins; chestnut bark → high in condensed tannins. |
| Growth Stage | Tannin levels fluctuate with time of year and plant development. | Young leaves may differ significantly from mature foliage. |
| Extraction – Temperature | Higher temperatures improve solubility and extraction rates, but very high heat can degrade sensitive tannins. | Hydrolyzable tannins degrade at high temperatures. |
| Extraction – Time | Longer extraction increases yield but may also extract undesirable compounds. Optimizing time ensures quality. | Risk of leaching sugars or lignin if too long. |
| pH | Solvent pH affects tannin solubility. | Alkaline → better for hydrolyzable tannins; Acidic → better for condensed tannins. |
| Concentration | Solvent-to-plant ratio affects yield. | Higher solvent concentration → higher extraction yield. |

The extraction of tannins for pigment production involves various methods, each with distinct advantages and challenges. Traditional aqueous extraction remains widely used due to its simplicity and cost-effectiveness, while solvent-based extraction offers higher yields and efficiency. Supercritical fluid extraction is a cutting-edge technique that provides high-quality, environmentally friendly extracts, but it comes with high costs and technical complexity. Optimizing the extraction method, based on factors like plant material, temperature, pH, and time, is key to obtaining high-quality tannins for use in various applications, particularly in sustainable pigment production.

## Advanced Extraction Techniques

### *Microwave-Assisted Extraction (MAE)*

Microwave-assisted extraction uses microwave radiation to heat the solvent and plant material. This method significantly accelerates the process compared to traditional heating, enhancing the solubility of tannins in the solvent.

- *Working Principle:* Microwaves generate rapid, localized heating, resulting in the rupture of plant cells, which helps release the tannins more efficiently.
- *Applications*: MAE is used for dyeing, cosmetics, bioactive compounds extraction, and biofuels production. Studies show that MAE extracts higher amounts of tannins from galls of plants like oak and sumac, making it useful for industries focused on eco-friendly products (Mullen et al., 2019).

### *Ultrasound-Assisted Extraction (UAE)*

UAE uses high-frequency sound waves to create cavitation bubbles in the solvent, which implode and create microscopic shockwaves, breaking open plant cells and enhancing the release of tannins. It is more energy-efficient than traditional methods.

- *Advantages:* UAE is faster, requires less solvent, and can be applied at ambient temperatures, making it more sustainable compared to traditional extraction methods.
- *Applications*: The technique has been applied for extracting tannins from tea leaves, mango peels, oak galls and grape seeds, enhancing sustainability in food and cosmetic industries (Zuo et al., 2020; Nisar et al., 2021; Dhiman et al., 2021).

### *Enzyme-Assisted Extraction (EAE)*

Enzymes such as cellulase and pectinase are used to break down the cell wall material in plant tissues, releasing tannins into the solvent. This method is mild and efficient, and typically performed at lower temperatures, preventing tannin degradation.

- *Mechanism:* Enzymes hydrolyze complex polysaccharides and lignin, making it easier to extract tannins, particularly from woody tissues like bark.
- *Applications*: EAE is increasingly used for the eco-friendly extraction of tannins from plant species such as mango and citrus (Hashem et al., 2020). It is valuable for industries that require high-quality, non-degraded tannins for cosmetics and pharmaceutical applications.

## Role of Mordants in Enhancing Tannin-Based Pigments

**Table 6.** Types of mordants in tannin-based dyeing

| Category | Details |
|---|---|
| Types of Mordants | Metal Mordants: Iron (Fe), aluminum (Al), and copper (Cu) salts are commonly used. When combined with tannins, these metals form coordination complexes that enhance the color intensity.<br>Iron: Forms a blue-black hue with tannins, widely used in leather dyeing and textile applications.<br>Aluminum: Forms lighter colors such as yellow or green, useful in more delicate fabric dyeing.<br>Copper: Produces blue to greenish shades, commonly used in textile and art industries.<br>Natural Mordants: Plant-based substances such as sumac, which contains tannic acid, are used as mordants. These natural mordants offer a sustainable option for dyeing. |
| Mechanism of Mordanting | The mordant reacts with the tannin molecule, forming a chelate complex with the dye, ensuring it binds to the fiber.<br>The metal ions from the mordant bridge the tannin molecules to the fibers, improving dye adhesion and making the color more resistant to fading or washing. |
| Advantages of Using Mordants | Color enhancement: Mordants significantly increase the depth and richness of colors produced by tannin-based dyes.<br>Stability: The complex formed with tannins and mordants enhances color fastness to light, washing, and rubbing.<br>Sustainability: Mordants like iron and aluminum are easily accessible and non-toxic, making them suitable for eco-friendly dyeing. |
| References | Sabaris et al., 2020; Zohra et al., 2019 |

Mordants are substances used to fix dyes on fibers. They interact with tannins to form a stable complex, improving the color fastness, vibrancy, and longevity of the dye. Table 6 provides information on the types of mordants, their mechanisms and advantages.

## Measuring Tannins and Black Pigments

Measuring tannins and the resulting black pigments is essential in determining their quality, consistency, and performance in various applications such as textile dyeing, cosmetics, pharmaceuticals, and automotive coatings. The properties of black pigments, such as their intensity, undertone, and stability, significantly impact their visual appearance and functional use. The following section discusses methods to quantify tannins and evaluate the blackness and jetness of black pigments derived from tannin-based sources.

## Measurement of Tannin Content

Quantifying tannins in plant extracts is crucial for assessing their potential for pigment production. Several methods are employed to evaluate total phenolic content and tannin concentration.

### *Folin-Ciocalteu Assay*

The Folin-Ciocalteu (FC) assay is one of the most widely recognized and established methods for determining the total phenolic content in plant extracts, including those rich in tannins. This colorimetric method is based on the ability of phenolic compounds, such as tannins, to reduce the phosphomolybdate/phosphotungstate complexes present in the Folin-Ciocalteu reagent, resulting in the formation of a blue-colored complex. The intensity of this blue color is proportional to the concentration of phenolic compounds in the sample and is measured spectrophotometrically at a wavelength of 765 nm.

Table 7 provides a clear and concise summary of the process, advantages, and limitations of the Folin-Ciocalteu assay for measuring phenolic compounds, particularly tannins as detailed by Singleton et al. (1999) and Tan et al. (2021).

**Table 7.** Summary of the Folin-ciocalteu assay for quantifying total phenolic content in plant extracts

| Category | Details |
|---|---|
| Process | |
| Sample Preparation | Grind or macerate plant material and dissolve in a suitable solvent (water or ethanol). |
| Reaction | Mix extract with Folin-Ciocalteu reagent (phosphomolybdate, phosphotungstate ions, $Na_2CO_3$). Phenolic compounds react with the reagent to form a blue-colored solution. |
| Measurement | Measure absorbance at 765 nm using a spectrophotometer. The intensity of the blue color correlates with phenolic concentration. |
| Quantification | Compare sample absorbance with a standard curve (using gallic acid). Quantify total phenolic content as gallic acid equivalents (GAE). |
| Advantages | |
| Simplicity | Straightforward method, ideal for routine and high-throughput applications. |
| Cost-effective | Inexpensive reagents and no need for costly consumables, suitable for labs with limited resources. |
| Wide Applicability | Can analyze tannin-rich extracts, fruits, leaves, bark, galls, and other plant materials with phenolic compounds. |
| Quantitative | Provides an accurate measure of total phenolic content, useful for assessing antioxidant and pigmenting properties in tannin-based dyes. |
| Limitations | |
| Non-specificity | Measures total phenolic content, not just tannins. Cannot specifically quantify tannins without further analysis. |

The method is sensitive to interference from other substances in the plant extract that may also react with the Folin-Ciocalteu reagent, potentially skewing results. To mitigate this, proper sample preparation and controls must be used (Singleton et al., 1999; Tan et al., 2021).

## Butanol-HCl Assay for Condensed Tannins (Proanthocyanidins)

The Butanol-HCl assay is a well-established method for quantifying condensed tannins, also known as proanthocyanidins (Porter et al., 1986; Ghosh et al., 2019). Condensed tannins are a group of polyphenolic compounds found in various plant materials, and they are particularly important due to their role in plant defense and their use as natural colorants in industries like textile dyeing and cosmetics. Unlike hydrolyzable tannins, which are easily hydrolyzed by water, condensed tannins are more stable and have a more complex structure.

**Table 8.** Butanol-HCl assay for condensed tannin quantification: Process, advantages, limitations, and applications

| Step/Factor | Details |
|---|---|
| Process | |
| Sample Preparation | Grind plant material and extract with solvent (e.g., ethanol or methanol) to prepare an extract rich in condensed tannins. |
| Reaction with Butanol-HCl Reagent | Mix the plant extract with butanol-HCl reagent, forming a colored complex due to the interaction between phenolic rings of condensed tannins and the reagent. The complex appears red or purple. |
| Heating | Heat the mixture to accelerate the reaction and enhance the formation of the colored complex, improving tannin extraction efficiency and color intensity. |
| Absorbance Measurement | Measure absorbance at 550 nm using a spectrophotometer. The wavelength corresponds to the maximum absorbance of the colored complex formed between condensed tannins and the reagent. |
| Calculation of Tannin Concentration | Compare the absorbance of the sample to a standard curve (e.g., catechin or epicatechin). The tannin concentration is calculated from this comparison. |
| Advantages | |
| Highly Sensitive | The assay is highly sensitive to condensed tannins, allowing for accurate measurement even at low concentrations. |
| Specific to Condensed Tannins | Unlike other methods (e.g., Folin-Ciocalteu), this assay specifically targets condensed tannins, making it ideal for studying proanthocyanidins. |
| Simplicity | The procedure is straightforward and requires only a spectrophotometer for absorbance measurement, making it accessible to various research labs. |
| Quantitative Results | The assay provides reliable, quantitative results based on a standard curve, facilitating precise determination of tannin concentrations in plant extracts. |
| Limitations | |
| Interference | Other phenolic compounds or colored substances may interfere with absorbance measurement, leading to inaccuracies. Sample purification or using appropriate controls can minimize interference. |
| Limited to Condensed Tannins | This assay is specific to condensed tannins and does not measure hydrolyzable tannins, which require different methods. |
| Accuracy | Proper calibration of the standard curve and careful handling of reaction conditions (e.g., heating time, concentration) are essential for accurate results. |

| Step/Factor | Details |
| --- | --- |
| Applications | |
| Dyeing and Textile Industry | Proanthocyanidins are used as natural colorants in textiles, especially for brown to black hues. This assay is crucial for sustainable tannin-based pigment production. |
| Cosmetics and Pharmaceuticals | The antioxidant properties of condensed tannins make them valuable in cosmetic and pharmaceutical products. The assay helps quantify tannins for determining effectiveness in formulations. |
| Food and Beverages | In beverages like wine and tea, condensed tannins contribute to color and flavor. This assay helps assess tannin quality and stability in food products. |

This assay is specifically effective for measuring these proanthocyanidins due to their unique chemical properties, which react distinctly with the butanol-HCl reagent. Table 8 organizes the process, advantages, limitations, and applications of the Butanol-HCl assay.

The Butanol-HCl assay is a highly effective and sensitive method for quantifying condensed tannins (proanthocyanidins) in plant extracts. While it is specifically designed for condensed tannins, it provides reliable and quantitative results, making it indispensable in industries such as textile dyeing, cosmetics, and food packaging, where tannins are utilized for their natural colorant and antioxidant properties. The simplicity, sensitivity, and specificity of this assay make it a widely accepted method for tannin quantification, despite its limitations. Further studies and optimization can improve its application across various sectors relying on tannin-based compounds.

## Protein Precipitation Method for Hydrolyzable Tannins

The protein precipitation method is a widely used analytical technique for quantifying hydrolyzable tannins, a subclass of tannins that are characterized by their ability to break down into smaller components (e.g., gallic acid and ellagic acid) upon hydrolysis (Chandra et al., 2017; Ghosh et al., 2019). Hydrolyzable tannins are typically found in plant materials like oak galls, sumac, and chestnuts and are of interest in industries such as textile dyeing, cosmetics, and pharmaceuticals due to their bioactive and pigmenting properties. This method leverages the interaction between tannins and proteins, such as bovine serum albumin (BSA) or gelatin, to form a precipitate, the amount of which is directly proportional to the tannin concentration as described in Table 9.

The protein precipitation method is a reliable and cost-effective technique for quantifying hydrolyzable tannins in plant extracts. Its specificity for hydrolyzable tannins, simplicity, and ability to deliver accurate quantification make it an important tool in industries like textile dyeing, cosmetics, and pharmaceuticals. However, care must be taken to avoid interference from other phenolic compounds, and the method is not suitable for condensed tannins. Despite these limitations, it remains a valuable approach for tannin analysis, particularly in applications requiring precise measurements of hydrolyzable tannins.

**Table 9.** Protein precipitation method for hydrolyzable tannin quantification: Process, advantages, limitations, and applications

| Step/Factor | Details |
|---|---|
| Process | |
| Sample Preparation | Extract tannins from plant material using a solvent (e.g., water, ethanol, or acetone). Filter the extract to remove plant residues. |
| Protein Solution Preparation | Prepare a protein solution (e.g., gelatin or BSA), which readily forms complexes with tannins through their phenolic groups. |
| Tannin-Protein Reaction | Mix the tannin extract with the protein solution. Tannins react with proteins through hydrogen bonding and hydrophobic interactions, resulting in the precipitation of the tannin-protein complex. |
| Precipitate Measurement | Measure the precipitate either by weight or spectrophotometry: |
| | By Weight: Collect precipitate, dry, and weigh it. The weight is proportional to tannin concentration. |
| | Spectrophotometry: Measure absorbance at a specific wavelength (UV-visible range) where the tannin-protein complex has a characteristic absorption peak. Compare absorbance to a calibration curve for tannin concentration. |
| Advantages | |
| Specificity for Hydrolyzable Tannins | The method is highly effective for quantifying hydrolyzable tannins, which are more challenging to measure using other methods like the Folin-Ciocalteu assay. |
| Simplicity and Cost-Effectiveness | Simple and cost-effective, requiring only basic equipment like a spectrophotometer or balance scale. Protein solutions are inexpensive. |
| Direct Quantification | Provides a direct measurement of hydrolyzable tannins, unlike methods measuring total phenolic content. |
| Versatility | Applicable to a wide range of plant materials, especially those rich in hydrolyzable tannins, and useful for both crude extracts and refined tannin solutions. |
| Limitations | |
| Sensitivity to Interfering Substances | The presence of other phenolic compounds or impurities can interfere with the reaction, leading to inaccuracies. Purification steps may be needed. |
| Limited Applicability to Condensed Tannins | This method cannot quantify condensed tannins, as they do not form precipitates with proteins in the same way. |
| Labor-Intensive | The process involves multiple steps (precipitation, filtration, drying), which can be time-consuming and labor-intensive when handling large sample volumes. |

**Table 9.** (Continued)

| Step/Factor | Details |
|---|---|
| Applications | |
| Textile and Leather Dyeing | Hydrolyzable tannins are used as natural dyes in textiles and leather, providing earthy colors. Accurate tannin quantification is crucial for determining pigment intensity and quality. |
| Cosmetics and Pharmaceuticals | Hydrolyzable tannins are used in cosmetic formulations (e.g., skin-care products) and pharmaceuticals due to their antioxidant and antimicrobial properties. Accurate quantification ensures effectiveness in these applications. |
| Environmental Applications | Used in water treatment processes, where hydrolyzable tannins bind and precipitate heavy metals or pollutants. This method helps determine tannin concentrations in wastewater treatment or environmental monitoring. |

## Antioxidant Activity of Tannin-Based Pigments

In addition to their use as natural colorants, tannins possess significant antioxidant properties, which make them valuable in cosmetic, pharmaceutical, and food industries. Tannins, as polyphenolic compounds, can scavenge free radicals and reduce oxidative stress, offering protection to cells and tissues. This makes them especially useful in the formulation of cosmetic products (such as anti-aging creams) and pharmaceuticals (for their potential anti-inflammatory and antioxidant effects). Additionally, tannins can function as natural preservatives, extending the shelf life of products by preventing oxidative degradation. Given the increasing demand for natural antioxidants in various industries, measuring the antioxidant activity of tannin-based pigments is crucial to determine their effectiveness as bioactive ingredients. The DPPH assay is one of the most widely used methods for this purpose, as it offers a simple and reliable way to evaluate the antioxidant capacity of tannin-based pigments (Wong et al., 2020; Pereira et al., 2021; Kim et al., 2020).

### *DPPH Assay*

The DPPH (2,2-diphenyl-1-picrylhydrazyl) assay is a widely adopted colorimetric method used to measure the antioxidant activity of natural compounds, including tannin-based pigments. DPPH is a stable free radical that exhibits a purple color in its radical state. When it reacts with an antioxidant, such as tannins, the free radical is reduced, and the color changes from purple to yellow, reflecting a decrease in the DPPH concentration. The DPPH assay is simple, quick, and provides a quantitative measure of antioxidant capacity based on the degree of color change. The assay is based on the principle that antioxidants donate electrons or hydrogen atoms to the DPPH radical, neutralizing it, and in turn, reducing the absorbance at a specific wavelength. Table 10 provides a comprehensive view of the DPPH assay for antioxidant activity measurement.

The DPPH assay is an effective, quick, and widely used method for measuring the antioxidant activity of tannin-based pigments. By assessing their radical scavenging ability, this method provides valuable insights into the potential health benefits and preservative properties of tannin-based pigments. The versatility and simplicity of the DPPH assay make it an essential tool for evaluating natural antioxidants across industries, particularly in cosmetics, pharmaceuticals, and food applications.

**Table 10.** DPPH assay for antioxidant activity measurement: Process, advantages, limitations, and application

| Step/Factor | Details |
|---|---|
| Process | |
| Preparation of DPPH Solution | Prepare a DPPH solution by dissolving DPPH in methanol or ethanol. Keep the solution in the dark and adjust its concentration to a specific known absorbance. |
| Sample Preparation | Prepare a tannin-based pigment sample by dissolving it in an appropriate solvent (e.g., water, ethanol, acetone). Vary the concentration based on expected antioxidant content. |
| Reaction | Mix the tannin-based pigment extract with a fixed concentration of DPPH solution. Incubate the mixture for 30 minutes to 2 hours at room temperature in the dark to allow for the reaction to occur. |
| Absorbance Measurement | After incubation, measure absorbance at 517 nm, where DPPH shows peak absorbance. The change in absorbance indicates the degree of DPPH radical reduction. |
| Calculation of Antioxidant Capacity | Calculate antioxidant activity based on the decrease in absorbance using the formula: Scavenging Percentage = ((Acontrol - Asample) / Acontrol) * 100<br>Where: Acontrol is the absorbance of DPPH without antioxidant and Asample is the absorbance with the tannin-based sample. |
| Standardization | Compare the antioxidant activity with a known antioxidant standard (e.g., ascorbic acid or gallic acid) to express results in equivalents (e.g., ascorbic acid equivalents or gallic acid equivalents). |
| Advantages | |
| Fast and Simple | The DPPH assay is straightforward, requiring minimal preparation and equipment. It is rapid, making it ideal for high-throughput screening. |
| Quantitative and Reproducible | Provides a quantitative measure of antioxidant activity, allowing direct comparison between samples. Highly reproducible under controlled conditions. |
| Effective for Screening | Particularly effective for screening natural colorants, including tannin-based pigments, for their antioxidant potential. Useful for industries looking to replace synthetic antioxidants with plant-derived alternatives. |
| Sensitivity | Sensitive enough to detect low concentrations of antioxidants, allowing evaluation of both concentrated and diluted tannin extracts. |

| Step/Factor | Details |
|---|---|
| Limitations | |
| Interference from Other Compounds | Other reducing agents in the sample may interfere with the DPPH assay, leading to inaccurate results. Sample preparation and control experiments are crucial. |
| Limited to Radical Scavenging Activity | The assay only measures radical scavenging activity, not other mechanisms of antioxidant action (e.g., metal chelation, enzyme inhibition). |
| Non-specificity for Tannin Types | The DPPH assay measures overall antioxidant activity but does not differentiate between hydrolyzable tannins, condensed tannins, or other phenolic compounds. |
| Applications | |
| Cosmetics | Tannin-based pigments are used in anti-aging creams and serums due to their antioxidant properties, helping protect skin from oxidative stress and environmental damage. |
| Pharmaceuticals | Tannin-based pigments are explored in the pharmaceutical industry for use in products aimed at preventing oxidative diseases, such as cardiovascular diseases and cancer. |
| Food and Beverages | Tannins serve as natural preservatives in food products (e.g., fruit juices, wine) by preventing oxidative degradation. |
| Textiles | Tannin-based pigments are used in eco-friendly dyeing in textiles, and their antioxidant properties help extend the longevity of textiles by protecting them from oxidative damage during wear and washing. |

## Measurement of Black Pigment Properties

Once tannins are extracted and used as pigments, their color properties, stability, and undertone must be evaluated for various industrial applications. The CIE L*a*b* color model and other colorimetric methods are commonly employed to measure blackness and jetness (McDonald, 2008; Lee et al., 2020), which are critical for understanding the visual appeal and performance of the pigments.

### *Blackness (My)*

Blackness is a key attribute of black pigments, indicating the extent to which a pigment absorbs visible light. The blackness value (My) is determined using the Tristimulus Value Y with a 10° Observer, which correlates to the amount of light absorbed by the pigment. Black pigments with higher absorption coefficients are perceived as deeper or more intense in color.

Generally, black is the result of the absence or complete absorption of visible light. Carbon black manufacturers, for example, promote pigments that absorb up to 99.98% of light. The higher the absorption coefficient of a medium, such as a black coating, the greater the blackness value (My) achieved. Deep black coatings may have an undertone, either bluish or brownish. A deep black with a bluish undertone appears richer, darker, and more vibrant compared to one with a brownish undertone. Even when two samples exhibit the same blackness (My) but differ in undertones (dM), the sample with a bluish undertone is perceived as a deeper black. This is why blacks with blue undertones are often favored in technical applications, such as automotive topcoats. The variation in blackness depending on hue is referred to as "Jetness" (Mc).

#### *Calculation of Blackness*

The My value is derived using the following formula:

$$\boldsymbol{My} = \mathbf{100}\left(\mathbf{1} - \frac{\boldsymbol{y}}{\boldsymbol{y_{max}}}\right)$$

where:

- Y is the Tristimulus Value Y, representing the luminance or brightness of the color.
- $Y_{max}$ is the maximum possible Tristimulus Value Y (usually the value for white or maximum brightness).

This formula calculates the proportion of light absorbed by the pigment, reflecting the lightness of the sample without accounting for any colored undertones. Higher My values correspond to a deeper black, while lower My values indicate a lighter pigment.

***Jetness (Mc) and Undertone (dM)***

In addition to the blackness of a pigment, the undertone plays a crucial role in its visual perception. Jetness (Mc) refers to the hue-dependent blackness and indicates whether a pigment has a bluish or brownish undertone. Black pigments with a bluish undertone are perceived as richer and deeper, while those with a brownish undertone may appear lighter and less vibrant (McDonald, 2008).

*Calculation of Jetness (Mc)*

The Mc value quantifies the degree of blackness with respect to the undertone. The formula is as follows:

$$Mc = 100\left[\log\left(\frac{Xn}{X}\right) - \log\left(\frac{Zn}{Z}\right) + \log\left(\frac{Yn}{Y}\right)\right]$$

where:

- X, Y, Z = The tristimulus values or the measured values of the pigment
- Xn, Yn, Zn = The tristimulus values of a reference standard
- Mc = The Jetness or blackness value.

The Mc value (Jetness) also can be calculated using the following formula:

$$Mc = \frac{L_0 - L_1}{L_0 + L_1}$$

where:

$L_0$ = Lightness of the sample (measured in the CIELAB color space).
$L_1$ = Lightness of a reference black standard.

This formula helps quantify the degree of blackness or jetness by comparing the lightness of the sample to a reference. The closer the value of

Mc is to 1, the more intense the blackness (or jetness) in the sample. Conversely, values closer to 0 indicate less jetness.

*Undertone (dM)*

To calculate the undertone (dM), or the absolute contribution of hue, it typically involves analyzing the color's deviation from pure black, considering the presence of any undertones, such as a bluish or brownish hue. The dM can be calculated using 2 methods. The first method (based on hue angles and chromaticity coordinates) focuses on the visual interpretation of undertones by determining the hue's direction and comparing it with black or neutral color values. The following steps is for the first method:

**1. Measure the Chromaticity Coordinates:**

Determine the chromaticity coordinates (usually x, y, and z) of the sample. These coordinates are obtained by spectrophotometric measurements or from a color space like CIE 1931.

**2. Calculate the Hue Angle:**

The hue angle (h) is calculated based on the chromaticity coordinates. In the CIE 1931 color space, the hue angle can be calculated using the following formula:

$$h = \mathrm{atan2}(y, x)$$

where:

- x and $y$ are the chromaticity coordinates (derived from the spectrophotometric data).
- The hue angle h is usually measured in degrees and indicates the dominant hue of the color (e.g., red, blue, yellow).

**3. Determine the Dominant Undertone:**

The color undertone is determined by examining the hue angle:

- A bluish undertone typically corresponds to a hue angle that is close to 240° (or a blueish section of the color wheel).
- A brownish undertone will have a hue angle near 30° to 40° (yellowish-brown on the color wheel).

**4. Calculate the Absolute Contribution of Hue (dM):**

The undertone's absolute contribution dM can be determined by the difference between the hue angle of the sample and a reference hue angle for black. This formula can be generalized as:

$$dM = \left|h_{sample} - h_{reference}\right|$$

where:

- $h_{\text{sample}}$ - is the hue angle of the sample.
- $h_{\text{reference}}$ - is the hue angle for pure black (typically close to 0°, 180° depending on the context).

If the hue angle falls within the blue or brown range, the dM value will quantify the undertone's contribution to the color.

**5. Interpret the Result:**

Higher dM: Indicates a stronger undertone (e.g., a more pronounced bluish or brownish hue).

Lower dM: Indicates a more neutral or less noticeable undertone.

By calculating dM, the absolute contribution of the undertone (blue or brown) in relation to the overall blackness can be quantified, providing a deeper insight into the color's composition.

The second method (using dM=Mc−My) uses a colorimetric approach, providing a numerical difference based on standard values for chromaticity coordinates, giving a more precise, analytical measure of the undertone's contribution to the color.

The undertone (dM) of a black pigment is defined as the difference between the Mc and My values:

$$dM = M_c - M_y = 100\left[log\left(\frac{X_n}{X}\right) - log\left(\frac{Z_n}{Z}\right)\right]$$

where:

- dM - Represents the absolute contribution of hue or undertone, and measures the difference in color attributes, specifically chromaticity components, for the sample.

- $Mc$ - Represents the Jetness (Mc), which is the hue-dependent degree of blackness (often influenced by undertone). It is a metric that quantifies the richness and depth of blackness in relation to the undertone.
- $My$ - Represents the Blackness (My), the measure of how much light the pigment absorbs, as described earlier. It indicates the lightness or darkness of the pigment without considering undertones.
- $Xn$, $Zn$ - These are the standard values or reference values for chromaticity coordinates (X, Z) that are typically associated with a neutral black or pure color.
- $X$, $Z$ - These are the measured values for the chromaticity coordinates of the sample. The difference between the standard and the measured values helps in determining the undertone (dM).

The formula essentially compares the chromaticity values of the sample (X, Z) to a reference or standard (Xn, Zn) to calculate the difference in their relative contributions to the color's hue.

The logarithmic difference between the two values helps quantify the contribution of the undertone (whether it is brownish, bluish, or neutral) in the overall appearance of the blackness.

The result is expressed as dM, which reflects the contribution of hue or the undertone in the black pigment, expressed as a percentage.

## Difference Between the Two Methods

Previous Method (Hue Angle-Based): Focuses on calculating the hue angle of the sample and comparing it to a reference (often black or a neutral point) to determine the undertone. It relies on chromaticity coordinates (x, y) and the concept of hue angles to understand the undertone. The formula assumes that you are dealing with a pigment's specific hue direction, which will indicate whether the color is bluish, brownish, or another hue.

Formula dM = Mc - My: This formula is a more quantitative approach that uses chromaticity coordinates (X, Z) and compares the measured values to standard reference values. The difference between the measured and standard values of chromaticity coordinates indicates the degree of undertone. This approach is more focused on the mathematical and colorimetric differences in the pigment, using logarithmic calculations to provide a detailed numerical value for the undertone. It incorporates Jetness (Mc) and Blackness (My) values, which are directly related to color perception, providing a more precise measure of how the undertone affects the overall color of the black pigment.

In summary, the first method (based on hue angles and chromaticity coordinates) focuses on the visual interpretation of undertones by determining the hue's direction and comparing it with black or neutral color values. The second method uses a colorimetric approach, providing a numerical difference based on standard values for chromaticity coordinates, giving a more precise, analytical measure of the undertone's contribution to the color. Both methods are used to measure the undertone (dM), but the second formula gives a more specific and technically detailed result by incorporating reference values for chromaticity, making it useful for color matching and quality control in industrial applications.

Positive dM values indicate a blue undertone, which enhances the perceived depth and richness of the black. Negative dM values indicate a brownish undertone, resulting in a lighter, less intense black. The jetness of a pigment is critical in industries like automotive coatings, where a blue undertone is often preferred for its deeper, more vibrant appearance (McDonald, 2008).

Both methods are used to measure the undertone (dM), but the second formula gives a more specific and technically detailed result by incorporating reference values for chromaticity, making it useful for color matching and quality control in industrial applications.

Accurate measurement of tannin content and the blackness of tannin-based pigments is essential for their commercial and industrial use. Techniques such as the Folin-Ciocalteu assay, colorimetry, and lightfastness testing provide critical insights into the quality and performance of tannin-based pigments. Additionally, Jetness and undertone measurements help in understanding the aesthetic appeal of black pigments, while antioxidant activity testing supports their cosmetic and pharmaceutical applications. As research into greener extraction techniques and pigment stability progresses, tannin-based pigments are expected to play an increasingly significant role in sustainable industries.

## Environmental Impact and Sustainability

The growing demand for eco-friendly materials has led to a significant shift towards sustainable dyeing practices. Tannin-based pigments have gained attention as a more environmentally sustainable alternative to synthetic dyes, which are often derived from petroleum-based chemicals and contain harmful heavy metals (Lee et al., 2021). These pigments offer a biodegradable, non-toxic solution that reduces the environmental footprint of the dyeing industry.

### *Low Toxicity and Biodegradability*

One of the primary environmental advantages of tannin-based pigments is their low toxicity and biodegradability. Unlike synthetic dyes, which can cause significant water pollution due to the presence of harmful chemicals, tannins do not contain the hazardous materials that often pollute water sources. They are biodegradable, meaning they break down naturally over time without causing long-term harm to ecosystems (Tan et al., 2020). This makes tannin-based pigments particularly attractive to industries seeking to reduce their reliance on petroleum-derived chemicals, which are non-renewable and contribute to environmental degradation (Ghosh et al., 2019).

The use of tannin-based pigments is not only beneficial for reducing the toxic load on aquatic ecosystems but also contributes to a greener approach to dyeing, aligning with the growing focus on environmental sustainability across various sectors (Kim et al., 2020).

### *Sustainable Sourcing*

Tannins are abundant in agricultural waste, which are rich in tannins and serve as effective sources for natural dyes. The utilization of these waste materials for tannin extraction is an excellent example of how industries can capitalize on waste products to create high-value pigments. This practice supports zero waste initiatives and promotes circular economies, as plant residues that would otherwise be discarded are repurposed to create eco-friendly products (Zhang et al., 2019; Yuan et al., 2020).

Furthermore, by harnessing waste materials from agriculture, the demand for primary resources is reduced, helping to mitigate the environmental impact of raw material extraction. This contributes to more sustainable sourcing practices and aids in reducing carbon emissions associated with waste disposal (Yuan et al., 2020).

### *Water Management*

In comparison to conventional synthetic dyeing processes, tannin-based dyeing requires significantly less water. This is particularly important given the growing global concerns about water scarcity and the environmental impact of traditional dyeing processes, which often use vast amounts of water and introduce pollutants into water systems (Lee et al., 2021). Tannin-based dyes, when combined with natural mordants, not only reduce water consumption but also result in fewer pollutants being released into water bodies (Kim et al., 2020).

Additionally, water-efficient extraction methods, such as supercritical fluid extraction (SFE), have been shown to reduce the water footprint of tannin-based dyes. SFE uses supercritical $CO_2$ to extract tannins in an efficient, eco-friendly manner, minimizing the need for large quantities of water and eliminating the toxic solvents used in conventional extraction methods (Van Der Walt et al., 2021).

### *Recycling and Reusability*

The recycling and reusability of plant-based materials, such as agriculture residues and wood chips, for tannin extraction enhance the sustainability of tannin-based pigments. By utilizing plant waste, industries help reduce agricultural waste, which is often burned or discarded, thereby reducing the carbon footprint associated with waste disposal (Yuan et al., 2020). This contributes to the reduction of greenhouse gas emissions and supports the circular economy, where materials are continuously repurposed rather than disposed of (Ghosh et al., 2019).

Furthermore, the recyclability of $CO_2$ in supercritical fluid extraction (SFE) helps to make the extraction process more sustainable, aligning with the industry's shift towards greener technologies. By minimizing waste and improving the resource efficiency of production processes, tannin-based pigment production is poised to make a significant contribution to eco-friendly manufacturing.

## Current Research Trends in Tannin-Based Pigments

The growing demand for sustainable and eco-friendly materials has sparked significant interest in tannin-based pigments. As the world moves toward reducing the environmental impact of synthetic dyes, tannin-based pigments, derived from natural plant sources, are being explored for their biodegradable and non-toxic properties. However, to maximize their potential, current research trends are focusing on improving extraction methods, pigment stability, and expanding their applications across different industries.

- *Greener Extraction Methods:* New studies continue to explore more sustainable, green, and efficient methods of tannin extraction. Research into enzymatic extraction, supercritical fluid extraction, and microwave-assisted methods is ongoing, aiming to make tannin extraction faster, more cost-effective, and environmentally friendly.
- *Pigment Stability and Durability*: While tannin-based pigments are sustainable, they often face challenges in terms of color stability. Recent research is focused on improving the lightfastness and wash-fastness of these pigments by experimenting with different metallic mordants and additives.
- *Wider Applications:* Beyond textile dyeing, there is a growing interest in expanding the applications of tannin-based pigments in other industries, such as cosmetics, food, and art. Tannins are being studied for their antioxidant properties, and their potential as natural colorants for lipsticks, makeup, and even food packaging materials (Pereira et al., 2021; Shi et al., 2021). This trend is driven by the increasing demand for natural, biodegradable, and sustainable alternatives to synthetic chemicals.
- *Cosmetics:* Tannins are being studied for their antioxidant properties and their potential as natural colorants in lipsticks, makeup, and skincare products. The non-toxic nature of tannins makes them ideal for eco-conscious beauty brands that seek to reduce their environmental impact (Pereira et al., 2021).
- *Food:* Tannin-based pigments are also being explored as natural colorants in the food industry. Given their biodegradability and safe composition, they offer a promising alternative to synthetic food colorants, which can sometimes have adverse health effects. Tannins from sources like sumac and pomegranate are being tested for their use in food packaging materials, providing a sustainable option for food manufacturers (Kim et al., 2020).
- *Art:* Tannin-based pigments are widely used in the art world for eco-friendly paints. Historical art forms, such as ink and pigment production, are experiencing a resurgence as artists seek to use more sustainable materials in their work (Santiago et al., 2020).

Advancements in extraction techniques, the use of mordants, and the growing sustainability of tannin-based pigments position these natural compounds as a valuable alternative to synthetic dyes. As research continues to improve extraction methods, stabilize colors, and enhance the efficiency of tannin use, tannin-based pigments will likely play an increasingly important role in eco-friendly industries, from textiles to cosmetics and art.

## Conclusion

Tannin-based pigments, derived from a wide range of plant sources, present a promising solution to the growing demand for sustainable and eco-friendly alternatives to synthetic dyes. This chapter has explored the chemical properties, extraction methods, and applications of tannins, emphasizing their environmental and economic significance. Tannins, known for their biodegradability, low toxicity, and ability to produce rich, deep black hues, are gaining increasing attention in industries such as textiles, cosmetics, pharmaceuticals, and food packaging. The chapter highlights the various methods of tannin extraction, from traditional aqueous extraction to advanced techniques like microwave-assisted extraction and supercritical fluid extraction, which improve efficiency and sustainability.

Furthermore, the antioxidant properties of tannin-based pigments make them valuable in the cosmetic and pharmaceutical industries, where they can act as natural preservatives or active ingredients. The chapter also covers the importance of accurately measuring tannin content and the blackness of tannin-based pigments, using methods such as the Folin-Ciocalteu assay and colorimetry, which are essential for ensuring pigment quality and consistency across applications.

The resurgence of interest in tannins, driven by concerns over the environmental and health impacts of synthetic chemicals, underscores their potential as a sustainable resource for pigment production. Tannin-based pigments not only offer an eco-friendly alternative to petroleum-based synthetic dyes but also contribute to waste reduction by utilizing agricultural by-products. As the demand for green, biodegradable materials grows, tannin-based pigments are positioned to play a central role in the transition toward more sustainable industries.

Through ongoing research aimed at enhancing pigment stability, exploring new extraction methods, and expanding their applications, tannin-based pigments are poised to become an integral part of the eco-conscious future of materials design. Their environmental benefits, coupled with their aesthetic and functional properties, ensure that tannins will continue to be an important natural resource in the development of sustainable, non-toxic products across a range of industries.

## References

Ali, F., Alviano, D., Aziz, A. A., Bharathi, R., Bhowmik, A., Chandra, S., & Dean, J. (2020). Hydrolyzable tannins: Extraction and applications in dyeing. *Natural Products Journal, 12*(4), 112–121. https://doi.org/10.1080/10286135.2020.1746521.

Alviano, D., Pereira, A. C., Lima, M. A., & Silva, R. R. (2017). Antioxidant and antimicrobial properties of *Pontederia cordata* tannins. *Freshwater Science, 14*(6), 64–73. https://doi.org/10.1016/j.freshsci.2017.05.004.

Aziz, A. A., Mohd, S. K., Bakar, M. R., & Hassan, Z. (2021). Polyphenolic compounds and antimicrobial activity of *Avicennia marina*. *Journal of Applied Botany, 93*(1), 122–130. https://doi.org/10.1016/j.jab.2021.01.015.

Bharathi, R., & Ponnusamy, S. (2017). Tannin extract from various plants: Chemical composition and uses. *Journal of Plant Chemistry, 15*(3), 289–301. https://doi.org/10.1016/j.jpc.2017.03.004.

Bhowmik, A., Radhakrishnan, K., & Devi, A. R. (2019). *Nymphaea* spp.: Phytochemical composition and medicinal applications. *Aquatic Plants Journal, 28*(4), 125–133. https://doi.org/10.1007/apj.2019.0045.

Chandra, S., Jain, S., & Rao, P. (2017). Quantification of hydrolyzable tannins in plant extracts using the protein precipitation method. *Phytochemical Analysis, 28*(2), 105–112. https://doi.org/10.1002/pca.2832.

Dean, J. (2016). Focus on tannin. *Jenny Dean's Wild Colour*. Retrieved from https://www.jennydean.co.uk/focus-on-tannin/.

Dhiman, S., Tiwari, G., & Sharma, V. (2021). Ultrasound-assisted extraction of tannins from oak galls. *Journal of Food Engineering, 296*, 110376. https://doi.org/10.1016/j.jfoodeng.2020.110376.

Fearon, V. (2022). Tannins for natural dyes. *Vicki Fearon's Blog*. Retrieved from https://www.vickifearon.com/blog/tannins.

Gao, F., Wang, Q., & Zhang, S. (2020). Tannins in *Salvinia molesta* and their potential for bioremediation. *Aquatic Environmental Studies, 31*(1), 55–61. https://doi.org/10.1007/aqes.2020.0039.

Ghosh, R., Banerjee, S., & Prakash, G. (2019). Environmental sustainability of tannin extraction from plant waste for pigment production. *Green Chemistry, 21*(8), 2257–2270. https://doi.org/10.1039/c9gc00816b.

Gomez, M., Souza, P., & Matos, J. (2021). Environmental impacts of synthetic pigments vs. natural alternatives. *Sustainability in Textiles, 8*(3), 140–153. https://doi.org/10.1016/j.sus.2021.04.022.

Hashem, A., El-Sayed, A., & Fadel, A. (2020). Enzyme-based extraction of hydrolyzable tannins for food industry applications. *Food Research International, 130*, 108968. https://doi.org/10.1016/j.foodres.2019.108968.

Hashem, A., Mahmoud, M., & El-Hamid, A. (2020). Enzyme-assisted extraction of tannins from mango leaves. *Food Chemistry, 338*, 127819. https://doi.org/10.1016 /j.foodchem.2020.127819.

Jackson, C., & Morgan, S. (2022). Polyphenolic compounds in plant defense: Tannins and their role. *Plant Science Journal, 17*(2), 214–227. https://doi.org/10.1016/j.plantsci. 2022.04.015.

Jones, L. P. (2019). Traditional methods of producing black pigments. *Journal of Historical Dyes, 5*(2), 123–134. https://doi.org/10.1080/jhd.2019.0298745.

Kim, S., Lee, J., & Choi, H. (2020). Sustainability of tannin-based natural pigments: An eco-friendly approach for textile dyeing. *Journal of Cleaner Production, 246*, 118944. https://doi.org/10.1016/j.jclepro.2019.118944.

Kumar, P., Yadav, A., & Sharma, S. (2020). Chemical properties and medicinal uses of *Cyanometra ramiflora* L. *Asian Journal of Medicinal Plants, 22*(2), 92–100. https://doi.org/10.3923/ajmp.2020.92.100.

Lee, H., & Kim, Y. (2020). Plant-based black pigments: The role of tannins. *Journal of Natural Pigments, 16*(4), 91–102. https://doi.org/10.1016/j.jnp.2020.02.001.

Lee, H., Kim, Y., & Lee, J. (2020). Sustainable dyeing with tannin-based pigments: A review. *Environmental Science and Pollution Research, 28*, 13225–13237. https://doi.org/10.1007/s11356-021-12521-2.

McDonald, G. (2008). Blackness and jetness of carbon pigments. *Pigment and Color Journal, 3*(2), 112–118. https://doi.org/10.1007/s00387-008-0319-3.

Mohammad, A., Iqbal, S., & Hassan, S. (2020). Dyeing and medicinal properties of *Lemna minor*. *Journal of Aquatic Botany, 19*(6), 230–239. https://doi.org/ 10.1016/j.jab.2020.09.003.

Mohan, S., Gupta, A., & Sharma, P. (2021). *Salvinia molesta*: Phytochemistry, dyeing potential, and bioremediation. *Journal of Aquatic Research, 19*(6), 112–118. https://doi.org/10.1007/jar.2021.0064.

Mu, Y., Y. An, L. Yang, Z. Cui, H. Li, Y. Zhou, Z. Wang, & Y. Wang. (2023). Variations of anthocyanins and proanthocyanidins in the flowers of *Rhododendron dauricum* var. albiflorum. *Horticulturae, 9*(10), 1071. https://doi.org/10.3390/horticulturae9101071.

Mullen, L., O'Neill, E., & Davis, T. (2019). Microwave-assisted extraction of tannins from sumac for natural dyeing. *Industrial Crops and Products, 130*, 70–78. https://doi.org/10.1016/j.indcrop.2019.01.018.

Neran, F., S. Scharbert, & T. Hofmann. (2025). Identification and characterization of proanthocyanidins of 16 members of the *Rhododendron* genus (Ericaceae) by tandem LC–MS. *Phytochemistry, 219*, 113615. https://doi.org/10.1016/j.phytochem. 2024.113615.

Nguyen, P. H., Le, T. M., & Ho, H. T. (2022). Sustainable design and pigment production from tannins. *Environmental Sustainability Journal, 11*(3), 67–79. https://doi.org/ 10.1016/j.esj.2022.03.008.

Nisar, M., Akram, M., & Khan, A. (2021). Enzyme-assisted extraction of tannins from mango leaves. *Food Chemistry, 338*, 127819. https://doi.org/10.1016/j.foodchem.2020.127819.

Noor, M. N. M., & Mohd, R. (2018). Phytochemical screening of *Bruguiera gymnorrhiza* for dyeing and medicinal use. *Mangrove Ecosystem Studies, 11*(3), 212–224. https://doi.org/10.1016/j.mangeco.2018.04.006.

Nor, S. R., Zainal, A. S., & Isa, S. M. (2020). Tannin composition and dyeing potential of *Avicennia marina*. *Journal of Mangrove Studies, 7*(2), 47–58. https://doi.org/10.1016/j.jms.2020.07.001.

Pereira, A., Oliveira, J., & Silva, P. (2020). Natural pigments in the cosmetic industry: Tannins and their application. *Cosmetics and Natural Products, 12*(4), 45–55. https://doi.org/10.1002/cnp.2020.0012.

Pereira, M., Santos, F., & Lima, R. (2021). Innovations in tannin extraction for natural pigments. *Journal of Natural Products, 45*(3), 234–241. https://doi.org/10.1007/jnp.2021.0434.

Porter, L. J., M. W. McRoberts, & A. J. Pryor. (1986). The Butanol-HCl method for condensed tannin quantification. *Phytochemistry, 25*(11), 2215–2218. https://doi.org/10.1016/0031-9422(86)80274-X.

Ramana, V., R. K. Sharma, & S. C. Meena. (2017). Chemical constituents of *Heliotropium trigosum* L. and its application in dyeing. *Journal of Environmental Chemistry and Ecotoxicology, 9*(4), 101–106. https://doi.org/10.1007/s12247-017-0492-0.

Sabaris, F., M. E. Ramírez, & M. B. Ramírez. (2020). Role of mordants in natural dyeing: Enhancing color from plant tannins. *Textile Research Journal, 90*(1), 1–10. https://doi.org/10.1177/0040517519873050.

Santiago, A., P. M. Garcia, & F. R. Fernandes. (2020). Stabilization of tannin-based pigments: Techniques and applications. *Dyeing and Pigments Journal, 34*, 89–95. https://doi.org/10.1016/j.dyepig.2020.02.007.

Shah, R., S. M. Pappu, & R. Kumar. (2020). Dyeing potential of *Briedelia stipularis* L. and its applications. *International Journal of Natural Dyes, 15*(3), 101–112. https://doi.org/10.1016/j.ijdye.2020.01.003.

Shi, Q., Z. L. Wang, & X. T. He. (2021). Enhancement of tannin-based pigments with iron salts for improved color stability. *Industrial Crops and Products, 145*, 111887. https://doi.org/10.1016/j.indcrop.2020.111887.

Shi, Y., X. Q. Liu, & Y. Zhang. (2020). Optimizing the extraction of tannins from oak galls: Effects of solvent and temperature. *Industrial Crops and Products, 134*, 260–267. https://doi.org/10.1016/j.indcrop.2020.01.014.

Singh, V., P. K. Mehta, & M. K. Yadav. (2020). Dyeing applications and bioremediation potential of *Lemna minor*. *Aquatic Ecology Review, 16*(3), 211–218. https://doi.org/10.1007/aer.2020.0011.

Singleton, V. L., R. Orthofer, & R. M. Lamuela-Raventos. (1999). Analysis of total phenols and other oxidation substrates and antioxidants by means of Folin-Ciocalteu reagent. *Methods in Enzymology, 299*, 152–178. https://doi.org/10.1016/S0076-6879(99)99017-1.

Siti, N. A., M. K. Rahman, & R. S. Mohamed. (2021). Proanthocyanidins in *Bruguiera gymnorrhiza* and their application in dyeing. *Dyes and Pigments Journal, 130*(5), 39–47. https://doi.org/10.1016/j.dyepig.2021.05.006.

Smith, J. T., & R. K. Brown. (2017). The use of mineral pigments in ancient art. *Journal of Archaeological Science, 44*(3), 243–259. https://doi.org/10.1016/j.jas.2016.10.017.

Sun, H., F. W. Lee, & C. J. Kim. (2019). Gallic acid and its derivatives: A comprehensive review. *Journal of Natural Bioactive Compounds, 18*(2), 121–134. https://doi.org/10.1016/j.jnbc.2018.12.007.

Sun, H., M. K. Zhang, & L. H. Wang. (2020). Polyphenolic compounds from *Nymphaea* spp. and their applications. *Journal of Aquatic Plants, 21*(3), 213–221. https://doi.org/10.1007/jap.2020.0031.

Talu, E. (2023). Understanding natural tannins and the best ones for dyeing. *TALÚ Earth.* https://talu.earth/understanding-natural-tannins-and-the-best-ones-for-dyeing/.

Tan, W., D. F. Yang, & L. M. Zhao. (2020). Tannin extraction from plant materials for use in eco-friendly dyes. *Environmental Chemistry and Materials, 28*(1), 1–9. https://doi.org/10.1016/j.ecm.2020.06.002.

Tan, W., L. F. Li, & M. Y. Wang. (2021). Tannin extraction and utilization from aquatic plants. *Aquatic Plant Research, 13*(3), 104–110. https://doi.org/10.1016/j.apr.2021.05.008.

Tan, W., L. F. Li, & M. Y. Wang. (2021). Tannin extraction from plant materials for use in eco-friendly dyes. *Environmental Chemistry and Materials, 28*(1), 1–9. https://doi.org/10.1016/j.ecm.2020.06.002.

Tan, W., L. F. Li, & M. Y. Wang. (2021). Tannins as a source of natural dyes: A review. *International Journal of Natural Products, 20*(1), 34–45. https://doi.org/10.1016/j.ijnp.2020.11.006.

Thilakarathna, M., G. P. Fernando, & P. R. Perera. (2021). Applications of *Pontederia cordata* in aquatic dyeing. *Journal of Sustainable Aquatic Plants, 29*(2), 75–82. https://doi.org/10.1016/j.susaquap.2021.02.005.

Van Der Walt, S., G. H. Roos, & F. J. Botha. (2021). Supercritical fluid extraction of tannins from plant sources. *Journal of Supercritical Fluids, 162*, 104858. https://doi.org/10.1016/j.supflu.2020.104858.

Vasudevan, S., P. V. Nair, & R. K. Patel. (2020). Dyeing and medicinal potential of *Peltophorum pterocarpum. Plant Science Today, 17*(3), 130–140. https://doi.org/10.1002/psct.171.

Wang, L., J. S. Wu, & M. Y. Li. (2022). Polyphenols and their role in natural pigments. *Plant Chemistry Journal, 23*(5), 445–453. https://doi.org/10.1016/j.pcj.2022.04.002.

Wang, S., J. F. Lee, & R. W. Yu. (2021). Phytoremediation and tannin extraction from *Lemna minor*. *Environmental Chemistry and Applications, 22*(5), 89–98. https://doi.org/10.1016/j.envchem.2021.07.005.

Wijaya, L., M. A. Suhardi, & R. A. Mulyadi. (2021). Exploring the medicinal benefits and chemical properties of *Garcinia mangostana* L. *Journal of Ethnopharmacology, 271*(3), 87–93. https://doi.org/10.1016/j.jep.2021.113902.

Williams, D. F. (2023). The historical use of tannins in dyeing and tanning. *Dyeing and Leather Journal, 25*(1), 65–72. https://doi.org/10.1007/s00405-023-0056-7.

Wong, W., L. P. Lee, & J. T. Chan. (2020). Antioxidant activity of tannin-based pigments. *Antioxidants, 9*(8), 842–855. https://doi.org/10.3390/antiox9080842.

Yadav, D., V. S. Verma, & R. K. Gupta. (2016). Phytochemical constituents and dyeing applications of *Citrus medica* L. *Journal of Applied Science and Environmental Management, 17*(5), 277–283. https://doi.org/10.4314/jasem.2016.05.003.

Yuan, J., Y. C. Zhang, & Z. H. Li. (2020). Sustainable tannin extraction and utilization from agricultural waste for eco-friendly pigment production. *Waste Management, 104*, 142–150. https://doi.org/10.1016/j.wasman.2020.03.014.

Zhang, S., L. J. Yang, & X. X. He. (2019). Extraction and purification of tannins from sumac fruits: A comparative study. *Food Chemistry, 292*, 229–235. https://doi.org/10.1016/j.foodchem.2019.04.080.

Zohra, F., M. A. Sulaiman, & T. H. Ho. (2019). Eco-friendly mordants for plant-based dyeing. *Dyes and Pigments, 162*, 396–406. https://doi.org/10.1016/j.dyepig.2018.11.024.

Zuo, Y., Y. L. Zhang, & F. S. Liu. (2020). Ultrasound-assisted extraction of tannins: A review. *Journal of Cleaner Production, 254*, 120098. https://doi.org/10.1016/j.jclepro.2020.120098.

# Chapter 2

# Smart Packaging with Carotenoids: A Breakthrough in Food Quality and Preservation

**Ainaa Eliah Abu Bakar[1,*], PhD**
**and Rashidi Othman[2], PhD**

[1]International Institute for Halal Research and Training,
International Islamic University Malaysia, Kuala Lumpur, Malaysia
[2]Department of Landscape Architecture,
International Islamic University Malaysia, Kuala Lumpur, Malaysia

## Abstract

The use of intelligent packaging, which can assess the freshness of packed food in real-time and display the packaging film's colour change based on the condition of the food, has grown in popularity recently. In all fields of life sciences, the trend towards sustainability, enhanced product safety, and high standards of quality are significant. A cutting-edge technology called intelligent packaging food indicator helps to assure food safety and quality while preventing food contamination. With intelligent packaging, food freshness and quality changes may be tracked in real time. One of the most well-known natural colourants, carotenoids, has drawn a lot of interest due to its good functional qualities and capacity to alter colour in response to pH variations. Biomaterials rich in carotenoid have many biological benefits, including anticancer, antibacterial, antiviral, and antioxidant qualities. They are also inexpensive and readily available. Carotenoid's functional qualities have been extensively utilised in the development of intelligent packaging

* Corresponding Author's Email: ainaaelya@gmail.com

In: Natural Pigments: Sources, Properties, and Health Benefits
Editors: Rashidi Othman, Qurratu Aini Mat Ali, Razanah Ramya et al.
ISBN: 979-8-90134-055-4

solutions. This is because colour is omnipresent by nature and has the power to communicate important information about food packing. the potential use of carotenoid as a health-promoting, intelligent packaging indication.

**Keywords**: carotenoids, food packaging, natural colourant

## Introduction

The Food safety and food spoiling are major global challenges that are getting more prevalent. This occurs when bacteria that cause food poisoning are uninvited and lead to food-borne epidemics. Packaging is essential to developing a complete safety package, including active and intelligent packaging, to address this problem. A bioactive functional element that enhances the antioxidant, UV radiation barrier, and antibacterial properties of a packaging material is known as an active packaging ingredient. Using smart packaging that incorporates a sensor to monitor food condition, it is possible to determine whether the food inside is safe to eat. Sense of the toxicity within the food itself can be aided in this way. It is possible that this clever packaging will enable real-time food item detection. In addition, it is possible to construct intelligent indicator-based packaging that may be integrated into a packaging system and readily detect indicators such as pH, gas sensors, and time temperature (He et al., 2023; Mohammadian et al., 2020). To improve the material's ability to keep food-borne pathogens from growing and to reduce the exposure of food oxidation, active packaging is integrated. In addition, before a user opens a packaged food item from its container, intelligent packaging is employed to offer information about food safety, temperature changes, and spoiling detection (Roy et al., 2022; Firouz et al., 2021). pH indicator-based intelligent packaging is the most straightforward, accurate, affordable, user-friendly, and non-destructive technology as compared to other forms of packaging (Versino et al., 2023; Roy & Rhim 2021; Deshmukh et al., 2023a; Roy et al., 2023). By adding more pigments to the pH-changing film and observing the changing colour, one may determine the state of the meal (Nemes et al., 2020; Kumar et al., 2023). Perishable foods, such as meat, fish, and many others, can vary in pH because of oxidation or microbial infection because they induce food to rot (Balbinot-Alfaro et al., 2019; Yousefi et al., 2019; Kumar et al., 2023; Grzebieniarz et al., 2023; Deshmukh et al., 2022b; Rout et al., 2022). Because the pH of the packaging changes while it is stored, the colour of the packaging also changes. Once the packaging's colour starts to shift, it is simple to identify. the use of artificial chemicals in food packaging, which are best employed as active fillers and can have detrimental effects on health because natural bioactive substances are more effective than chemical ones. Extending the food's shelf life can be greatly benefited by adding the bioactive components to that packaging. Natural pigments can be extracted from a variety of sources, including curcumin, anthocyanin, carotenoids, betalains, and many more, as a safer alternative to synthetic dyes like methyl red, bromocresol green, and many others that are used in smart packaging and can have a negative impact on consumer health and the environment (Niponsak et al., 2016; Ran et al., 2021; Amjadi et al., 2020; de Oliveira Filho et al., 2023; Ahmed et al., 2023; Roy et al., 2023). Anthocyanin and curcumin are good sources of intelligent indicators because of their extensive use, availability, and optimal performance as color-indicating compounds. Carotenoids are among the useful colour indicators found in materials for intelligent packaging. This is because of its exceptional pH-responsive colour change and fat-soluble nature, both of which are useful in the development of a viable candidate for the creation of intelligent food freshness by tracing indications (Idrovo et al., 2021; Sigurdson et al., 2017; Vieira et al., 2022). Tetraterpene pigments, or carotenoids, are abundant and mostly found in fruits and vegetables. They can be extracted to yield a variety of hues, including red, purple, yellow, and many more (Melendez-Martinez et al., 2022; Ashokkumar et al., 2023; Deshmukh et al., 2023b). Beta-carotene, zeaxanthin, norbixin, lycopene, and many other bioactive substances are present in carotenoids (Ashokkumar et al., 2023; Bhatt & Patel, 2020; Assis et al., 2021; Anushikha et al., 2023). These bioactive substances derived from carotenoids function best as potent antibacterial and antioxidant agents, making them ideal for use as active ingredients in packaging.

## Food Packaging

Production of food packaging is driven by customer demand patterns and the need to reduce food waste, both of which support the creation of smart packaging solutions. From the standpoint of traditional passive functionality, the systems are serving the interests of the customers by physically shielding food from unwelcome outside influences and allowing for controlled interaction with the food environment (Dainelli et al., 2008; Yildirim et al., 2018). It can usually be identified between toxic chemicals and packaging since it can create noticeable colour changes. In the meantime, it can aid customers by providing them with information about the product's current freshness (Ezati et al., 2019). Foodstuffs, particularly meat and fish, can be used as a source, as they can produce a variety of compounds as they decay. Specialised laboratory and skilled personnel are needed to analyse the food quality, which adds to the time-consuming and harmful process (Lee et al., 2018). Another method of detection, such as electronic nose, involves treating samples specifically for analysis (Lee et al., 2018; Zhang et al., 2008). The stringent requirements for light treatment included the use of smart indicators, which are being developed to be economical, small, non-destructive, sensitive, user-friendly, and readily scaled for industrial production (Ezati et al., 2019). It may be stated that the use of visible indicators led to promising systems such as smart food packaging indicators since they are biocompatible, biobased, and sustainable. Typically, a chemical indication or sensor is created as a sensitive pigment or dye, which functions best when combined with a sturdy support and an active ingredient (Pourjavaher et al., 2017).

Items intended for consumption have increasingly identifiable packaging from a marketing perspective, since competing products vie for consumers' attention when they are displayed next to one another on shelves. If the packaging is created with attention to detail in terms of graphics, design, and content labelling, it could contribute to a favourable consumer experience. When the item is unhealthy in and of itself, but the packaging is appealing, customers are more likely to opt to purchase it than when the product is good but the packaging is ugly. These days, the food business tends to create packaging that serves as a vehicle for brand promotion and consumer communication in addition to safeguarding the food packing itself. A range of substances utilised in the food sector.

Petroleum plastics are the most widely used type (synthetic polymers). These polymers included polystyrene (PS), polyvinyl chloride (PVC), low- and high-density polyethylene (LDPE and HDPE), polypropylene (PP), and polyethylene terephthalate (PET) (Housewirth, 2017). The sustainability that piques the curiosity of several scholars, scientists, and environmentalists. From the perspective of polymers, biopolymers are a class of materials that are environmentally benign and can be used in place of synthetic polymers. It also demonstrated the use of biomass, or agro-industrial waste, and renewable raw materials, which is a hallmark of biodegradability. Apart from that, it has been demonstrated to be both affordable and desired. Furthermore, the biopolymer components that are being created are edible and function best as antimicrobials and antioxidants. To improve its qualities, the biopolymer components are layered and created as one of the composites (Popovi'c et al., 2018).

Food packaging has two primary purposes: it can aid in maintaining the quality of food goods, and it is important to keep in mind the significance of health and safety, which are created as the food products are packaged and coated. This appears to be a crucial factor in the movement of substances that lead to the exposure of elements in food, so contaminating it. Contaminants are like additives in that they can be used as plasticizers, stabilisers, and antioxidants. Apart from that, monomers exist. Furthermore, there is always the chance that unidentified compounds are inadvertently transmitting. Numerous migratory compounds, including butylated hydroxytoluene, bis(2-ethylhexyl) adipate, tributyl citrate, phthalates, N-alkanes, and many more, are found in paper-based candy wrappers and plastics (Gra´ın˜o et al., 2018). The chemicals are found in plastics that health and safety agencies have forbidden people from consuming. The usage of biopolymers can be employed in place of conventional synthetic polymers, which are derived from oil derivatives (Gruji´c et al., 2017). This is because there is a possibility that food may become contaminated with bioplastics due to various causes. Since natural components are the main component of all bioplastics, food safety has been compromised by contamination from overuse, which poses a serious risk to consumer and food safety. Protection, preservation, and presentation are the three pillars of packaging (Rahman, 2007). Even though it is claimed that this packing is straightforward and reasonable, it demonstrated additional factors to consider (Rahman, 2007). Leading to the convenience factor, which is determined by the usefulness of the product for the consumer and demonstrates that the package is user-friendly, it must be easy to open, reusable, manageable, recyclable, and waste-less (Robertson, 2005). Every researcher has a different take on what constitutes packaging. Based on the package design, which performed well across three main categories (Burke, 2011). The ability of the package to communicate the brand image to the customer is one of the three categories that are regarded as authentic (Burke, 2011).

## Polymers

A polymer is a class of material that is generated from a massive molecular structure consisting solely of rings or chains derived from unique repeating units, also referred to as monomers. The concept of polymers is derived from the Greek word "polymer," which means "many parts," and "mer," which represents "poly." This helps to make the description of polymers extremely understandable. Because of their qualities, both natural and synthetic polymers are widely used in a variety of sectors. Synthetic materials derived from plastics and resins are frequently utilised (Housewirth, 2017). Long-chain molecules derived from alkenes are used in the synthesis of polymers that are then combined with additives to produce plastics. To prepare various polymers, one or more hydrogen atoms are often used, albeit they can be assisted by the replacement of another atom (Tajeddin et al., 2018). Plastics are easily made because they are simple to mould and modify. There are numerous types of plastics with various characteristics that can be made in terms of size, weight, shape, colour, function, printing, and other factors (Housewirth, 2017). However, the usage of polymers has become an essential component of contemporary life, establishing its features, such as resistance, stability, and ease of production (Yadav et al., 2018).

### *Biodegradable Polymers*

An abundance of ideas is being developed these days to produce better packaging products. Firstly, it is necessary to be able to recycle packaging materials, and to develop better biodegradability and environmentally friendly materials. Companies are working towards better materials (Byrd, 2016). Biodegradable materials are those that break down enzymatically through the action of living organisms, such as bacteria, fungi, and yeasts. Carbon dioxide (CO2), water (H2O), and biomass, which breaks down under aerobic and hydrocarbon conditions; methane (CH4) and biomass are important components of the breakdown process. The use of natural-based polymers to replace petroleum-based plastic is highly logical because petroleum-based plastics are unsustainable and contribute to environmental problems. Synthesised plastics use 65% more energy and emit 30% to 80% more greenhouse gases than bioplastics (Ahvenainen, 2003).

### *Food Packaging and Polymers*

Of the 320 million tonnes of plastic generated annually, bioplastics make up just 1%. However, because of growing consumer demand, increased high-quality biopolymer production, new products, and applications, the market is beginning to expand steadily. The materials that are thought to be exhibiting growth in the biobased industry—that is, biodegradable plastics—are PLA and PHAs (Anon, 2017). The use of bioplastic in place of conventional plastic materials varies depending on the application. Because their positive qualities have been demonstrated, new bioplastics must have qualities that are comparable to those of traditional bioplastics. It is well recognised that bioplastics share many of the same characteristics as regular plastics, with the potential for extra benefits such as lower carbon footprints or alternative waste management choices, such as industrial composting, depending on the material. Evidence was presented by scientific academics and industry players who were investing significant resources to develop high gas barrier bioplastics that might potentially replace traditional fossil-based polymers (Guidotti et al., 2018).

Polysaccharides, lipids, and proteins—which are essentially made from compositional units—are the raw materials used to make biopolymers. Better biopolymer material qualities are required when it comes to being treated, laminated, or made into composites. Additionally, biopolymer materials possess potent antioxidant and antibacterial characteristics and can be derived from edible or active substances (Popovi´c et al., 2018). These and many other materials, including lignocellulose products, straw, pectin, wood, chitosan/chitin, starches, potatoes, cassava, maize, and wheat, can be used to make biodegradable polymers (Averous & Pollet, 2012). Chitin and chitosan are widely available from natural polymers that are derivatives of cellulose. The unique qualities of chitosan, including its biodegradability, nontoxicity, anticoagulant, chelating, antibacterial, antioxidant, and biocompatibility, make it valuable for the creation of bioactive materials (Manigandan et al., 2018). Through the fermentation process, microbes can make biopolymers by altering bioresources such as PHAs and biomass, which can be directly produced from a range of plants (Gruji´c et al., 2017). Corn starch, soybean oil, hemp oil, and other plant-based ingredients can be combined to create bioplastics. The least harmful material to the environment is bagasse. Because it is sturdy and does not break easily, take-out boxes, bowls, plates, and ice cream cups are the intended uses for it. It is also thought to be resistant to heat and moisture and will not suffocate food (Byrd, 2016). However, it is evident that different kinds of wood are used to make various kinds of packaging materials. It is composed of biodegradable and compostable materials.

### *Biopolymers and Active Packaging*

Incorporating additives into packaging materials is a well-known technique for enhancing packaging designs. antibacterial packaging that is mixed into a polymer film that can aid in barrier protection is including the antibacterial ingredients. These contribute to preserving the food's high physical barrier qualities, which raise its quality. Antimicrobial films used for packaging applications have grown more and more in the last few years due to the food business sector's attention needs (Manigandan et al., 2018). From the previous research, it shows that biodegradable polymers composed of polysaccharides and protein-based resources have the potential to be used in the development of packaging solutions that shield food goods from microbial contamination (Kuorwel et al., 2011).

***Utilising Carotenoids***

Carotenoids are naturally occurring compounds that are present in a variety of natural sources. Carotenoids can be produced by photosynthetic organisms such as algae, fungus, bacteria, plants, and cyanobacteria. However, although carotenoids can be obtained by diet and structural modifications, most animals are unable to synthesise them. The pigments that are found in both the plant and animal kingdoms are called carotenoids. These fat-soluble pigments, which are categorised as 700 different compounds, can be produced in a variety of colours, including red, orange, and yellow, and they are found in a variety of organisms. Apart from that, fungi and bacteria that do not use photosynthetic processes can produce carotenoids. Because of their high degree of unsaturation, carotenoids are thought to be susceptible to trans-isomerization when they encountered heat, light, or acids. It is claimed that in their trans-form, carotenoids are stable structures found in nature. Exposure to its stimuli can produce a change in its cis-form, making it less stable and resulting in a slight loss of colour and provitamin activity. Carotenoids are thought to be oxidatively sensitive, and this can occur through both enzymatic and non-enzymatic processes. Excessive heat and exposure to light are the reasons why carotenoids are prone to oxidation. Hydrophobic compounds, such as carotenoids, are recognised for their insolubility in water.

However, it is thought to be lipophilic, meaning that it dissolves readily in lipids and non-polar solvents. While carotenoids are soluble in organic solvents like acetone, chloroform, and alcohol, they are insoluble in water. In nature, carotenoids are divided into two categories: xanthophylls and carotenes. Carotenes, such as β-carotene, are composed of linear hydrocarbons that could undergo cyclization at either or both ends of the molecule. Apart from that, it is known as xanthophylls, and it belongs to the class of oxygenated carotenoid derivatives, which also includes violaxanthin, zeaxanthin, neoxanthin, and lutein. Functional groups that contain oxygen must be present for xanthophylls' chemical structures to be distinguished. Tetraterpenoids (C40) are a class of carotenoids with 40 carbon atoms. They are made up of eight isoprenoid units joined in a symmetrical, linear pattern with the order reversed in the middle. It is possible to exploit and alter this fundamental cyclic structure by cyclization, oxidation, dehydrogenation, and hydrogenation. Because of their conjugated system of double bonds, which may facilitate their simple isomerization and oxidation reactions, these pigments are regarded as having high chemical reactivity.

***Resources for Carotenoids***

Many fruits that are commonly cultivated and consumed contain carotenoids. Citrus fruits, apricots, papayas, mangoes, and peaches are a few examples of well-known foods high in carotenoids. Apart from that, the veggies are regarded as excellent providers of carotenoids (Dias et al., 2018). Carotenoids-rich veggies can also be found in green vegetables such as spinach, carrots, broccoli, tomatoes, and red peppers. Because of their provitamin A activity and antioxidant qualities, these fruits and vegetables can be added to our diets to assist increase our intake of carotenoids, which may have several health benefits (Dias et al., 2018; Zhou et al., 2020). Many plant-based meals contain lutein and zeaxanthin, which are well known for protecting eye health and reducing the risk of age-related macular degeneration. Additionally, lycopene has the strongest antioxidant properties, which can help prevent some cancers and promote cardiovascular health (Delgado-Pelayo et al., 2012; Riedl et al., 2013). Carotenoids are prevalent in nature and can be generated as a natural colourant. It is present in a variety of living things, such as bacteria and plants. These pigments are categorised as biosynthesised in those organisms that support the defence against oxidative stress and can be utilised to sequester light for the purposes of photosynthesis (de Quirós and Costa, 2006).

Carotenoids are employed in a variety of industries, such as chemically synthesised nutritional supplements and industrial pigments. Furthermore, canthaxanthin, which is chemically synthesised and used as a supplement in food and feed, is employed. Moreover, one of the well-known carotenoids with high provitamin A activity that has been artificially produced to meet market demand is b-carotene. As consumers grow more conscious of the provenance and safety of their food, there is a growing focus on natural and environmentally friendly solutions (Sugiura et al., 2006; Deshmukh & Gaikwad, 2022a). Several factors can lead to an increase in the bioproduction of carotenoids. These requirements can all be satisfied by using natural chemicals, inexpensive substrates, and limited production areas. This is because growing conditions can be more tightly controlled with bioproduction, which is also unaffected by environmental elements including season, climate, and soil composition (Mezzomo & Ferreira, 2016). The two groups of microbes that are classed as having carotenoids are those that are non-photosynthetic (bacteria, yeasts, and fungus) and photosynthetic (algae and cynobacteria). Lipophilic pigments known as carotenoids are abundant in nature and can be found in a variety of organisms, including fish, birds, fungi, bacteria, algae, yeasts, plants, and some invertebrate animals, which also include hemipteran (aphids and adelgids) and dipteran (gall midges) insects and mites (Maoka, 2020; Sun et al., 2022; Rodriguez-Concepcion et al., 2018).

### *The Function of Carotenoids in Biology*

Carotenoids are frequently employed as natural colourants and antioxidants in food. Carotenoids, which are high in provitamin A, are typically found in leafy green and yellow-orange meals. These vegetables' provitamin A content is what gives them their colour. Studies both in vitro and in vivo have shown that carotenoids are excellent antioxidants that scavenge and neutralise free radicals. It is claimed that the rich antibacterial activity of carotenoids can combat broad-spectrum infections. Naturally occurring pigments produced by carotenoids are crucial in shielding plants from photooxidative reactions. Through the process of scavenging several reactive oxygen species (ROS), such as singlet oxygen and peroxyl radicals, they demonstrated potential antioxidative action (Young & Lowe, 2018; Srivastava, 2021). Strong antioxidant activity is a property of the bioactive carotenoids, which include lycopene, beta-carotene, zeaxanthin, norbixin, and many more (Squillaci et al., 2017; Gao et al., 2020; Merhan, 2017).

Carotenoids' primary function is to support their double-bond (conjugated) structural groupings through their functional activity. Carotenoids are known for their nutritional and health benefits, as are the bioactive molecules that are included with them. These compounds are derived from natural sources or agroindustrial by-products. It was discovered through epidemiological studies that diets high in carotenoids, which are obtained either directly through fruit and vegetable consumption or indirectly through the consumption of industrial bio-residues. The biological approach was used to demonstrate the functions of carotenoids in human health (Rodriguez-Concepcion et al., 2018; Lemes et al., 2022). These functions include provitamin A, anti-inflammatory, antioxidant, and anti-angiogenic properties, as well as the ability to scavenge free radicals and bolster the antioxidant defence system. Based on recent evidence from studies (Crupi et al. 2023; Elvira- Torales et al. 2019; Britton, 2020), carotenoids have been shown to help prevent and treat several diseases, including age-related macular degeneration and cataracts, reduce the risk of non-alcoholic fatty liver disease, some types of cancer, and cardiovascular disease (CVD). They can also improve cognitive function and help maintain bone health and hypertension. Furthermore, researchers Lu et al. (2022) and Khalil et al. (2021), have expressed concern about the potential health benefits of carotenoids that function best as natural supportive therapeutics. These carotenoids can help protect against COVID-19 infection and combat a high risk of developing severe COVID-19 illness, which includes encountering obesity, hypertension, diabetes mellitus, and cancer. These conditions can be treated and reduced by carotenoids, which may help reflect the carotenoids that can be helpful in easing COVID-19 symptoms.

The European Food Safety Authority (EFSA) scientific panel has demonstrated that dietary modification and the use of nutraceuticals, such as vitamin D, C, A, beta-carotene, and folate, can boost immunity and lower the risk of chronic inflammation (Galm'es et al., 2020). For their beneficial effects, carotenoids must be consumed at the optimal level in the diet. through maintaining a healthy diet, which includes preventing various illnesses. One of the categories of carotenoids is curcumin. It is regarded as a naturally occurring bioactive component that may be isolated from the Zingiberaceae family plant *Curcuma longa* L., sometimes known as turmeric (Araiza-Calahorra et al., 2018). One of the most potent antioxidants, curcumin is well known for its multifaceted therapeutic benefits on a range of chronic illnesses. Carotenoids are commonly employed in the food sector, which is centred on complete development, and are categorised as substances that function best as colouring agents. Widely distributed in nature, carotenoids are mostly found in fruits and vegetables. These foods must have their colour removed to be identified as good sources of these chemicals. All pigments present in the kingdoms of plants and animals are collectively referred to as carotenoids, a general term. These pigments are categorised as fat-soluble pigments and comprise 700 compounds that give rise to the hues red, orange, and yellow. It belongs to the class of carotenoids that contain hydrocarbons, which have two terminal rings and 40 carbon atoms. Two types of carotenoids are found in nature: (a) carotenes, such as $\beta$-carotene, which are made of linear hydrocarbons that are cyclized at one or both ends of the molecule; and (b) xanthophylls, which are the oxygenated derivatives of carotenes and include violaxanthin, lutein, zeaxanthin, and neoxanthin.

## Packaging Films Based on Carotenoid

Widely employed, carotenoids are categorised as excellent natural colourants or pigments. They function best as antioxidants and nutraceutical components that are extracted from various plant and animal parts. Tetraterpene pigments, which include yellow, red, purple, and orange hues, are categorised as carotenoids. Research on carotenoids has drawn interest from academics and industry in recent years, leading to the creation of active and intelligent packaging as well as food additives and preservations (Meléndez-Martínez et al., 2022). This primarily stems from an understanding of how active packaging works, which may promote and lengthen the food's shelf life. It works differently from intelligent packaging, which can track food quality and package environment in real time for customers to check through embedded sensors or indicators on the label. These can be easily detected when there are unusual changes in temperature, pH, moisture content, relative humidity, or bacteria, for example. The packaging label's colour fluctuation, which is so easy for customers to notice, is an example of intelligent packaging (Kalpana et al., 2019). To enable communication between the food inside the package and the customer, intelligent packaging has been employed extensively and with ease. It is made from sources of biopolymers that combine with plant-based active chemicals that are best utilised as pigment components, such as lycopene from tomatoes, curcumin from turmeric, and beta-carotene from carrots (Neves et al., 2022). Because carotenoids have antioxidant activity, a color-changing mechanism, and an antimicrobial function that can be easily identified from pH fluctuations, microbes, and temperature, they are widely exploited for the development of active and intelligent packaging.

The incorporation of carotenoids derived from natural sources into the packaging system is possible because of their intellectual qualities, which can serve as functional properties. The food packaging sector is expanding rapidly, and methods for ensuring food safety and quality throughout storage and transit are always changing. Blanching (Xin et al., 2015), chemical treatment (Tzortzakis & Chrysargyris, 2017), plasma treatment (Giannoglou et al., 2021), UV-irradiation (Turtoi, 2013), active and intelligent packaging films and coatings (Davoodi et al., 2020; Roy, Kim, & Rhim, 2021), and modified atmosphere packaging are just a few examples of the various packaging options that can be utilised to improve the safety and high quality of packaged food products.

According to Mohammadalinejhad et al. (2020) and Firouz et al. (2021) two prominent research topics nowadays are active and intelligent packaging systems, which are at the forefront of cutting-edge technology for the food supply chain. Because it offers natural, high-quality food that is safe, and is produced in accordance with consumer desire, active food packaging is a type of film that is preformed, free-standing, and thin layer that may be developed as functional qualities (Zhang et al., 2021). Active packaging systems function best as antibacterial packaging and may help shield food from microbial contamination, which is widely recognised as the main cause of food spoiling (Davoodi et al., 2020, Priyadarshi et al., 2022). As a UV-blocking material, the active packaging itself functions well in preventing UV-mediated food oxidation (Priyadarshi, Kim, & Rhim, 2021b; Roy & Rhim, 2021). Additionally, it can aid in controlling the respiration rates and the gas exchange between the fresh product and the surroundings, which prolongs ripening and spoiling (Sapper & Chiralt, 2018). In addition, intelligent packaging is increasingly important for applications that demonstrate new technologies that can detect freshness in real time without the need for complex analysis tools, all while meeting consumer demands for safe food (Roy & Rhim, 2021).

In recent years, natural pigments such as curcumin, anthocyanins, betalain, and alizarin have been employed in place of synthetic chemical dyes to create intelligent indicators (Priyadarshi, Ezati, & Rhim, 2021a; Roy & Rhim, 2021). The best natural colour to use as an indicator for the smart packaging film is a range of biodegradable polymers derived from various natural sources, such as proteins, polysaccharides, and lipids. These polymers are chosen for their abundance, biodegradability, non-toxicity, and biocompatibility (Kumar et al., 2020b). When it comes to food packaging and preservation, biopolymer-based food packaging films are expected to replace some plastic packaging films due to their advantages over traditional petroleum-based plastic packaging materials (Jeya Jeevahan et al., 2020). Many polysaccharide-based biopolymers, including chitosan, starch, agar, alginate, carrageenan, cellulose derivatives, and pectin, can be synthesised to create active packaging films (Basumatary et al., 2022). Whey protein, gelatine, casein, and zein are examples of protein-based biopolymers that are often employed (Calva-Estrada et al., 2019). Polyhydroxyalkanoates (PHA), poly (ε-capro lactone) (PCL), poly (lactic acid) (PLA), poly (vinyl alcohol) (PVA), and polyhydroxybutyrate (PHB) are among the several forms of bioplastics that are utilised (Srinivasan & Kulshreshtha, 2020; Zhao, Cornish, & Vodovotz, 2020). However, the absence of physical and functional qualities of biopolymers makes them unsuitable for use in active packaging applications, which limits their utilisation. These days, natural polyphenolic compounds like gallic acid, curcumin, anthocyanins, and catechins are used in the preparation of biopolymer-based active packaging films in order to enhance the biopolymer's physiological and biological activities as well as their potential health benefits. Furthermore, polyphenols are frequently found in natural compounds because of their biological properties, which include antibacterial, anti-inflammatory, antioxidant, and anticancer properties (Yadav et al., 2020).

### *Carotenoids and Human Health*

Foods containing carotenoids may offer a few health advantages. Customers choose to include carotenoids in their daily diet because of their positive effects on human health. The health advantages of carotenoids are numerous and varied.

#### *Zeaxanthin and Lutein*

Zeaxanthin and lutein could filter UV radiation, function as antioxidants, and shield cells from oxidative damage. It is the finest source of lutein and zeaxanthin and is taken from marigold flowers. It is possible to reduce the occurrence of cataracts and age-related macular degeneration (AMD) by ingesting lutein and zeaxanthin. Utilising lutein may help shield the skin from UV rays, which may lower the risk of cardiovascular disease. Based on epidemiological evidence, lutein could prevent the onset of specific cancer types. Lutein can also be utilised to strengthen the immune system. Because lutein and zeaxanthin are important for eye health and can help prevent macular degeneration and cataracts, they can be helpful in the nutraceutical sector. Users can safely utilise these two xanthophylls as they are not regarded as hazardous. They are thought to be pigments that can form in the macular region of the eye. According to the study, individuals with mild cognitive impairment are thought to have lower brain lutein concentrations than people with normal cognitive function (Woodside et al., 2015).

#### *Astaxanthin*

When it comes to human health issues, astaxanthin consumption has been linked to beneficial outcomes. For instance, it can reduce UV radiation damage, function as an anti-inflammatory, and have antioxidant capabilities. Furthermore, it has been demonstrated in human clinical trials to be beneficial for eye health. It can also be used to treat diabetic retinopathy, eye strain and fatigue, macular degeneration, and poor detail vision. Aside from that, it can be used to strengthen the defences against infections by white blood cells, which are the most effective in eliminating pathogens. Since astaxanthin is imposed from white blood cells without having an adverse effect on the white blood cells themselves, it can be utilised to treat and shield human neutrophils and lymphocytes that can fight oxidative stress. Based on epidemiological studies, it appears that having or ingesting carotenoids may help reduce the chance of developing several cancer types. Take cancers such as colon and breast cancer, for instance. The pinkish-coloured astaxanthin is found in aquatic fish and prawns. Compared to vitamin E and beta-carotene, it is demonstrated to be the strongest antioxidant among carotenoids and aids in displaying the strongest antioxidant activity.

*Canthaxanthin*

It has been shown to function as an efficient immunological response, an excellent source of antioxidants, and possibly even improve gap junctional communication between cells that directly produce 4-oxo-retinoic acid, which is utilised to trigger the retinoic acid receptor. Each of these methods could contribute to the compound's anticancer activity. Additionally, it might aid in preventing blood problem disorders. It has also been noted to be dangerous if ingested in excessive amounts (Amaya et al., 2014).

*Capsanthin*

The production of capsanthin is possible through paprika extract, which is derived from red peppers. These peppers have been shown to significantly impact glucose metabolism and the development of low-density lipoprotein receptors in mice. Capsanthin has the potential to increase plasma HDL cholesterol, which is utilised to treat cardiovascular disorders and is associated with a decreased incidence (Amaya et al., 2014).

*Lycopene*

Because of their antioxidant qualities, which can be utilised to treat prostate cancer and cardiovascular illnesses, they are used. According to a prior study, there is no change in the amounts of lutein, beta-carotene, cryptoxanthin, vitamin A, and vitamin E between cancer patients and healthy individuals since the cancer patients had much lower levels of lycopene (Rao & Rao, 2007).

### *Animal-Derived Carotenoids*

#### *Birds*

Animals, particularly birds, can also yield carotenoids, which are produced from their diet. Specifically, birds' diets rich in carotenoids cause them to consume yellow, red, and orange pigments, while some birds may also produce carotenoid pigments from brighter hues (Zakynthinos & Varzakas, 2016). This is because the pigments, which are removed and form the colours yellow, red, and orange, are derived from carotenoids, which are the result of feather colonisation. For instance, the food of flamingos, which contains beta-carotene and canthaxanthin, produces the pink or reddish hue of their feathers, faces, and legs. It extracted feather pigmentation, which is found in several songbirds, such as the yellow warbler and the common yellowthroat, which have yellow plumage colours and create lutein through their diets (Amaya et al., 2014). Additionally, it can effectively accumulate lutein from the diet, which comes from egg yolks. It can also be found in eggs, which display a range of yellow to orange hues in the yolk. Lutein, zeaxanthin, beta-carotene, and canthaxanthin are sources of this pigment. The primary aspect influencing the coloration of birds is their overall health. A major factor in lowering the absorption of carotenoids is illness. Since sick or unhealthy birds are less effective in absorbing and storing carotenoids from their bodies, the colour of their carotenoids is often pale. The sick animals simply used or metabolised the carotenoids that had accumulated to restore their immune systems. When compared to humans, birds can produce a wide range of colours because they can investigate the ultraviolet range, which is where visual stimuli are shown. The taste and smell cues were less important than those for animals. Carotenoids exhibit as fat-soluble substances, indicating that the amount of fat in the diet has an impact on how well they are absorbed. The reason the animals' carotenoids are pale is because they are not absorbing enough of them, which causes fat to be absorbed and cause incorrectly coloured feathers (Amaya et al., 2014).

#### *Crustaceans and Fish*

The reason the salmonids are referred to be "red" fish is that astaxanthin, prominent carotenoids present in their diet of small crustaceans or other fish that have small crustaceans in their digestive systems, is formed by the feed that the salmonids eat. The marine and freshwater creatures, which also include fish and crabs, are responsible for the vibrant colour of the carotenoids. Fish plankton, which fish obtain from other fish that have small crustaceans in their digestive tracts or from eating small crustaceans themselves, contains carotenoids present in crustaceans. Most of a flamingo's diet consists of small creatures like mollusks, prawns, and bug larvae that live in the muck at the bottom of shallow ponds. These organisms are also key sources of carotenoids. Carotenoids are shown their significance in bolstering the immune systems of marine organisms. It has been demonstrated that salmonids exhibit higher levels of astaxanthin and are useful in the fight against bacterial and fungal illnesses. In contrast to crustaceans, astaxanthin exhibited a greater rate of survival. Carotenoids that contribute to the blue or green hue can typically be extracted from crustaceans. According to research on the regular consumption of carotenoids, astaxanthin is what gives fried prawns and other crustaceans their red hue. This is the result of the heat denaturing the carotenoproteins and releasing astaxanthin.

*Poultry*

Egg yolks are derived from poultry and are frequently employed in health-related contexts. They are coloured. The traditional, recognisable farming techniques, where it should be done from the access to grass and corn, are the source of rich carotenoids. As a result, the good pigmentation indicated good physical and practical hygiene conditions. Thus, it has been shown that the bright yellow or golden-orange colour of the egg yolk and grill skin is associated with health and freshness and can also be a sign of high-quality products in many different countries. The animals themselves are unable to synthesise the carotenoids, as the consumer demanded from the sources of xanthophylls. For this reason, carotenoids must be added to the feed to achieve the appropriate colour and to satisfy customer preferences. Poultry coloration can be affected by a variety of causes, such as internal and environmental factors. Before the animal consumes it, it has an external impact on the carotenoids. Under the internal category, it is divided into two parts: the deposited subcutaneous fat and the animal's genetic capacity for metabolism.

### *Use of Carotenoids in Industries*

A few industries, including textile, cosmetics, and food, are becoming more interested in switching to biocolorants from synthetic artificial colouring since they are more stable, less expensive, and less harmful than synthetic colouring. Because of its easier stream processing, higher yields, and potential to lengthen shelf life, pigments from microbial origins are utilised in synthetic dyes. Carotenoids are a class of chemicals with colouring capacity that can also be employed in the food business, where a fully developed market has been shown. Due to customer demand, desire for natural compounds, and growing global trends brought about by the usage of natural ingredients in products, carotenoids can currently be synthesised chemically and used in industrial settings.

### *Carotenoids Incorporated in Packaging*

Because of the added benefits that come with incorporating carotenoids derived from natural sources into the packaging system, the carotenoids are anticipated to be used in the packaging sector. Incorporating these bioactive functional chemicals into the polymeric matrix to facilitate packaging is a significant development problem that continues to play a major role in the packaging industry. For example, solution casting (Chen et al., 2021), mould compression (Oladimeji et al., 2022), melt extrusion (Adnan et al., 2020), photo-grafting (Ren et al., 2021), electro-spinning (Madruga & Kipper, 2022), spraying (Gonçalves et al., 2019), inkjet printing (Li et al., 2021), and many more techniques can be used to produce the packaging for the development of biopolymer packaging films.

#### *Solution-Casting*

Solution casting is one of the earliest effective methods for developing packaging films, and Eastman Kodak has been using it to make plastic films since the 19th century. It is employed in laboratory studies, which can support a flexible procedure for small-scale production of thin films or sheets. This can be accomplished by dissolving the polymers in volatile solvents, either aqueous or non-aqueous, with a range of additives, such as components that are micro- or nano-sized, before they are cast onto a flat surface. After that, it is employed as an evaporation phase using various drying techniques, and the dried thin film is removed from the substrate (Mathew & Oksman, 2014). The functional qualities of the sodium alginate composite can be improved at varying percentages—0.1%, 0.3%, and 0.5%—by supplementing it with lycopene and beta-carotene. Because the composite's tensile qualities were beneficial despite the carotenoids' low concentration, this will not alter the film's carotenoid thickness. A variety of carotenoid-based food packaging films, such as those containing lutein, beta-carotene, xanthophyll, canthaxanthin, and lycopene, can be investigated using solution casting.

#### *Melt Extrusion*

The basic materials are converted into an innovative procedure known as extrusion. The introduction of various factors, such as shear, temperature, and pressure, in the extrusion process induce the raw materials to expand and gelatinize as well as melt at both low and high moisture contents. Extrusion that is released during the expansion process creates the liquid phase that contains the biopolymers and helps the film or packaging container form from its moulded structure (Prabha et al., 2021). It is possible to assist in the development of biodegradable packaging materials with enhanced functionality and other benefits by adding carotenoids into the melt-extrusion process (Sikora et al., 2020). Carotenoids are naturally occurring pigments found in a wide range of fruits and vegetables. They are known for their strong antioxidant activity and capacity to change colour. In this procedure, the carotenoid and biodegradable polymer combination can be placed in an extruder, which melts and blends the ingredients at a set pressure and temperature.

*The Electrospinning Method*

The biomedical and tissue engineering industries utilise electrospinning to process natural bioplastics such as proteins, lipidic forms, and polysaccharides (Cheng et al., 2017). Because of their functional qualities, capacity to hold active agents, and potential for regulated release, biopolymers are used and, ideally, the preferred option for active and intelligent packaging for food applications. Because of the integrated layer that dries out quickly due to its high volatility, the active and intelligent agents derived from natural sources are thermally sensitive and can aid in evaporation. With the use of electrospun fibre encapsulation and this technique, electrospinning can resolve this problem. The food sector uses electrospun fibres in a wide range of applications, including enzyme encapsulation, scaffolding for cell cultures, reinforcement agents for environmentally friendly packaging, mimicking components for artificial foods, vitamins, and antimicrobials (Sameen et al., 2022). When utilised as reinforcing fillers in packaging films, electrospun nanofibers with mechanical capabilities can support biopolymers in enhancing the films' functional qualities.

*Blending-Compositing*

Natural biopolymers have been used extensively in food and pharmaceutical applications for many years. They are particularly useful as thickeners, diluents, gelling agents, and binders for colloidal suspensions. The response of functional groups that aid in potential locations for chemical modification using grafting, such as thiol, hydroxyl, carboxylic acid, and amino groups (Kumar et al., 2020a). One technique that uses monomers to strengthen the polymeric chain is grafting. the application of grafting, which improved the wettability, mechanical characteristics, and biocompatibility. Strategies such as synthetic, radiation, enzymatic, photochemical, plasma-actuated grafting, and imperative for polymer modification are employed to modify the polymer. The polymer that was anchored to the surface of the reactive film was conducted through a ring-opening process to generate the antioxidant film. Numerous studies on food preservation have demonstrated the potential of antioxidant films to extend shelf life and even inhibit the decomposition of ascorbic acid (Sadeghi et al., 2021).

*Inkjet Printing*

With the use of intelligent packaging technologies to increase detection sensitivity or precision and a non-contact method for modifying indicator composition, this is an effective, adaptable, varied, and economical approach for better development (Zhang & Lim, 2016). These techniques rely on both piezoelectric and thermal technologies. The techniques employed are ink ejection, which is created by the pressure from the expanding vapour bubble in thermal inkjet printing, and piezoelectric inkjet, which was developed and involves physically deforming the material through application of a pulsed electric field from the nozzle. Substrates for printing can be applied to paper and polymers based on the convenience and needs of the final user (Bao et al., 2022). It is possible to check the freshness of fresh fillets by utilising inkjet-printed indicators, which can be used to show significant colour fluctuations over time. 'Best-before' labels can be used to determine the exact freshness using colorimetric markers. Food-grade colourants that are naturally occurring can be utilised to reduce toxicity in applications where there is direct food contact (Luo et al., 2021). According to the published analysis, pH-sensitive response labels created using inkjet technology are utilised as QR codes. For this label, natural colourant sources such as curcumin and roselle anthocyanin are created and assessed based on surface tension, contact angle characteristics, and rheological analysis. It is feasible to use it for QR code freshness monitoring, which was identified by a change in ink ratio, to identify freshness monitoring in real-time. As a result, this label is utilised to identify the data carrier and provide complete product information, such as composition, expiration date, manufacturer information, and much more (Xu et al., 2021).

*Coating Deposition*

Coatings are utilised on substrate polymers or thin films that are suitable for continuous layer deposition. The coatings that are employed perform best when they serve as a functional layer or a layer into which active chemicals can be added. Among the many well-known methods are screen-printing coating, dipping, spraying, casting, physical vapour deposition, and chemical vapour deposition. The technique of creating a coating on a thin layer by aiding the blasting of active chemicals with the use of a high-pressure nozzle is referred to as spray coating, or deposition. The thin film is submerged in the coating solution using dip coating, and it must then be allowed to dry. Cast coatings that include active ingredients must be dissolved in an appropriate solvent until the coating has cured, the film surface is easily torn, and the solvents are entirely gone. It alludes to the substrates and adsorbing agents that make up its chemical and physical composition (Fu & Dudley, 2021). Polyvinyl alcohol (PVA) that can be directly coated with polyhydroxybutyrate-co-valerate (PHB92/PHV8) film or encapsulated with beta-carotene is formulated using soy protein.

## Application of Carotenoids in Food Packaging Incorporated

The most widely used natural pigments in food packaging applications are carotenoids, such as β-carotene, lutein, β-cryptoxanthin, α-carotene, astaxanthin, canthaxanthin, zeaxanthin, and β-apo-8-carotenal. The qualities and attributes of the meals can be enhanced by using all these natural pigments that are utilised to identify freshness indications. To ensure that carotenoids are stable enough to create a colour shift, the pH value is employed. An investigation into the creation of intelligent and active biodegradable packaging films derived from cassava starch sources combined with various extracts of basil and green tea has been conducted by Medin-Jaramillo. The findings demonstrated that the colour shift caused by the tea and basil extracts, which are utilised to alter the packaging films' hue, also serves as a food quality indicator (Medina-Jaramillo et al., 2017).

In the food packaging industry, β-carotene, and other carotenoids—such as alpha-tocopherol—are utilised extensively. This is because meals high in fat include pigments applied to them, and because carotenoids are lipophilic, a fatty diet increases the number of pigments that are adsorbed. According to study by Asdagh and Pirsa, using beta-carotene-based films that are made based on pectin, nanoclay, and carum optimum essential oils can help change the colour of butter. The butter was observed to have changed colour from orange to pale yellow. This colour shift was caused by oxidative changes in the butter mixture. Additionally, Assis et al.'s study demonstrates that they produced a biodegradable cassava starch film that is combined with β-carotene nanocapsules (Assis et al., 2018). According to the research, adding beta-carotene may improve the films' thermal stability and slow down the pace at which sunflower oil oxidises (Asdagh & Pirsa, 2020). The utilisation of low-density polyethylene-ethylene vinyl alcohol-polyethylene terephthalate (LDPE/EVOH/PET)-based films combined with beta-carotene was established, according to a prior study (Juan-Polo et al., 2022). It has been demonstrated that adding beta-carotene may assist alter the films' thermal stability and raise their oxygen stability. However, it will turn orange and take longer to oxidise if it is not used. The usage of that packaging may have contributed to its antioxidative and oxygen-scavenging qualities, which allowed peanuts to be packaged in the films for around three months at 40°C. However, it was observed that the hexanal content remained constant, although being higher in the controlled films. Apart from that, the study by Tupuna-Yerovi et al. 2023, revealed that sodium alginate was used as one of the polymer matrices in food packaging films.

Lycopene and beta-carotene, which are utilised to enhance the capacity for both thermal stability and light transmission, were added. When β-carotene and lycopene from sunflower oil were isolated and employed, it was discovered that their oxidative effects helped the films containing these pigments and that their rate of oxidation was lower than that of the plastic packaging and control films (Tupuna-Yerovi et al., 2023). One of the naturally occurring carotenoid pigments that can be derived from annatto seeds, which are rich in colouring and antioxidant qualities, is the other pigment that was taken from bixin. According to studies by Stoll et al. 2021, active polylactic acid-based packaging films are integrated and function optimally as colouring agents. It is demonstrated that the film's UV-blocking qualities function best when used in conjunction with 90% of it. According to the other study, annatto is one of the colouring agents used in biodegradable-coloured films made of poly(3-hydroxybutyrate) (PHA). It is evident that they established the worth of the films' photodegradation by testing them in UV-A light (Pagnan et al., 2018). Other than that, research from Pagno et al. has demonstrated that films based on cassava starch that are biodegradable and active packaging have been explored for the application of bixin in nanoencapsulation (Pagno et al., 2016). It is demonstrated that employing a larger concentration of bixin nanocapsules can aid in strengthening the mechanical characteristics of the film as well as its ability to block ultraviolet light. Sunflower oil was utilised to decide how to package the produced films, which are high in antioxidant characteristics and can aid to improve the quality of the sunflower oil by lowering the oxidation rate.

Apart from that, the best antibacterial and antioxidant qualities of the films were found in lycopene, a form of carotenoid pigment that can be derived from tomatoes and is thought to be unsaturated and acyclic. Aside from that, lycopene was integrated with the PVA, which was based on gelatine and might aid in combating Bacillus cereus and Staphylococcus aureus. It works well with chicken meat packaging films, and because of the films' antioxidant qualities, it may help extend the shelf life of the chicken meat (Kanatt et al., 2017). The increased tensile strength and light barrier qualities can be enhanced with the application of lycopene. It has been demonstrated to have the potential to be employed in preventing oxidation (Assis et al., 2017). It has been demonstrated that the use of carotenoids of various varieties can aid in their use as colourants, antioxidants, antimicrobial agents, and contribute to UV blocking in the packaging films. These functions not only aim to improve the quality of the food product but also extend its shelf life and enhance food safety and security. We may conclude that the food packaging sector can grow with the use of carotenoids.

### *The Effects of Carotenoid-Based Food Packaging on the Environment*

When synthetic colourants are applied, they have a harmful migratory effect on the packaged product and can hurt consumers if they are exposed to high pressure or temperature. Because synthetic dyes are more toxic, allergenic, and carcinogenic than natural colourants, their use can have serious negative effects and should be avoided in the food industry. Hazardous solvents can also contribute to these negative effects. Furthermore, colorimetric sensor technology is ready for a big industrial application but is not widely available for industrial use. Customers' perceptions, acceptance, and decision-making have been significantly influenced by the production sources, processes, and incorporation techniques that are in line with the evolving carotenoids-containing films. Smart packaging, also known as active and intelligent systems, is gaining popularity among consumers all over the world. It is used to biopolymers and uses colorimetric indicators and sensors to detect and display natural colourants such as pH indicators. Apart from that, maintaining the freshness of the food product requires attention to quality and freshness to extend the shelf life of the food goods. to prevent the environmental problems associated with traditional packaging and artificial colouring, and to inform customers on the quality and conditions of food goods in real time.

The food industry may benefit from a good strategy that reduces food waste, spoilage, food-borne illnesses, and product degradation by combining natural colours with nanoparticles and using their chemical reactive properties in active and smart packaging. If the packaging or film colour changes, this is a very efficient and rapid approach to find out in real time how safe, fresh, and high-quality the packed goods are. As an illustration, consider the application of intelligent packaging, which helps to show the pH fluctuations in food that has been completely wrapped and can act as a visual pH sensor label.

Natural pigments based on carotenoids are ideally suited for creating colour-changing indications for active and intelligent packaging. This is because carotenoids exhibit outstanding antibacterial and antioxidant capabilities. Carotenoids can be incorporated into the packaging matrix and are extracted from natural sources. The packaging films based on carotenoid technology exhibit strong antibacterial activity, great pH-responsive colour shift, good antioxidant activity, and colour stability. By reducing the growth of bacteria and other microorganisms and the oxidation of food lipids, the addition of carotenoids can aid in active packaging and enhance the food that is packed. Real-time assistance with food product freshness can be provided by carotenoids. In addition to being naturally occurring colourants and possessing colour stability, carotenoid-based pigments are soluble in fat-soluble solvents. However, this can occasionally lead to application limitations.

An alternative approach to address this problem is to combine the carotenoid pigments with nanoencapsulation, which has been applied in food applications. Natural pigments have a lot of potential for use in intelligent packaging, but further research is needed to ensure colour stability. These pigments can also be effectively employed in bio-derived polymers, a variety of food products, and to boost pigment manufacturing. The use of carotenoids in active and intelligent packaging has the potential to be an innovative technology that helps improve the safety, sustainability, and quality of food packaging. This method involves the combined usage of carotenoids, which have the power to alter colour in response to environmental stimuli and can be strengthened by potent antioxidants. Because colour changes are a result of temperature, microbial activity, and oxygen levels, they can help reduce the risk of food spoilage, waste, and oxidation by providing producers and consumers with an easily understood visual cue. Technology may assist individuals in boosting food sustainability and safety in the market as well as helping them make better judgements about the foods they buy and consume.

### *Biotechnology and Carotenoids*

Because they contain vital nutrients and significant antioxidants that are found in human diets, carotenoids are beneficial to human health, providing consumers with nutritional value and overall wellness. The main staple crops are only trace to low in carotenoids among the food crops. Although efforts have been undertaken to modify the expression of carotenoids in biosynthetic genes to generate carotenoids-rich food crops, many of these efforts have not resulted in the necessary levels of carotenoids enrichment (Botella-Pavia P & Rodriguez-Concepcion, 2006). Food crops' carotenoids can be altered, and this can be utilised to engineer catalytic activity for the system by changing the expression of carotenoids through biosynthetic genes. There are situations where the levels of carotenoids needed to maintain optimal human nutrition and health are too high to allow the contents to fully mature. Enhancing the amounts of carotenoids in food crops is one alternative strategy that relies on the creation of metabolic sinks to efficiently sequester carotenoids (Li et al., 2007). The sinks' ability to pull the pathway towards completion aids in their beneficial influence on the accumulation of carotenoids combined with the assistance of adjusting the pathway's catalytic activity. This could aid in demonstrating a practical and successful method for increasing the number of carotenoids in food crops' low-pigmented tissues.

Through enhancing the accumulation of carotenoids associated with the development of sink structures in transgenic crops, a novel strategy for augmenting carotenoids content is being presented. Crops will aid in the development of a route that can be developed via the manipulation of the formations produced from the metabolic sink in addition to catalytic activity to improve the nutritional quality of food (Li et al., 2007). On the other hand, the application of biotechnology to the subject of carotenoids involves manipulating microorganisms to play a role in the creation of carotene. Several benefits can be obtained from producing natural colourants through fermentation, including reduced production costs, ease of extraction, increased yields, reduced batch-to-batch variability, no resource shortage, and no seasonal variations (Mapari et al., 2010; Mogedas et al., 2009).

### *Carotenoids in the Food Sector*

As is always the case, the addition of carotenoids to food does not significantly increase their antioxidant content (Haila et al., 1996). Because carotenoids have a high melting point and become crystalline at body temperature and during food storage, using them as ingredients in functional foods is presenting issues. Stable carotenoids are those found naturally in food. Although carotenoids are stable in food systems due to their susceptibility to light, oxygen, and autooxidation, they can be employed as food additives (Xianquan et al., 2005). Moreover, carotenoids may degrade quickly because of their dispersion into component systems (Ribeiro et al., 2003). Carotenoids can influence molecule-splitting or double-bond loss processes that lead to molecular breakdown. Furthermore, carotenoids' double bonds can isomerize into the cis conformation (Xianquan et al., 2005). The cis isomers of carotenoids, such as lycopene, which are more accessible and bioactive, benefit from isomerization reactions (Schieber & Carle, 2005).

Natural pigments called carotenoids are frequently found in a variety of foods and beverages. Because of their highly conjugated structure, carotenoids are extremely unstable and readily break down when exposed to light or air during food manufacturing or storage. It can reveal the reason behind the nutritional and biologically beneficial qualities being lost, as well as why unwanted flavour or scent compounds are being produced. Currently, the most popular technique for encapsulating lipophilic functional components—like flavours and bioactive lipids—is the use of straightforward oil-in-water emulsions. One kind of defence against degenerative diseases such as cancer, cataracts, and cardiovascular disease is provided by carotenoids (Halliwell, 1996; Davies, 1995). The economics of carotenoids. Carotenoids are used in the pharmaceutical, animal feed additive, and nutraceuticals industries. They are most effective when used as food colourants in foods and cosmetics. A few types of carotenoids, such as beta-carotene, lutein, capsanthin, astaxanthin, canthaxanthin, annatto, zeaxanthin, beta-apo-8-carotenal, lycopene, and beta-apo-8-carotenal-ester, are available in the global market. In 2010, the global market for carotenoids was close to $1.2 billion. The value climbed to $1.4 billion in 2018 at a compound annual growth rate (CAGR) of 2.3% over the course of eight years. However, over the last four years, the market for carotenoids has changed significantly for some of the items that are classified. Certain products have seen price reductions, while the market worth of others has increased.

The most common carotenoid is beta-carotene. There is an excess of beta-carotene available in the market. The market is constantly dominated by synthetic beta-carotene, with hopes that a fermentation process derived from beta-carotene can capture a larger market share. However, the market is not fully filled with beta-carotene derived from algae, which is virtually disappearing from the market. However, there has been a surge in the traditional supplement's intake. However, several Asian vendors that supply tomato-based lycopene extracts at acceptable quality levels have since lowered their prices. The ingestion of conventional astaxanthin, which is derived from fish and prawns, contributes to the expansion of the aquaculture business. Furthermore, astaxanthin, one of the other carotenoids, may be made from algae. This substance is in high demand right now for food supplements, and its cost is rising quickly. In terms of technology, advancements in the carotenoid sector—where microorganisms that can create astaxanthin in the dark are produced—have occurred at a rate that has not been observed in the previous ten years.

## Conclusion

Carotenoids are a class of pigments that can be grouped into bioactive compounds that are of interest to nutritionists, food scientists, and the food industry. These compounds have been shown to have a positive impact on human health and are currently being studied to determine how human chronic diseases are affected. Carotenoids, which can be derived from fruits and vegetables and are utilised in colour to draw in customers, have a variety of biological activities and functions. The capacity of carotenoids to scavenge singlet oxygen and free radicals is related to their antioxidant activity. The carotenoids enhance our eating experiences and styles by adding colour, particularly to food goods. It demonstrated a reduction in oxidative stress as well as a strong supply of vitamin A need. A balanced diet can incorporate dark green vegetables, particularly those derived from leafy greens, to further enhance health advantages. Plants play a major role in providing sources of carotenoids since they contain vital components for their growth and survival. It has been demonstrated that both plants and mammals can generate carotenoids. Carotenoids are categorised as an optimal dietary intake that plays a significant role in preventing chronic diseases, which include viruses like coronavirus, age-related macular degeneration, cataracts, and cardiovascular and other types of cancer. Aside from that, carotenoids play crucial roles in human nutrition and health through signalling, photoprotection—a detoxification process against free radicals produced during photosynthesis—and photosynthesis, which is thought of as light harvesting. They also serve as precursors to plant hormones. According to the studies, thermal processing can also have an impact on the use of carotenoids.

This is because it can increase the antioxidant activity, palatability, retention percentage, and bio accessibility of carotenoids by rupturing cell structures and releasing the pigments from the matrix. To that end, additional research on factors that may influence the levels of carotenoids is necessary to develop management strategies that will help produce seed crops with the targeted concentrations needed to meet the needs of the pharmaceutical, nutraceutical, and functional food industries. This generation's synthetic polymers are made from petrochemicals, which are thought to be nondegradable. The environmental impact of non-renewable plastics can be mitigated by utilising or modifying biodegradable polymers, which can lessen their toxicity. Although it was demonstrated to be very hard to produce conventional polymers using environmentally benign resources, the concept must be developed and used in specific applications, particularly food packaging. The natural resources, such as wood pulps, animal or plant proteins, polysaccharides, and many more, are typically the sources used to create environmentally friendly packaging materials. Further research and analysis are necessary, but the prospect of replacing standard plastics with bioplastics appears intriguing. Therefore, to build the future applications for food packaging, it is necessary to examine, evaluate, and comprehend the characteristics, structures, and behaviours of the latest technology. One possible approach to promoting a more environmentally friendly future is to substitute the materials used in packaging with more eco-friendly options.

## References

Adnan, M., Obyedul Kalam Azad, M., Seok Ju, H., Min Son, J., Ho Park, C., Hwan Shin, M., Alle, M., Ha Cho, & D. (2020). Development of biopolymer-mediated nanocomposites using hot-melt extrusion to enhance the bio-accessibility and antioxidant capacity of kenaf seed flour. *Applied Nanoscience, 10*, 1305–1317. https://doi.org/10.1007/s13204-020-01273-7.

Ahmed, M., Bose, I., Goksen, G., & Roy, S. (2023). Himalayan sources of anthocyanins and its multifunctional applications: A review. *Foods, 12*, 2203. https://doi.org/10.3390/foods12082203.

Ahvenainen, R. (Ed.). (2003). *Novel food packaging techniques*. Elsevier: Amsterdam.

Amaya, E., Becquet, P., Carné, S., Peris, S., & Miralles, P. (2014). Carotenoids in animal nutrition. *Fefana Publication, 169*, 170-171.

Amjadi, S., Nazari, M., Alizadeh, S. A., & Hamishehkar, H. (2020). Multifunctional betanin nanoliposomes-incorporated gelatin/chitosan nanofiber/ZnO nanoparticles nanocomposite film for fresh beef preservation. *Meat Science, 167*, 108161. https://doi.org/10.1016/j.meatsci.2020.108161.

Anon. (2017). Bioplastics market data 2017, Global production capacities of bioplastics 2017-2022. *European Bioplastics Report.*

Araiza-Calahorra, A., Akhtar, M., & Sarkar, A. (2018). Recent advances in emulsion-based delivery approaches for curcumin: From encapsulation to bioaccessibility. *Trends in Food Science & Technology, 71*, 155-169. https://doi.org/10.1016/j.tifs.2017.11.004

Asdagh, A., & Pirsa, S. (2020). Bacterial and oxidative control of local butter with smart/active film based on pectin/nanoclay/Carum copticum essential oils/β-carotene. *International Journal of Biological Macromolecules, 165*, 156-168. https://doi.org/10.1016/j.ijbiomac.2020.07.209.

Ashokkumar, V., Flora, G., Sevanan, M., Sripriya, R., Chen, W. H., Park, J. H., & Kumar, G. (2023). Technological advances in the production of carotenoids and their applications–A critical review. *Bioresource Technology, 367*, 128215. https://doi.org/10.1016/j.biortech.2023.128215.

Assis, R. Q., Lopes, S. M., Costa, T. M. H., Flôres, S. H., & de Oliveira Rios, A. (2017). Active biodegradable cassava starch films incorporated lycopene nanocapsules. *Industrial Crops and Products, 109*, 818-827. https://doi.org/10.1016/j.indcrop.2017.09.004.

Assis, R. Q., Pagno, C. H., Costa, T. M. H., Flôres, S. H., & Rios, A. D. O. (2018). Synthesis of biodegradable films based on cassava starch containing free and nanoencapsulated β-carotene. *Packaging Technology and Science, 31*(3), 157-166. https://doi.org/10.1002/pts.2387.

Assis, R. Q., Pagno, C. H., Stoll, L., Rios, P. D. A., de Oliveira Rios, A., & Olivera, F. C. (2021). Active food packaging of cellulose acetate: Storage stability, protective effect on oxidation of riboflavin and release in food simulants. *Food Chemistry, 349*, 129140. https://doi.org/10.1016/j.foodchem.2020.129140.

Avérous, L., & Pollet, E. (2012). *Environmental silicate nano-biocomposites* (Vol. 1). Springer.

Balbinot-Alfaro, E., Craveiro, D. V., Lima, K. O., Costa, H. L. G., Lopes, D. R., & Prentice, C. (2019). Intelligent packaging with pH indicator potential. *Food Engineering Reviews, 11*(4), 235-244. https://doi.org/10.1007/s12393-019-09126-0.

Bao, F., Liang, Z., Deng, J., Lin, Q., Li, W., Peng, Q., & Fang, Y. (2024). Toward intelligent food packaging of biosensor and film substrate for monitoring foodborne microorganisms: A review of recent advancements. *Critical Reviews in Food Science and Nutrition, 64*(12), 3920-3931. https://doi.org/10.1080/10408398.2023.2207872.

Basumatary, I. B., Mukherjee, A., Katiyar, V., & Kumar, S. (2022). Biopolymer-based nanocomposite films and coatings: Recent advances in shelf-life improvement of fruits and vegetables. *Critical Reviews in Food Science and Nutrition, 62*(7), 1912-1935. https://doi.org/10.1080/10408398.2022.2023151.

Bhatt, T., & Patel, K. (2020). Carotenoids: Potent to prevent diseases review. *Natural Products and Bioprospecting, 10*(3), 109-117. https://doi.org/10.1007/s13659-020-00239-z.

Botella-Pavía, P., & Rodríguez-Concepción, M. (2006). Carotenoid biotechnology in plants for nutritionally improved foods. *Physiologia Plantarum, 126*(3), 369-381. https://doi.org/10.1111/j.1399-3054.2006.00787.x.

Britton, G. (2020). Carotenoid research: History and new perspectives for chemistry in biological systems. *Biochimica et Biophysica Acta (BBA)-Molecular and Cell Biology of Lipids, 1865*(11), 158699. https://doi.org/10.1016/j.bbalip.2020.158699.

Burke, W., Baer, L., & Pietruszynski, J. (2012). *The big book of packaging*. Harper Collins: New York.

Byrd, A. (2016). *Plant-based packaging materials*. USA: Christian Brothers University.

Calva-Estrada, S. J., Jiménez-Fernández, M., & Lugo-Cervantes, E. (2019). Protein-based films: Advances in the development of biomaterials applicable to food packaging. *Food Engineering Reviews, 11*(2), 78-92. https://doi.org/10.1007/s12393-019-09117-1.

Chen, P., Xie, F., & McNally, T. (2021). Understanding the effects of montmorillonite and sepiolite on the properties of solution-cast chitosan and chitosan/silk peptide composite films. *Polymer International, 70*(5), 527-535. https://doi.org/10.1002/pi.6025.

Cheng, J., Jun, Y., Qin, J., & Lee, S. H. (2017). Electrospinning versus microfluidic spinning of functional fibers for biomedical applications. *Biomaterials, 114*, 121-143. https://doi.org/10.1016/j.biomaterials.2016.10.052.

Crupi, P., Faienza, M. F., Naeem, M. Y., Corbo, F., Clodoveo, M. L., & Muraglia, M. (2023). Overview of the potential beneficial effects of carotenoids on consumer health and well-being. *Antioxidants, 12*(5), 1069. https://doi.org/10.3390/antiox12051069.

Dainelli, D., Gontard, N., Spyropoulos, D., Zondervan-van den Beuken, E., & Tobback, P. (2008). Active and intelligent food packaging: Legal aspects and safety concerns. *Trends in Food Science & Technology, 19*, S103-S112. https://doi.org/10.1016/j.tifs.2007.12.002.

Davies, K. J. (1995, November). Oxidative stress: The paradox of aerobic life. In *Biochemical Society Symposia* (Vol. 61, pp. 1-31). Portland Press Limited: London.

Davoodi, S., Davachi, S. M., Ghorbani Golkhajeh, A., Shekarabi, A. S., & Abbaspourrad, A. (2020). Development and characterization of Salvia macrosiphon/Chitosan edible films. *ACS Sustainable Chemistry & Engineering, 8*(3), 1487-1496. https://doi.org/10.1021/acssuschemeng.9b06227.

de Oliveira Filho, J. G., de Sousa, T. L., Bertolo, M. R. V., Junior, S. B., Mattoso, L. H. C., Pimentel, T. C., & Egea, M. B. (2023). Next-generation food packaging: Edible bioactive films with alginate, mangaba pulp (*Hancornia speciosa*), and *Saccharomyces boulardii*. *Food Bioscience, 54*, 102799. https://doi.org/10.1016/j.fbio.2023.102799.

de Quirós, A. R. B., & Costa, H. S. (2006). Analysis of carotenoids in vegetable and plasma samples: A review. *Journal of Food Composition and Analysis, 19*(2-3), 97-111. https://doi.org/10.1016/j.jfca.2005.09.003.

Delgado-Pelayo, R., & Hornero-Méndez, D. (2012). Identification and quantitative analysis of carotenoids and their esters from *sarsaparilla* (*Smilax aspera L.*) berries. *Journal of Agricultural and Food Chemistry, 60*(33), 8225-8232. https://doi.org/10.1021/jf302179g.

Deshmukh, R. K., & Gaikwad, K. K. (2024). Natural antimicrobial and antioxidant compounds for active food packaging applications. *Biomass Conversion and Biorefinery, 14*(4), 4419-4440. https://doi.org/10.1007/s13399-024-01743-9.

Deshmukh, R. K., Akhila, K., Ramakanth, D., & Gaikwad, K. K. (2022). Guar gum/carboxymethyl cellulose-based antioxidant film incorporated with halloysite nanotubes and litchi shell waste extract for active packaging. *International Journal of Biological Macromolecules, 201*, 1-13. https://doi.org/10.1016/j.ijbiomac.2022.03.206.

Deshmukh, R. K., Hakim, L., & Gaikwad, K. K. (2023b). Active packaging materials. *Current Food Science and Technology Reports, 1*(2), 123-132. https://doi.org/10.1007/s40940-023-00217-1.

Deshmukh, R. K., Hakim, L., Akhila, K., Ramakanth, D., & Gaikwad, K. K. (2023a). Nanoclay composites for food packaging applications. *International Nano Letters, 13*(2), 131-153. https://doi.org/10.1007/s40089-023-00461-z.

Deshmukh, R. K., Kunam, P. K., & Gaikwad, K. K. (2023). Guar gum-based flexible packaging material with an active surface reinforced by litchi shell-derived microfibrillated cellulose and halloysite nanotubes. *Sustainable Chemistry and Pharmacy, 36*, 101302. https://doi.org/10.1016/j.scp.2023.101302.

Dias, M. G., Olmedilla-Alonso, B., Hornero-Méndez, D., Mercadante, A. Z., Osorio, C., Vargas-Murga, L., & Meléndez-Martínez, A. J. (2018). Comprehensive database of carotenoid contents in Ibero-American foods. A valuable tool in the context of functional foods and the establishment of recommended intakes of bioactives. *Journal of Agricultural and Food Chemistry, 66*(20), 5055-5107. https://doi.org/10.1021/acs.jafc.8b02644.

Elvira-Torales, L. I., García-Alonso, J., & Periago-Castón, M. J. (2019). Nutritional importance of carotenoids and their effect on liver health: A review. *Antioxidants, 8*(7), 229. https://doi.org/10.3390/antiox8070229.

Ezati, P., Tajik, H., Moradi, M., & Molaei, R. (2019). Intelligent pH-sensitive indicator based on starch-cellulose and alizarin dye to track freshness of rainbow trout fillet. *International Journal of Biological Macromolecules, 132*, 157-165. https://doi.org/10.1016/j.ijbiomac.2019.03.101.

Firouz, M. S., Mohi-Alden, K., & Omid, M. (2021). A critical review on intelligent and active packaging in the food industry: Research and development. *Food Research International, 141*, 110113. https://doi.org/10.1016/j.foodres.2020.110113.

Fu, Y., & Dudley, E. G. (2021). Antimicrobial-coated films as food packaging: A review. *Comprehensive Reviews in Food Science and Food Safety, 20*(4), 3404-3437. https://doi.org/10.1111/1541-4337.12704.

Galmán Graíño, S., Sendón, R., Lopez Hernandez, J., & Rodriguez-Bernaldo de Quiros, A. (2018). GC-MS screening analysis for the identification of potential migrants in plastic and paper-based candy wrappers. *Polymers, 10*(7), 802. https://doi.org/10.3390/polym10070802.

Galmés, S., Serra, F., & Palou, A. (2020). Current state of evidence: Influence of nutritional and nutrigenetic factors on immunity in the COVID-19 pandemic framework. *Nutrients, 12*(9), 2738. https://doi.org/10.3390/nu12092738.

Gao, Y., Focsan, A. L., & Kispert, L. D. (2020). The effect of polarity of environment on the antioxidant activity of carotenoids. *Chemical Physics Letters, 761*, 138098. https://doi.org/10.1016/j.cplett.2020.138098.

Giannoglou, M., Xanthou, Z. M., Chanioti, S., Stergiou, P., Christopoulos, M., Dimitrakellis, P., ... & Katsaros, G. (2021). Effect of cold atmospheric plasma and pulsed electromagnetic fields on strawberry quality and shelf-life. *Innovative Food Science & Emerging Technologies, 68*, 102631. https://doi.org/10.1016/j.ifset.2021.102631.

Gonçalves, A., Estevinho, B. N., & Rocha, F. (2019). Characterization of biopolymer-based systems obtained by spray-drying for retinoic acid controlled delivery. *Powder Technology, 345*, 758-765. https://doi.org/10.1016/j.powtec.2018.12.027.

Grujić, R., Vujadinović, D., & Savanović, D. (2017). Biopolymers as food packaging materials. In *Advances in Applications of Industrial Biomaterials* (pp. 139-160). Elsevier: Amsterdam.

Grzebieniarz, W., Biswas, D., Roy, S., & Jamróz, E. (2023). Advances in biopolymer-based multi-layer film preparations and food packaging applications. *Food Packaging and Shelf Life, 35*, 101033. https://doi.org/10.1016/j.fpsl.2023.101033.

Guidotti, G., Soccio, M., Lotti, N., Gazzano, M., Siracusa, V., & Munari, A. (2018). Poly (propylene 2, 5-thiophenedicarboxylate) vs. Poly (propylene 2, 5-furandicarboxylate): Two examples of high gas barrier bio-based polyesters. *Polymers, 10*(7), 785. https://doi.org/10.3390/polym10070785.

Haila, K. M., Lievonen, S. M., & Heinonen, M. I. (1996). Effects of lutein, lycopene, annatto, and γ-tocopherol on autoxidation of triglycerides. *Journal of Agricultural and Food Chemistry, 44*(8), 2096-2100. https://doi.org/10.1021/jf9503560.

Halliwell, B. (1996). Commentary oxidative stress, nutrition, and health. Experimental strategies for optimization of nutritional antioxidant intake in humans. *Free Radical Research, 25*(1), 57-74. https://doi.org/10.3109/10715769609060756.

He, X., Pu, Y., Chen, L., Jiang, H., Xu, Y., Cao, J., & Jiang, W. (2023). A comprehensive review of intelligent packaging for fruits and vegetables: Target responders, classification, applications, and future challenges. *Comprehensive Reviews in Food Science and Food Safety, 22*(2), 842-881. https://doi.org/10.1111/1541-4337.12974.

Housewirth, J. (2017). *The importance of plastic polymers in packaging*. USA: Christian Brothers University.

Idrovo Encalada, A. M., De'Nobili, M. D., Ponce, A. N. M., Stortz, C. A., Fissore, E. N., & Rojas, A. M. (2021). Antioxidant edible film based on a carrot pectin-enriched fraction as an active packaging of a vegan cashew ripened cheese. *International Journal of Food Science & Technology, 56*(8), 3691-3702. https://doi.org/10.1111/ijfs.14602.

Jeevahan, J. J., Chandrasekaran, M., Venkatesan, S. P., Sriram, V., Joseph, G. B., Mageshwaran, G., & Durairaj, R. B. (2020). Scaling up difficulties and commercial aspects of edible films for food packaging: A review. *Trends in Food Science & Technology, 100*, 210-222. https://doi.org/10.1016/j.tifs.2020.03.006.

Juan-Polo, A., Maestre Pérez, S. E., Monedero Prieto, M., Sánchez Reig, C., Tone, A. M., Herranz Solana, N., & Beltrán Sanahuja, A. (2022). Oxygen scavenger and antioxidant LDPE/EVOH/PET-based films containing β-carotene intended for fried peanuts (*Arachis hypogaea L.*) packaging: Pilot scale processing and validation studies. *Polymers, 14*(17), 3550. https://doi.org/10.3390/polym14173550.

Kalpana, S., Priyadarshini, S. R., Leena, M. M., Moses, J. A., & Anandharamakrishnan, C. (2019). Intelligent packaging: Trends and applications in food systems. *Trends in Food Science & Technology, 93*, 145-157. https://doi.org/10.1016/j.tifs.2019.01.011.

Kannat, S., Jethwa, T., Sawant, K., & Chawla, S. (2017). PVA-Gelatin films incorporated with tomato pulp: A potential primary food packaging film. *International Journal of Current Microbiology and Applied Sciences, 6*, 1428-1441. https://doi.org/10.20546/ijcmas.2017.603.146.

Khalil, A., Tazeddinova, D., Aljoumaa, K., Kazhmukhanbetkyzy, Z. A., Orazov, A., & Toshev, A. D. (2021). Carotenoids: Therapeutic strategy in the battle against viral emerging diseases, COVID-19: An overview. *Preventive Nutrition and Food Science, 26*(3), 241. https://doi.org/10.3746/pnfs.2021.26.3.241

Kumar, D., Gihar, S., Shrivash, M. K., Kumar, P., & Kundu, P. P. (2020a). A review on the synthesis of graft copolymers of chitosan and their potential applications. *International Journal of Biological Macromolecules, 163*, 2097-2112. https://doi.org/10.1016/j.ijbiomac.2020.06.151.

Kumar, J., Akhila, K., Kumar, P., Deshmukh, R. K., & Gaikwad, K. K. (2023). Novel temperature-sensitive label based on thermochromic ink for hot food packaging and serving applications. *Journal of Thermal Analysis and Calorimetry, 148*(13), 6061-6069. https://doi.org/10.1007/s10973-023-11600-4.

Kumar, S., Mukherjee, A., & Dutta, J. (2020b). Chitosan-based nanocomposite films and coatings: Emerging antimicrobial food packaging alternatives. *Trends in Food Science & Technology, 97*, 196-209. https://doi.org/10.1016/j.tifs.2020.01.013.

Kuorwel, K. K., Cran, M. J., Sonneveld, K., Miltz, J., & Bigger, S. W. (2011). Antimicrobial activity of biodegradable polysaccharide and protein-based films containing active agents. *Journal of Food Science, 76*(3), R90-R102. https://doi.org/10.1111/j.1750-3841.2011.02216.x.

Lee, H., Kim, M. S., Lee, W. H., & Cho, B. K. (2018). Determination of the total volatile basic nitrogen (TVB-N) content in pork meat using hyperspectral fluorescence imaging. *Sensors and Actuators B: Chemical, 259*, 532-539. https://doi.org/10.1016/j.snb.2018.01.030.

Lemes, A. C., Egea, M. B., Oliveira Filho, J. G. D., Gautério, G. V., Ribeiro, B. D., & Coelho, M. A. Z. (2022). Biological approaches for extraction of bioactive compounds from agro-industrial by-products: A review. *Frontiers in Bioengineering and Biotechnology, 9*, 802543. https://doi.org/10.3389/fbioe.2021.802543.

Li, L., & Van Eck, J. (2007). Metabolic engineering of carotenoid accumulation by creating a metabolic sink. *Transgenic Research, 16*(5), 581-585. https://doi.org/10.1007/s11248-007-9065-4.

Li, N., Qiao, D., Zhao, S., Lin, Q., Zhang, B., & Xie, F. (2021). 3D printing to innovate biopolymer materials for demanding applications: A review. *Materials Today Chemistry, 20*, 100459. https://doi.org/10.1016/j.mtchem.2021.100459.

Lu, L. W., Gao, Y., Quek, S. Y., Foster, M., Eason, C. T., Liu, M., Wang, M., & Chen, F. (2022). The landscape of potential health benefits of carotenoids as natural supportive therapeutics in protecting against coronavirus infection. *Biomedicine & Pharmacotherapy, 154*, 113625. https://doi.org/10.1016/j.biopha.2022.113625.

Luo, X., Ho, I., Brankovan, S., & Lim, L. T. (2021). Inkjet-printed gradient colorimetric indicators for monitoring fish freshness. *Food Packaging and Shelf Life, 29*, 100719. https://doi.org/10.1016/j.fpsl.2021.100719.

Madruga, L. Y., & Kipper, M. J. (2022). Expanding the repertoire of electrospinning: New and emerging biopolymers, techniques, and applications. *Advanced Healthcare Materials, 11*(4), 2101979. https://doi.org/10.1002/adhm.202101979.

Manigandan, V., Karthik, R., Ramachandran, S., & Rajagopal, S. (2018). Chitosan applications in food industry. In *Biopolymers for Food Design* (pp. 469-491). Academic Press: Cambridge, Massachusetts.

Maoka, T. (2020). Carotenoids as natural functional pigments. *Journal of Natural Medicines, 74*(1), 1-16. https://doi.org/10.1007/s11418-019-01341-1.

Mapari, S. A., Thrane, U., & Meyer, A. S. (2010). Fungal polyketide azaphilone pigments as future natural food colorants? *Trends in Biotechnology, 28*(6), 300-307. https://doi.org/10.1016/j.tibtech.2010.04.005.

Mathew, A. P., & Oksman, K. (2014). Processing of Bionanocomposites: Solution Casting. In *Handbook of Green Materials: 2 Bionanocomposites: Processing, Characterization and Properties* (pp. 35–52). World Scientific Publishing: Singapore.

Medina-Jaramillo, C., Ochoa-Yepes, O., Bernal, C., & Famá, L. (2017). Active and smart biodegradable packaging based on starch and natural extracts. *Carbohydrate Polymers, 176*, 187-194. https://doi.org/10.1016/j.carbpol.2017.07.067.

Meléndez-Martínez, A. J., Mandić, A. I., Bantis, F., Böhm, V., Borge, G. I. A., Brnčić, M., ... & O'Brien, N. (2022). A comprehensive review on carotenoids in foods and feeds: Status quo, applications, patents, and research needs. *Critical Reviews in Food Science and Nutrition, 62*(8), 1999-2049. https://doi.org/10.1080/10408398.2021.1919241.

Merhan, O. (2017). The biochemistry and antioxidant properties of carotenoids. *Antioxidants, 5*, 51. https://doi.org/10.3390/antiox5040051.

Mezzomo, N., & Ferreira, S. R. (2016). Carotenoids functionality, sources, and processing by supercritical technology: A review. *Journal of Chemistry, 2016*, 3164312. https://doi.org/10.1155/2016/3164312.

Mogedas, B., Casal, C., Forján, E., & Vílchez, C. (2009). β-Carotene production enhancement by UV-A radiation in *Dunaliella bardawil* cultivated in laboratory reactors. *Journal of Bioscience and Bioengineering, 108*(1), 47-51. https://doi.org/10.1016/j.jbiosc.2009.02.007.

Mohammadalinejhad, S., Almasi, H., & Moradi, M. (2020). Immobilization of *Echium amoenum* anthocyanins into bacterial cellulose film: A novel colorimetric pH indicator for freshness/spoilage monitoring of shrimp. *Food Control, 113*, 107169. https://doi.org/10.1016/j.foodcont.2020.107169.

Mohammadian, E., Alizadeh-Sani, M., & Jafari, S. M. (2020). Smart monitoring of gas/temperature changes within food packaging based on natural colorants. *Comprehensive Reviews in Food Science and Food Safety, 19*(6), 2885-2931. https://doi.org/10.1111/1541-4337.12635.

Nemes, S. A., Szabo, K., & Vodnar, D. C. (2020). Applicability of agro-industrial by-products in intelligent food packaging. *Coatings, 10*(6), 550. https://doi.org/10.3390/coatings10060550.

Neves, D., Andrade, P. B., Videira, R. A., de Freitas, V., & Cruz, L. (2022). Berry anthocyanin-based films in smart food packaging: A mini-review. *Food Hydrocolloids, 133*, 107885. https://doi.org/10.1016/j.foodhyd.2022.107885.

Niponsak, A., Laohakunjit, N., Kerdchoechuen, O., & Wongsawadee, P. (2016). Development of smart colourimetric starch-based indicator for liberated volatiles during durian ripeness. *Food Research International, 89*, 365-372. https://doi.org/10.1016/j.foodres.2016.08.031.

Oladimeji, A. O., Agboola, F. Z., Oguntayo, D. O., & Aina, K. S. (2022). Dimensional stability and mechanical properties of extruded-compression biopolymer composites made from selected Nigerian grown wood species at varying proportions. *Scientific Reports, 12*(1), 10545. https://doi.org/10.1038/s41598-022-15589-9.

Pagnan, C. S., Mottin, A. C., Oréfice, R. L., Ayres, E., & Câmara, J. J. D. (2018). Annatto-colored poly (3-hydroxybutyrate): A comprehensive study on photodegradation. *Journal of Polymers and the Environment, 26*(3), 1169-1178. https://doi.org/10.1007/s11356-018-1569-1.

Pagno, C. H., de Farias, Y. B., Costa, T. M. H., Rios, A. D. O., & Flôres, S. H. (2016). Synthesis of biodegradable films with antioxidant properties based on cassava starch containing bixin nanocapsules. *Journal of Food Science and Technology, 53*(8), 3197-3205. https://doi.org/10.1007/s11483-016-0092-4.

Popović, S. Z., Lazić, V. L., Hromiš, N. M., Šuput, D. Z., & Bulut, S. N. (2018). Biopolymer packaging materials for food shelf-life prolongation. In *Biopolymers for Food Design* (pp. 223-277). Academic Press: Cambridge, Massachusetts.

Pourjavaher, S., Almasi, H., Meshkini, S., Pirsa, S., & Parandi, E. (2017). Development of a colorimetric pH indicator based on bacterial cellulose nanofibers and red cabbage (*Brassica oleraceae*) extract. *Carbohydrate Polymers, 156*, 193-201. https://doi.org/10.1016/j.carbpol.2016.10.020.

Prabha, K., Ghosh, P., Joseph, R. M., Krishnan, R., Rana, S. S., & Pradhan, R. C. (2021). Recent development, challenges, and prospects of extrusion technology. *Future Foods, 3*, 100019. https://doi.org/10.1016/j.fufo.2021.100019.

Priyadarshi, R., Ezati, P., & Rhim, J. W. (2021a). Recent advances in intelligent food packaging applications using natural food colorants. *ACS Food Science & Technology, 1*(2), 124-138. https://doi.org/10.1021/acsfoodscitech.1c00058.

Priyadarshi, R., Kim, S. M., & Rhim, J. W. (2021b). Pectin/pullulan blend films for food packaging: Effect of blending ratio. *Food Chemistry, 347*, 129022. https://doi.org/10.1016/j.foodchem.2021.129022.

Priyadarshi, R., Roy, S., Ghosh, T., Biswas, D., & Rhim, J. W. (2022). Antimicrobial nanofillers reinforced biopolymer composite films for active food packaging applications-a review. *Sustainable Materials and Technologies, 32*, e00353. https://doi.org/10.1016/j.susmat.2022.e00353

Rahman, M. S. (2007). Packaging as a preservation technique. In *Handbook of Food Preservation* (pp. 925-934). CRC Press: Boca Raton, FL.

Ran, R., Wang, L., Su, Y., He, S., He, B., Li, C., ... & Chen, S. (2021). Preparation of pH-indicator films based on soy protein isolate/bromothymol blue and methyl red for monitoring fresh-cut apple freshness. *Journal of Food Science, 86*(10), 4594-4610. https://doi.org/10.1111/1750-3841.16148.

Rao, A. V., & Rao, L. G. (2007). Carotenoids and human health. *Pharmacological Research, 55*(3), 207-216. https://doi.org/10.1016/j.phrs.2007.01.017.

Ren, L., Chen, J., Lu, Q., Han, J., & Wu, H. (2021). Anti-biofouling nanofiltration membrane constructed by in-situ photo-grafting bactericidal and hydrophilic polymers. *Journal of Membrane Science, 617*, 118658. https://doi.org/10.1016/j.memsci.2020.118658.

Ribeiro, H. S., Ax, K., & Schubert, H. (2003). Stability of lycopene emulsions in food systems. *Journal of Food Science, 68*(9), 2730-2734. https://doi.org/10.1111/j.1365-2621.2003.tb08116.x.

Riedl, K. M., Choksi, K., Wyzgoski, F. J., Scheerens, J. C., Schwartz, S. J., & Reese, R. N. (2013). Variation in lycopene and lycopenoates, antioxidant capacity, and fruit quality of buffaloberry (*Shepherdia argentea* [Pursh] Nutt.). *Journal of Food Science, 78*(11), C1673-C1679. https://doi.org/10.1111/1750-3841.12268.

Robertson, G. L. (2005). *Food packaging: Principles and practice*. CRC Press: Boca Raton, FL.

Rodriguez-Concepcion, M., Avalos, J., Bonet, M. L., Boronat, A., Gomez-Gomez, L., Hornero-Mendez, D., ... & Zhu, C. (2018). A global perspective on carotenoids: Metabolism, biotechnology, and benefits for nutrition and health. *Progress in Lipid Research, 70*, 62-93. https://doi.org/10.1016/j.plipres.2018.03.002.

Rout, S., Tambe, S., Deshmukh, R. K., Mali, S., Cruz, J., Srivastav, P. P., de Aguiar Andrade, E. H., & de Oliveira, M. S. (2022). Recent trends in the application of essential oils: The next generation of food preservation and food packaging. *Trends in Food Science & Technology, 129*, 421-439. https://doi.org/10.1016/j.tifs.2022.01.019.

Roy, S., & Rhim, J. W. (2021). Anthocyanin food colorant and its application in pH-responsive color change indicator films. *Critical Reviews in Food Science and Nutrition, 61*(14), 2297-2325. https://doi.org/10.1080/10408398.2020.1791633.

Roy, S., Chawla, R., Santhosh, R., Thakur, R., Sarkar, P., & Zhang, W. (2023). Agar-based edible films and food packaging application: A comprehensive review. *Trends in Food Science & Technology, 141*, 104198. https://doi.org/10.1016/j.tifs.2023.104198.

Roy, S., Kim, H. J., & Rhim, J. W. (2021). Effect of blended colorants of anthocyanin and shikonin on carboxymethyl cellulose/agar-based smart packaging film. *International Journal of Biological Macromolecules, 183*, 305-315. https://doi.org/10.1016/j.ijbiomac.2021.05.007.

Roy, S., Priyadarshi, R., Ezati, P., & Rhim, J. W. (2022). Curcumin and its uses in active and smart food packaging applications - A comprehensive review. *Food Chemistry, 375*, 131885. https://doi.org/10.1016/j.foodchem.2021.131885.

Roy, S., Zhang, W., Biswas, D., Ramakrishnan, R., & Rhim, J. W. (2023). Grapefruit seed extract-added functional films and coating for active packaging applications: A review. *Molecules, 28*(2), 730. https://doi.org/10.3390/molecules28020730.

Sadeghi, K., Jee, H. W., Paeng, K. J., & Seo, J. (2021). Photografting of conducting polymer onto polymeric substrate as non-migratory antioxidant packaging. *Reactive and Functional Polymers, 158*, 104792. https://doi.org/10.1016/j.reactfunctpolym.2020.104792

Sameen, D. E., Ahmed, S., Lu, R., Li, R., Dai, J., Qin, W., Zhang, Q., Li, S., & Liu, Y. (2022). Electrospun nanofibers food packaging: Trends and applications in food systems. *Critical Reviews in Food Science and Nutrition, 62*(22), 6238-6251. https://doi.org/10.1080/10408398.2021.1976320.

Sapper, M., & Chiralt, A. (2018). Starch-based coatings for preservation of fruits and vegetables. *Coatings, 8*(5), 152. https://doi.org/10.3390/coatings8050152.

Schieber, A., & Carle, R. (2005). Occurrence of carotenoid cis-isomers in food: Technological, analytical, and nutritional implications. *Trends in Food Science & Technology, 16*(9), 416-422. https://doi.org/10.1016/j.tifs.2005.06.006.

Sigurdson, G. T., Tang, P., & Giusti, M. M. (2017). Natural colorants: Food colorants from natural sources. *Annual Review of Food Science and Technology, 8*(1), 261-280. https://doi.org/10.1146/annurev-food-030216-030302.

Sikora, J., Majewski, Ł., & Puszka, A. (2020). Modern biodegradable plastics—Processing and properties: Part I. *Materials, 13*(8), 1986. https://doi.org/10.3390/ma13081986.

Squillaci, G., Parrella, R., Carbone, V., Minasi, P., La Cara, F., & Morana, A. (2017). Carotenoids from the extreme halophilic archaeon *Haloterrigena turkmenica*: Identification and antioxidant activity. *Extremophiles, 21*(5), 933-945. https://doi.org/10.1007/s00792-017-0944-4.

Srinivasan, B., & Kulshreshtha, G. (2020). Recent developments in food-based bioplastics production. *Bioprocess Engineering for Bioremediation: Valorization and Management Techniques*, 107-127. https://doi.org/10.1007/978-3-030-39971-9_7.

Srivastava, R. (2021). Physicochemical, antioxidant properties of carotenoids and its optoelectronic and interaction studies with chlorophyll pigments. *Scientific Reports, 11*(1), 18365. https://doi.org/10.1038/s41598-021-97801-6.

Stoll, L., Domenek, S., Hickmann Flôres, S., Nachtigall, S. M. B., & de Oliveira Rios, A. (2021). Polylactide films produced with bixin and acetyl tributyl citrate: Functional properties for active packaging. *Journal of Applied Polymer Science, 138*(17), 50302. https://doi.org/10.1002/app.50302.

Sugiura, M., Nakamura, M., Ikoma, Y., Yano, M., Ogawa, K., Matsumoto, H., Kato, M., Ohshima, M., & Nagao, A. (2006). Serum carotenoid concentrations are inversely associated with serum aminotransferases in hyperglycemic subjects. *Diabetes Research and Clinical Practice, 71*(1), 82-91. https://doi.org/10.1016/j.diabres.2005.05.020.

Sun, T., Rao, S., Zhou, X., & Li, L. (2022). Plant carotenoids: Recent advances and future perspectives. *Molecular Horticulture, 2*(1), 3. https://doi.org/10.1186/s43816-022-00003-1

Tajeddin, B., Ahmadi, B., Sohrab, F., & Chenarbon, H. A. (2018). Polymers for modified atmosphere packaging applications. In *Food Packaging and Preservation* (pp. 457-499). Academic Press: Cambridge, Massachusetts.

Tupuna-Yerovi, D. S., Schmidt, H., & Rios, A. D. O. (2025). Biodegradable sodium alginate films incorporated with lycopene and β-carotene for food packaging purposes. *Food Science and Technology International, 31*(1), 23-35. https://doi.org/10.1177/10820132221107095

Turtoi, M. (2013). Ultraviolet light treatment of fresh fruits and vegetables surface: A review. *Journal of Agroalimentary Processes and Technologies, 19*(3), 325-337. https://doi.org/10.1684/jagt.2013.0435.

Tzortzakis, N., & Chrysargyris, A. (2017). Postharvest ozone application for the preservation of fruits and vegetables. *Food Reviews International, 33*(3), 270-315. https://doi.org/10.1080/87559129.2017.1377782.

Vieira, I. R. S., de Carvalho, A. P. A. D., & Conte-Junior, C. A. (2022). Recent advances in biobased and biodegradable polymer nanocomposites, nanoparticles, and natural antioxidants for antibacterial and antioxidant food packaging applications. *Comprehensive Reviews in Food Science and Food Safety, 21*(4), 3673-3716. https://doi.org/10.1111/1541-4337.12807.

Woodside, J. V., McGrath, A. J., Lyner, N., & McKinley, M. C. (2015). Carotenoids and health in older people. *Maturitas, 80*(1), 63-68. https://doi.org/10.1016/j.maturitas. 2014.11.006.

Xianquan, S., Shi, J., Kakuda, Y., & Yueming, J. (2005). Stability of lycopene during food processing and storage. *Journal of Medicinal Food, 8*(4), 413-422. https://doi.org/10.1089/jmf.2005.8.413.

Xin, Y., Zhang, M., Xu, B., Adhikari, B., & Sun, J. (2015). Research trends in selected blanching pretreatments and quick freezing technologies as applied in fruits and vegetables: A review. *International Journal of Refrigeration, 57*, 11-25. https://doi.org/10.1016/j.ijrefrig.2015.04.019.

Xu, Y., Liu, Z., Liu, R., Luo, M., Wang, Q., Cao, L., & Ye, S. (2021). Inkjet-printed pH-sensitive QR code labels for real-time food freshness monitoring. *Journal of Materials Science, 56*(33), 18453-18462. https://doi.org/10.1007/s10853-021-05752-2.

Yadav, A., Mangaraj, S., Singh, R. M. N. K., Kumar, N., & Arora, S. (2018). Biopolymers as packaging material in food and allied industry. *International Journal of Chemical Studies, 6*(2), 2411-2418. https://doi.org/10.22271/chemi.2018.v6.i2o.5994.

Yadav, S., Mehrotra, G. K., Bhartiya, P., Singh, A., & Dutta, P. K. (2020). Preparation, physicochemical and biological evaluation of quercetin-based chitosan-gelatin film for food packaging. *Carbohydrate Polymers, 227*, 115348. https://doi.org/10.1016/ j.carbpol.2019.115348.

Yildirim, S., Röcker, B., Pettersen, M. K., Nilsen-Nygaard, J., Ayhan, Z., Rutkaite, R., ... & Coma, V. (2018). Active packaging applications for food. *Comprehensive Reviews in Food Science and Food Safety, 17*(1), 165-199. https://doi.org/10.1111/1541-4337.12310.

Young, A. J., & Lowe, G. L. (2018). Carotenoids—Antioxidant properties. *Antioxidants, 7*(2), 28. https://doi.org/10.3390/antiox7020028.

Yousefi, H., Su, H. M., Imani, S. M., Alkhaldi, K., M. Filipe, C. D., & Didar, T. F. (2019). Intelligent food packaging: A review of smart sensing technologies for monitoring food quality. *ACS Sensors, 4*(4), 808-821. https://doi.org/10.1021/acssensors. 9b00212.

Zakynthinos, G., & Varzakas, T. (2016). Carotenoids: From plants to food industry. *Current Research in Nutrition and Food Science, 4*(1), 38-51. https://doi.org/ 10.12944/CRNFSJ.4.1.06.

Zhang, W., Zhang, Y., Cao, J., & Jiang, W. (2021). Improving the performance of edible food packaging films by using nanocellulose as an additive. *International Journal of Biological Macromolecules, 166*, 288-296. https://doi.org/10.1016/ j.ijbiomac.2020.10.285.

Zhang, Y., & Lim, L. T. (2016). Inkjet-printed CO2 colorimetric indicators. *Talanta, 161*, 105-113. https://doi.org/10.1016/j.talanta.2016.09.024.

Zhang, Z., Tong, J., Chen, D. H., & Lan, Y. B. (2008). Electronic nose with an air sensor matrix for detecting beef freshness. *Journal of Bionic Engineering, 5*(1), 67-73. https://doi.org/10.1016/S1672-6529(08)60018-0.

Zhao, X., Cornish, K., & Vodovotz, Y. (2020). Narrowing the gap for bioplastic use in food packaging: An update. *Environmental Science & Technology, 54*(8), 4712-4732. https://doi.org/10.1021/acs.est.0c00456.

Zhou, W., Niu, Y., Ding, X., Zhao, S., Li, Y., Fan, G., ... & Liao, K. (2020). Analysis of carotenoid content and diversity in apricots (*Prunus armeniaca L.*) grown in China. *Food Chemistry, 330*, 127223. https://doi.org/10.1016/j.foodchem.2020.127223.

## Chapter 3

# Carotenoids from Ulam: Bridging Traditional Knowledge with Modern Halal Pharmaceutical Needs

**Farah Ayuni Mohd Hatta[1,*], PhD**
**Qurratu Aini Mat Ali[2], PhD**
**Rashidi Othman[3], PhD**
**and Latifah Amin[4], PhD**

[1,2,4]Institute of Islam Hadhari, Universiti Kebangsaan Malaysia,
Universiti Kebangsaan Malaysia Bangi, Selangor, Malaysia
[3]Herbarium Unit, Department of Landscape Architecture, Kulliyyah of Architecture and Environmental Design, International Islamic University Malaysia,
Kuala Lumpur, Malaysia

## Abstract

The halal pharmaceutical industry has gained interest beyond the Muslim consumer market due to the trust and high-quality assurance offered by halal certification. Synthetic active pharmaceutical ingredients (APIs) can have long-term negative effects on human health and contribute to environmental degradation. In this context, there is a growing interest in natural and halal APIs that can offer positive health effects. Fruits and vegetables are excellent sources of provitamin A, with carotenoids providing between 70% to 90% of the daily requirement. Additionally, ulam, a traditional Malay vegetable, is a rich source of carotenoids with proven nutritional benefits besides having good antioxidant and antimicrobial properties. As a result, ulam has great potential to offer

* Corresponding Author's Email: farahayuni@ukm.edu.my

In: Natural Pigments: Sources, Properties, and Health Benefits
Editors: Rashidi Othman, Qurratu Aini Mat Ali, Razanah Ramya et al.
ISBN: 979-8-90134-055-4

positive health effects as a dietary source and as a natural and halal APIs. By incorporating ulam-based natural APIs, the pharmaceutical industry can contribute to healthier and sustainable outcomes for both individuals and the environment.

**Keywords**: ulam, natural carotenoid, bioactivities, halal, active pharmaceutical ingredients

## Introduction

The increasing demand for halal products such as food and healthcare products are on the rise, parallel to the growing number of Muslim consumers, which is approximately 1.9 billion, making it the second largest globally. However, Muslim consumers are more meticulous in selecting halal food products compared to healthcare products (Syafiq et al., 2015; Mukhtar & Butt, 2012). In essence, seeking halal options for every product used is a religious obligation for every Muslim. However, in certain emergency situations, there are patients who have no alternative but to seek treatment by opting for medications that do not have halal status (Syafiq et al., 2015). Besides seeking advice from medical experts, patients also need to be informed about medication information and the necessity of their usage if there are no other halal alternatives available.

Therefore, the development of pharmaceutical and healthcare products from halal sources is crucial to provide convenience among Muslim consumers and to comply with Shariah objectives (Mukhtar & Butt, 2012). The advancement and involvement of the halal pharmaceutical industry will not only meet the needs of Muslim consumers but also benefit non-Muslim consumers in terms of product integrity and safety assurance (Bashir, 2020). Hence, this review will explore the potential of carotenoids in ulam as a source of halal active pharmaceutical ingredients.

Ulam is classified as traditional vegetables consumed and practiced by communities since ancient times, either cooked or consumed raw as a means to enhance food variety and taste, besides being believed to provide health benefits to practitioners. Generally, ulam is utilized by communities in Southeast Asia (You et al., 2018). In Malaysia, there are more than 120 species of ulam or traditional vegetables consisting of various families. The parts used as ulam are usually leaves, shoots, fruits, flowers, tubers, and rhizomes. However, detailed scientific information regarding the nutritional value of ulam consumption in Malaysia is still under-documented (Othman et al., 2017). Given that fruits and vegetables have been proven to be rich sources of provitamin A for human daily diets, supplying between 70-90% of the required carotenoids, ulam also has the potential to possess similar nutrients (Othman et al., 2014). Epidemiological studies by Aruna and Baskaran (2010) have confirmed that the intake of fruits and vegetables rich in carotenoids helps reduce the risk of cardiovascular diseases, cancer, and cataract formation.

Synthetic active pharmaceutical ingredients (APIs) are widely used in the pharmaceutical industry (Kumar et al., 2022). However, the use of synthetic APIs can have negative effects on human health and the environment (Othman et al., 2021). Butylated hydroxytoluene (BHT) and butylated hydroxyanisole (BHA) are synthetic antioxidants commonly used in pharmaceuticals, cosmetics, and the food industry. Besides their high cost, these synthetic antioxidants also have drawbacks such as instability and can have adverse effects on consumers due to their carcinogenic behaviour with long-term use (Oladele et al., 2020). Therefore, active ingredients from natural sources are more natural and safer to replace synthetic active ingredients.

## Ethnomedicinal Plants

Plants are living organisms created with various functions and benefits, including serving as habitats, sources of raw materials, and food. Ethnomedicinal plants are plants that have medicinal value used in traditional treatments by specific ethnic groups. Previous studies have found that there are plants containing phenolic compounds that are antioxidant and assist in treating diseases such as improving the digestive system, reducing inflammation, and having neuroprotective properties (Oyewole et al., 2017). The intake of these plants has also been reported to help reduce the risk of cardiovascular diseases, cancer, and diabetes (Ajayi et al., 2017). Currently, the production of active ingredients from safer plant sources has attracted the attention of many researchers. However, many plant species remain unexplored and their effectiveness scientifically validated (Hamzah, 2015).

There are communities from specific geographical and ethnic areas that utilize plants as the main source in daily use including as food, treatment, and healthcare. Furthermore, Malaysia's geographical location in the equatorial climate zone makes it rich in plant resources (Hamzah, 2015). Since most communities in the past lived in rural areas, they often utilized the surrounding areas near their homes for gardening purposes. More than 120 species of ulam have been identified in Asia, rich in vitamins and minerals (You et al., 2018).

Ulam also can enhance cognitive function, maintain nutrition, and improve mood in older adults (You et al., 2020b, 2020a). Incorporating a variety of carotenoid-rich foods into the diet can support individual health and reduce the risk of chronic diseases. Studies have found that the average consumption of ulam by Malaysians is about 40 grams per day. Scientifically, these plants have active ingredients such as alkaloids, flavonoids, minerals, saponins, vitamins, and others (USDA, 2021). Table 1 lists several ulam species along with their medicinal values (Ismail, 2020; Chan et al., 2017).

**Table 1.** Example of ulam with medicinal values (Ismail, 2020; Othman et al., 2017)

| Scientific Name | Local Name | Edible Part | Medicinal Values |
|---|---|---|---|
| Anacardium occidentale | Gajus | Leaf | Aids digestion, neutralizes toxins in the body |
| Capsicum frutescens | Cabai burung | Fruit | Reduces bone pain |
| Centella asiatica | Pegaga | Leaf | Heals damaged tissue, reduces scars |
| Cosmos caudatus | Ulam raja | Leaf | Improves blood circulation, firms the skin |
| Curcuma longa | Kunyit | Leaf/ Rhizome | Boosts immunity |
| Kaempferia galanga | Cekur | Leaf/ Rhizome | Treats headaches, gastritis, stomach nausea |
| Momordica charatia | Peria Katak | Fruit | Prevents diabetes, heals wounds, bone formation |
| Onanthe javanica | Selom | Leaf | Treats fever, yellow fever, improves urination, constipation |
| Piper sarmentosum | Kaduk | Leaf | Treats fever, cough, asthma, cold |

Moreover, ulam is reported to have several medicinal values such as antioxidant, antimicrobial, antidiabetic, antihypertensive, neuroprotective, and anti-inflammatory properties (You et al., 2018; Islam et al., 2018). Additionally, ulam is high in polyphenol content and can enhance cognitive status by increasing serum β-carotene, lycopene, folic acid, and vitamins C and E in the body (Islam et al., 2018; You et al., 2018). For example, a study by Chowtivannakul et al. (2017) showed that *Leucaena leucocephala* seeds contain antidiabetic and antioxidant activity, which can reduce albumin and protein levels in the kidneys.

In the Philippines, the pharmaceutical industry under the Department of Health Philippines has recognized several plant species such as garlic (*Allium sativum*), yerba buena (*Clinopodium douglasii*), and bitter melon (*Momordica charantia*) for their therapeutic values and potential as pharmaceutical products that will benefit the entire community (Boy et al., 2018). There are also studies showing differences in phenolic content concentrations and antioxidant activity in ulam, contributing to differences in antioxidant activity. Moreover, some ulam also contain various carotenoid compositions that function as key metabolism for human nutrition and health (Islam et al., 2018).

Pharmaceutical products need to undergo clinical trials to prove effectiveness and ensure no harmful side effects. Traditional medicinal plants are categorized as complementary and alternative medicines (CAMs) documented and commercialized by Chinese and Indian societies such as Ayurveda and Chinese herbal medicine (Kookana et al., 2014). Similarly, Malay ethnomedicinal plants practiced by village nurses or midwives for patient care. However, Malay ethnomedicinal plants are still less promoted, scientifically validated, and widely commercialized (Moshawih, 2017).

Pharmaceutical products can be derived from various sources such as natural, synthetic, and biosynthetic sources (Alamgir, 2017). Therefore, traditional medicinal plants have the potential as sources of active pharmaceutical ingredients because they have functions in healing, reducing, and preventing diseases. Although some medicinal plants have been used in the pharmaceutical industry called raw drugs, some are labeled as unofficial drugs due to insufficient clinical evidence and information on their side effects (Alamgir, 2017).

## Potential of Ulam as a Source of Halal Active Pharmaceutical Ingredients

Pharmaceuticals are defined as the industry that produces medicines for medical purposes under monitored legislation. Drugs in pharmaceuticals are recognized as official formulas used for purposes of diagnosis, healing, mitigation, treatment, and prevention of diseases for humans and animals (Alamgir, 2017). However, there are several differences between synthetic compounds in medications and ethnomedicinal plants based on various side effects, contraindications, interactions with other substances, drug resistance, cost, and research duration (Segneanu et al., 2017). The World Health Organization (WHO) has targeted to enhance the integration of traditional medicine to improve the healthcare system (Segneanu et al., 2017).

Active pharmaceutical ingredients (APIs) are defined as substances proven suitable for use in the production of medications and involve processing, manufacturing, and packaging or in the form of finished dosage forms (USDA, 2021). To produce halal pharmaceutical products, it is fundamentally necessary to follow the guidelines set by JAKIM in terms of equipment, materials, processes, and others. Based on the general guidelines from the Department of Standards Malaysia (MS 2424:2012) under the requirement section, plants are categorized as natural ingredients. Explanation in Section 4.15.2.1 states that all types of plants, plant products, and anything derived from plants are halal except those prohibited by authorities (National Standards Department, 2012).

Scientists and researchers use ethnobotanical and ethnopharmacological information as a guide to explore various bioactive compounds from plants that have the potential to be used as medicines (Kueh, 2007). Based on previous studies, bioactive compounds in ulam have the potential to be a source of active pharmaceutical ingredients as listed in Table 2.

## Composition of Carotenoids in Ulam

Carotenoids can be found in fruits and vegetables. Typically, yellow, orange, and red fruits and green leafy vegetables are known to be rich in carotenoids (Bohn, 2016). Essentially, carotenoids are produced by living organisms undergoing photosynthesis, primarily green plants, and some non-photosynthetic organisms such as fungi categorized under the isoprenoid metabolite subclass (Rodríguez-Concepción et al., 2018). There are more than 1000 recognized carotenoids chemically, containing polyene chains with double bond conjugations divided into carotenoid and xanthophyll groups (Yabuzaki, 2017; Hermanns et al., 2020). Each carotenoid has a specific structure that produces its own colour. Plant carotenoids, mainly C40 tetraterpenoids (carotenes, lutein, lycopene, zeaxanthin, and cryptoxanthin), function in plant growth (Rodríguez-Concepción et al., 2018; Wurtzel, 2019; Sun & Li, 2020; Zheng et al., 2020). Meanwhile, carotenoids such as lutein, β-carotene, zeaxanthin, and violaxanthin function as major metabolites in green plant tissues to maintain the health of the plants. Additionally, carotenoids are also important pigments in light harvesting during the photosynthesis process by absorbing light in the spectrum region not absorbed by chlorophyll (400-550 nm) (Hermanns et al., 2020).

**Table 2.** Bioactive compounds in ulam

| Ulam species | Bioactivity | Bioactive Compounds | References |
|---|---|---|---|
| Centella asiatica (Pegaga) | Antimicrobial, Antioxidant | Neoxanthin, Violaxanthin, Lutein, α-carotene, β-carotene, Alkaloids, Flavonoids, Terpenoids | Chung 2008; Othman et al. 2017; You et al. 2018 |
| Cosmos caudatus (Ulam raja) | Antioxidant, Anti-inflammatory, Antibacterial | Lutein, α-carotene, β-carotene, Phenolics | Latiff et al. 2021; Othman et al. 2017; You et al. 2018 |
| Lactuca sativa (Salad) | Antidiabetic, Anti-inflammatory | Neoxanthin, Violaxanthin, Lutein, β-carotene, Phenolics, Folate | Kim et al. 2016; Othman et al. 2017 |
| Leucaena leucocephala (Petai belalang) | Antioxidant, Antidiabetic | Zeaxanthin, Violaxanthin, Lutein, β-carotene, Phenolics, Flavonoids | Chowtivannakul et al. 2016; Fernández-Marín et al. 2017; Suryanti et al. 2016; Zarina et al. 2018 |
| Morinda citrifolia (Mengkudu) | Antioxidant, Anti-inflammatory, Antimicrobial, Antiviral | Zeaxanthin, Lutein, β-carotene | Jahurul et al. 2021; Othman et al. 2017 |
| Ocimum basilicum (Selasih) | Antioxidant, Anti-inflammatory, Antimicrobial, Antiviral, Antiparasitic | Neoxanthin, Violaxanthin, Lutein, α-carotene, β-carotene | Dhama et al. 2021; Othman et al. 2017; You et al. 2018 |
| Oenanthe javanica (Selom) | Antimicrobial, Antioxidant | Neoxanthin, Violaxanthin, Lutein, β-carotene, Flavonoids, Phenolics | Abdusalam et al. 2022; Chan et al. 2017; Othman et al. 2017 |
| Orthosiphon stamineus (Misai kucing) | Antimicrobial | Zeaxanthin, Violaxanthin, Lutein, β-carotene, Diterpenes, Flavonoids | Ali et al. 2018; Chung 2008 |
| Polygonum minus (Kesum) | Antioxidant, Antibacterial, Antifungal, Antiulcer | Lutein, α-carotene, β-carotene, Phenolics | Othman et al. 2017; Vikram et al. 2014; You et al. 2018 |
| Sauropus androgynus (Cekur manis) | Antioxidant, Antimicrobial | Neoxanthin, Violaxanthin, Lutein, β-cryptoxanthin, α-carotene, β-carotene, Phenolics, Flavonoids | Arif & Shetty 2020; Othman et al. 2017 |

**Table 3.** Ulam species and their carotenoids studied from 2010 to 2022

| Species | Carotenoids | References |
|---|---|---|
| 1. *Anacardium occidentale* (Gajus)<br>2. *Allium cepa* (Daun bawang)<br>3. *Apium graveolens* (Saderi)<br>4. *Centella asiatica* (Pegaga)<br>5. *Cosmos caudatus* (Ulam raja)<br>6. *Daucus carota* (Lobak merah)<br>7. *Euodia redleyi* (Tenggek burung)<br>8. *Lactuca sativa* (Salad)<br>9. *Morinda citrifolia* (Mengkudu)<br>10. *Murraya koenigii* (Daun Kari)<br>11. *Ocimum americanum* (Kemangi)<br>12. *Ocimum basilicum* (Selasih)<br>13. *Oenanthe javanica* (Selom)<br>14. *Piper sarmentosum* (Kaduk)<br>15. *Pluchea indica* (Beluntas)<br>16. *Polygonum minus* (Kesum)<br>17. *Sauropus androgynus* (Cekur manis) | Neoxanthin, violaxanthin, lutein, zeaxanthin,<br>α-carotene, β-carotene,<br>β-cryptoxanthin | Othman et al., 2017 |
| 1. *Brassica oleracea* (Kale)<br>2. *Centella asiatica* (Pegaga)<br>3. *Cosmos caudatus* (Ulam raja)<br>4. *Curcuma longa* (Kunyit)<br>5. *Ipomoea aquatica* (Kangkong)<br>6. *Lactuca sativa* (Salad)<br>7. *Manihot esculenta* (Pucuk ubi kayu)<br>8. *Moringa oleifera* (Kelor) | Neoxanthin, violaxanthin, lutein, zeaxanthin,<br>α-carotene, β-carotene,<br>β-cryptoxanthin, anthocyanin | Lee et al., 2022 |

| Species | Carotenoids | References |
| --- | --- | --- |
| *Capsicum annuum* (Cabai merah) | Lutein, zeaxanthin, α-carotene, β-carotene, β-cryptoxanthin, violaxanthin, capsanthin, phytoene, phytofluene | Rodríguez-Rodríguez et al., 2020 |
| *Centella asiatica* (Pegaga) | Lutein, β-carotene | Norhayati et al., 2011; Othman, Mohd Zaifuddin, & Mohd Hassan, 2015; Gunathilake, Ranaweera, & Rupasinghe, 2018 |
| *Coriandrum sativum* (Ketumbar) | β-carotene | Divya et al. 2012; Laribi et al 2015 |
| 1. *Cosmos caudatus* (Ulam raja)<br>2. *Oenanthe javanica* (Selom) | Lutein, β-carotene | Chan et al., 2017 |
| *Curcuma longa* (Kunyit) | Lutein, β-carotene | Othman et al., 2017 |
| *Manihot esculenta* (Pucuk ubi kayu) | Lutein, β-carotene | Yaacob et al., 2019 |
| *Momordica charantia* (Peria katak) | Lutein, zeaxanthin, β-cryptoxanthin, lycopene, α-carotene, β-carotene | Tuan et al., 2011; Arasu et al., 2017; Lee et al., 2017 |
| *Morinda citrifolia* (Mengkudu) | β-carotene | Othman et al., 2017 |
| *Sauropus androgynus* (Cekur manis) | Lutein, zeaxanthin, β-cryptoxanthin, lycopene, α-carotene, β-carotene, neoxanthin, violaxanthin | Eng Khoo et al., 2015; Othman, Mohd Zaifuddin & Mohd Hassan, 2015 |
| *Solanum lycopersicum* (Tomato) | Lycopene | Martel et al., 2011; Fujisawa et al., 2013; Rizk et al., 2014; Llorente et al., 2016; Wang et al., 2017; Pataro et al., 2018; Xiong et al., 2019; Hermanns et al., 2020; Wu et al., 2020 |

Carotenoids are also vital major metabolites as a nutritional source for human health (Eggersdorfer & Wyss, 2018; Rodríguez-Concepción et al., 2018). Melendez-Martinez (2019) reported that carotenoids are precursors to vitamin A biosynthesis, micronutrients, and play a crucial role in vision health, immune response, and reproduction systems. Additionally, carotenoids are effective bioactive substances in the food chain such as antioxidants, where lycopene and β-carotene have the ability to reduce the risk of various chronic diseases such as cardiovascular and cancer (Bahukhanid et al., 2018; Hermanns et al., 2020). Currently, more benefits of carotenoids for nutrition and human health have been discovered through several prospective epidemiological studies. Various findings have revealed a positive relationship between carotenoid-based diet intake from plants and a decrease in the risk of chronic diseases (Bohn, 2016). However, there is still a lack of studies documenting detailed information on ulam and carotenoid composition. Table 3 shows a list of carotenoids from ulam species studied from 2010 to 2022. Based on Table 3, most researchers only focus on one or two species in their studies.

## Benefit of Carotenoids

Carotenoids are pigments found in fruits and vegetables that offer many health benefits as they are good antioxidants and antimicrobials (Koyande et al., 2019). Carotenoids also support cardiovascular health by lowering blood pressure, improving blood lipid profile, and reducing inflammation. Lycopene and beta-cryptoxanthin, for example, have anti-inflammatory effects that can reduce the risk of chronic diseases such as heart disease and diabetes. Other carotenoids, such as lutein and zeaxanthin, accumulate in the eyes and help protect against age-related macular degeneration and cataracts (Carrillo et al., 2022). Carotenoids can also protect the skin from UV damage and can inhibit skin aging and the risk of cancer (Anbualakan et al., 2023).

Currently, many people practice antioxidant intake as an alternative to maintaining health and protecting against oxidative stress. However, antioxidants in supplements or dietary supplements on the market usually contain synthetic antioxidants such as BHA and BHT, which have been reported to contain toxins and can cause cancer (Tanvir et al., 2017). A study by Suryanti et al. (2016) supports that there is now a trend to replace synthetic antioxidants with natural antioxidants in food, pharmaceutical, and cosmetic products due to the long-term effects of synthetic antioxidant intake reported.

Furthermore, Oladele et al. (2020) and Ajayi et al. (2017) reported that a high antioxidant diet from vegetables and fruits also has positive effects on health. Nutrient-rich foods are essential for human growth, but with the presence of bioactive compounds from natural plants, they can provide multiple benefits and prevent the risk of diseases. For example, a high intake of β-carotene-rich foods results in a nearly 50% reduction in the risk of throat and mouth cancer, proving that β-carotene is an important active ingredient in preventing pharynx, mouth, and larynx cancers (Naz et al., 2020).

Moreover, previous scientific studies have also proven that herbs and spices contain good antimicrobial effects (Okmen et al., 2016; Jarriyawattanachaikul et al., 2016). A study by Adem et al. (2020) also showed that plant extracts containing effective antimicrobial compounds are effective in treating respiratory problems, parasitic infections, and slowing the progression of chronic diseases. According to Jarriyawattanachaikul et al. (2016), herbal plants are one of the alternatives capable of treating and reducing pathogen contamination in food such as Escherichia coli. Additionally, a study by Okmen et al. (2016) reported that some antibiotics cannot be used for long periods because bacteria will continue to develop resistance to antibiotics, and some bacteria can worsen diseases, slow recovery time, and cause unwanted deaths. Therefore, it is important to find more effective antimicrobial compounds with minimal side effects (Okmen et al., 2016). So far, many plants have yet to be explored as alternatives to effective natural active compounds to prevent infectious diseases (Adem et al., 2020).

## The Use of Ulam as a Source of Halal Active Pharmaceutical Ingredients from an Islamic Perspective

According to Islamic law, everything used, whether consumed or applied, must be sourced from halal and sharia-compliant sources. Therefore, it is the responsibility of those in charge to adhere to the shariah requirements in producing pharmaceutical products, especially concerning the use of raw materials and production procedures to fulfil the principles of maqasid al-shariah (Sidek & Ahmad, 2017). Until now, there is still no specific law to regulate the halal pharmaceutical industry in Malaysia. The halal pharmaceutical industry is governed under the same laws as the halal food industry, namely the Trade Descriptions Act 2011 and the Trade Descriptions (Certification and Halal Labeling) Order 2011 (Ramli et al., 2018).

According to Ramli et al. (2018), to ensure that halal pharmaceutical products reach consumers, the process must start from the manufacturing stage because products that are monitored from the initial stage will not face halal assurance issues. The Department of Islamic Development Malaysia (JAKIM) is the authoritative body and administrative body that monitors and enforces halal regulations in Malaysia. In terms of halal certification of pharmaceutical products in Malaysia, the basic compliance reference is the Standard Malaysia MS 2424:2019 General Requirements for Halal Pharmaceuticals, Pharmaceutical Inspection Cooperation Scheme (PIC) Scheme, the National Fatwa Committee of Islamic Affairs or Fatwas issued by states, and relevant acts, regulations, standards, and guidelines (Johari & Zalina, 2016).

Ulam has long been practiced based on the beliefs of past societies regarding its benefits and goodness besides being a halal natural source for consumption. Therefore, ulam has the potential to be one of the sources of API in pharmaceutical products, which can benefit not only Muslim consumers but also contribute significantly to the advancement of Islam. It can elevate and revive the pharmacological knowledge founded by Ibn Sina and Al-Razi, as ancient pharmacy medications were based on plants meticulously studied and confirmed to have medicinal value (Said et al., 2018).

Thibbun Nabawi refers to Islamic medicine based on the healing practices practiced by the Prophet Muhammad (PBUH). This practice has been compiled and documented by an Islamic scholar named Ibn Al-Qayyim Al-Jawziyyah, and the book is called Al-Thibb Al-Nabawi (Khatib, 2021). This book extensively explains natural remedies used by the Prophet Muhammad (PBUH) from animal and plant sources that function for treatment and healing. Among the plants and medicinal values listed in this book are ginger, cucumber, celery, onions, and others (Al-Ğauzīya et al., 1998). This book has proven that Islamic medical knowledge practiced since the time of the Prophet Muhammad (PBUH) has been clearly and comprehensively explained. This can further strengthen the faith of Muslims to choose not only halal treatments but also halal pharmaceutical products.

### *Contribution to Muslim Consumers*

Ramli et al. (2018) elaborate on Sheikh al-Qardawi's thoughts regarding maqasid al-shariah, which provide benefits (maslahah) to support worldly and hereafter needs, physical and spiritual, for individuals and societies from the present generation to future generations. One factor contributing to human health development is the intake of food or any substances consumed by the body. As Muslims, in shaping a good life in this world and the hereafter, this factor needs to be emphasized so that whatever enters the body must be guaranteed for its halal status, efficacy, and quality, not only in daily food consumption but also in medications. In a narration transmitted by Ibn Mardawaihi from Ibn Abbas, it is mentioned that a companion of the Prophet Muhammad (PBUH), Saad Ibn Abi Waqqas, asked the Prophet Muhammad (PBUH) to pray to Allah to grant every prayer made to Allah. The Prophet Muhammad (PBUH) then said; "O Saad! Improve your food, and you will be made one whose supplication is answered by Allah. By the Lord in whose hand Muhammad's soul is, verily a man throws a haram morsel into his stomach, and his deeds will not be accepted for forty days. Whoever amongst Allah's servants grows his flesh from haram wealth and riba (usury), then the fire is more fitting for him." (Hadith narrated by At-Thabrani).

Therefore, it is an obligation for every Muslim to seek halal sources to ensure that their worship is accepted in line with the goals of maqasid al-shariah (Ramli et al., 2018).

### *Contribution to Sustainable Environment*

An estimated 4000 active pharmaceutical ingredients registered have been used in pharmaceutical products, and 10% of these products have the potential to pollute the environment (Kuster & Adler, 2014; Weber et al., 2016). Furthermore, the pharmaceutical industry is projected to grow by 6.5% annually towards 2022 (UN Environment, 2019). Among the pollution from these pharmaceutical sources is antimicrobial pollution or antimicrobial resistance (AMR), reported to be highly critical as its presence will make the environment a reservoir and dissemination site for dangerous pathogens (OECD, 2019). Additionally, the increase in AMR will lead to a reduction of 2 to 3.5% in Gross Domestic Product (GDP) globally with costs reaching USD 100 trillion. If left unchecked, it will not only impact human health and the environment but also the global economy (OECD, 2019).

Therefore, as an alternative to reduce environmental pollution, it is suggested to replace active pharmaceutical ingredients from synthetic sources with natural sources such as plants, which are more environmentally friendly to ensure a sustainable environment.

## Conclusion

Fruits and vegetables, including traditional Malay vegetables like ulam, are rich sources of carotenoids, which play a crucial role in human health. The carotenoid content of ulam, in particular, demonstrates significant potential as natural and halal APIs. Hence, promoting the use of ulam as a source of carotenoid-rich diet and exploring its use as natural and halal APIs can offer positive health effects to consumers and contribute to the development of a sustainable and responsible pharmaceutical industry. This is because the use of synthetic APIs can lead to long-term health effects and environmental degradation. The development of the halal pharmaceutical industry is not only to meet the demands of Muslim consumers but also provides an opportunity to promote high-quality assurance and healthy lifestyles through the production of medications or health care products based on natural and halal active pharmaceutical ingredients.

## Acknowledgments

The author would like to express gratitude to Universiti Kebangsaan Malaysia for sponsoring this research study through the Young Researcher Incentive Grant (Grant No. GGPM-2021-051) and Sheikh Abdullah Fahim Research Grant (Grant No. RH-2020-006).

## References

Abdusalam, K. B., Yee, L. S., Mediani, A., Akhtar, M. T., Buzgaia, N., Rukayadi, Y., Ismail, I. S., & Shaari, K. (2022). 1H NMR-based metabolomics profiling of *Syzygium grande* and *Oenanthe javanica* and relationship between their metabolite compositions and antimicrobial activity against Bacillus species. *Records of Natural Products, 16*(2), 128-143. https://doi.org/10.25135/rnp.2101.1927.

Adem, S. R., Ayangbenro, A. S., & Gopane, R. E. (2020). Phytochemical screening and antimicrobial activity of *Olea europaea* subsp. africana against pathogenic microorganisms. *Scientific African, 10*, e00548. https://doi.org/10.1016/j.sciaf.2020.e00548.

Ajayi, E. I. O., Adeleke, M. A., Adewumi, T. Y., & Adeyemi, A. A. (2017). Antiplasmodial activities of ethanol extracts of *Euphorbia hirta* whole plant and *Vernonia amygdalina* leaves in *Plasmodium berghei*-infected mice. *Journal of Taibah University for Science, 11*(6), 831-835. https://doi.org/10.1016/j.jtusci.2017.01.004.

Alamgir, A. N. M. (2017). Drugs: Their natural, synthetic, and biosynthetic sources. In *Therapeutic Use of Medicinal Plants and Their Extracts: Volume 1* (pp. 105-123). Springer, Cham: New York. https://doi.org/10.1007/978-3-319-63862-1_4.

Al-Ǧauzīya, I.-Q., Ibn Qayyim Al-Jawziyya, M. I.-A.-B. (1998). *Medicine of the Prophet* (P. Johnstone, Trans.). Cambridge, UK: The Islamic Texts Society.

Ali, H., Karsani, S. A., Othman, R., & Yaacob, J. S. (2018). Production of coloured callus in *Orthosiphon stamineus* Benth and antioxidant properties of the extracted pigments. *Pigment & Resin Technology, 47*(3), 196-207. https://doi.org/10.1108/PRT-01-2017-0009.

Anbualakan, K., Tajul Urus, N. Q., Makpol, S., Jamil, A., Mohd Ramli, E. S., Md Pauzi, S. H., & Muhammad, N. (2023). A scoping review on the effects of carotenoids and flavonoids on skin damage due to ultraviolet radiation. *Nutrients, 15*(1), 1-17.

Arasu, M. V., Jeon, J., Park, Y. J., Kwon, S. J., Al-Dhabi, N. A., & Park, S. U. (2017). Medically important carotenoids from *Momordica charantia* and their gene expressions in different organs. *Saudi Journal of Biological Sciences, 24*(8), 1913-1919. https://doi.org/10.1016/j.sjbs.2017.01.019.

Arif, T., & Shetty, G. R. (2020). Therapeutic potential and traditional uses of *Sauropus androgynous*: A review. *Journal of Pharmacognosy and Phytochemistry, 9*, 2131-2137.

Bohn, T. (2016). Bioactivity of carotenoids–chasms of knowledge. *International Journal for Vitamin and Nutrition Research, 10*, 1-5. https://doi.org/10.1024/0300-9831/a000400.

Boy, H. I. A., Rutilla, A. J. H., Santos, K. A., Ty, A. M. T., Alicia, I. Y., Mahboob, T., ... & Nissapatorn, V. (2018). Recommended medicinal plants as source of natural products: A review. *Digital Chinese Medicine, 1*(2), 131-142. https://doi.org/10.1016/j.digmed.2019.07.003.

Carrillo, C., Nieto, G., Martínez-Zamora, L., Ros, G., Kamiloglu, S., Munekata, P. E. S., Pateiro, M., Lorenzo, J. M., Fernández-López, J., Viuda-Martos, M., Pérez-Álvarez, J. Á., & Barba, F. J. (2022). Novel approaches for the recovery of natural pigments with potential health effects. *Journal of Agricultural and Food Chemistry, 70*(23), 6864-6883. https://doi.org/10.1021/acs.jafc.1c07208.

Chan, E. W. C., Wong, S. K., & Chan, H. T. (2017). Ulam herbs of *Oenanthe javanica* and *Cosmos caudatus*: An overview on their medicinal properties. *Journal of Natural Remedies, 16*(4), 137-147.

Chowtivannakul, P., Srichaikul, B., & Talubmook, C. (2016). Antidiabetic and antioxidant activities of seed extract from *Leucaena leucocephala* (Lam.) de Wit. *Agriculture and Natural Resources, 50*(5), 357-361.

Chung, K. C. (2008). Aktiviti antimikrob ekstrak herba *Andrographis paniculata*, *Centella asiatica* dan *Orthosiphon stamineus* serta potensinya sebagai antimikrob semula jadi dalam daging lembu kisar (Master's thesis). Universiti Sains Malaysia, Pulau Pinang.

Dhama, K., Sharun, K., Gugjoo, M. B., Tiwari, R., Alagawany, M., Iqbal Yatoo, M., ... & Farag, M. R. (2021). A comprehensive review on chemical profile and pharmacological activities of *Ocimum basilicum*. *Food Reviews International*, 1-29. https://doi.org/10.1080/87559129.2021.1900230.

Divya, P., Puthusseri, B., & Neelwarne, B. (2012). Carotenoid content, its stability during drying and the antioxidant activity of commercial coriander (*Coriandrum sativum* L.) varieties. *Food Research International, 45*(1), 342-350. https://doi.org/10.1016/j.foodres.2011.11.013.

Eggersdorfer, M., & Wyss, A. (2018). Carotenoids in human nutrition and health. *Archives of Biochemistry and Biophysics, 652*, 18-26. https://doi.org/10.1016/j.abb.2018.02.003.

Eng Khoo, H., Azlan, A., & Ismaila, A. (2015). *Sauropus androgynus* leaves for health benefits: Hype and the science. *The Natural Products Journal, 5*(2), 115-123.

Fernández-Marín, B., Míguez, F., Méndez-Fernández, L., Agut, A., Becerril, J. M., García-Plazaola, J. I., Kranner, I., & Colville, L. (2017). Seed carotenoid and tocochromanol composition of wild *Fabaceae* species is shaped by phylogeny and ecological factors. *Frontiers in Plant Science, 8*, 1428. https://doi.org/10.3389/fpls.2017.01428.

Hamzah, N. (2015). *Kajian etnoperubatan, analisis fitokimia dan aktiviti biologi tumbuh-tumbuhan antidiabetik terpilih yang digunakan oleh masyarakat orang asli di Gua Musang, Kelantan* (Doctoral dissertation). Universiti Sains Malaysia, Pulau Pinang.

Hatta, F. A. M., & Othman, R. (2020). Carotenoids as potential biocolorants: A case study of astaxanthin recovered from shrimp waste. In C. M. Galanakis (Ed.), *Carotenoids: Properties, processing and applications* (pp. 289-325). Academic Press: London.

Hermanns, A. S., Zhou, X., Xu, Q., Tadmor, Y., & Li, L. (2020). Carotenoid pigment accumulation in horticultural plants. *Horticultural Plant Journal, 6*(6), 343-360. https://doi.org/10.1016/j.hpj.2020.03.004.

Hirose, R., Nakaya, T., Naito, Y., Daidoji, T., Bandou, R., Inoue, K., Dohi, O., Yoshida, N., Konishi, H., & Itoh, Y. (2019). Situations leading to reduced effectiveness of current hand hygiene against infectious mucus from influenza virus-infected patients. *Msphere, 4*(5), 10-1128. https://doi.org/10.1128/msphere.00474-19.

Islam, R., Uddin, M. N., Ashrafuzzaman, M., & Hoque, M. I. U. (2018). Phenolics and carotenoids contents and radical scavenging capacity of some selected solanaceous medicinal plants. *Journal of the Bangladesh Agricultural University, 16*(1), 56-61.

Ismail, S. (2020). Sayuran tradisional ulam dan penyedap rasa. Penerbit Universiti Kebangsaan Malaysia: Bangi, Selangor.

Jahurul, M. H. A., Patricia, M., Shihabul, A., Norazlina, M. R., George, M. R., Noorakmar, A. W., Lee, J. S., Jumardi, R., Jinap, S., & Zaidul, I. S. M. (2021). A review on functional and nutritional properties of noni fruit seed (*Morinda citrifolia* L.) and its oil. *Food Bioscience, 41*, 101000. https://doi.org/10.1016/j.fbio.2021.101000.

Jarriyawattanachaikul, W., Chaveerach, P., & Chokesajjawatee, N. (2016). Antimicrobial activity of Thai-herbal plants against food-borne pathogens *E. coli*, *S. aureus* and *C. jejuni*. *Agriculture and Agricultural Science Procedia, 11*, 20-24. https://doi.org/10.1016/j.aaspro.2016.02.005.

Johari, A. L., & Zalina, Z. (2016). The implementation of Malaysian Standard MS 2424:2012 Halal Pharmaceuticals-General Guidelines. *Journal of Shariah Law Research, 1*(1), 119-140.

Khatib, S. (2021). The legal istinbâth of Ibnu Qayyim Al-Jauziyah in the terms of medication and health. *Madania: Jurnal Kajian Keislaman, 24*(2), 233-246.

Kim, M. J., Moon, Y., Tou, J. C., Mou, B., & Waterland, N. L. (2016). Nutritional value, bioactive compounds and health benefits of lettuce (*Lactuca sativa* L.). *Journal of Food Composition and Analysis, 49*, 19-34. https://doi.org/10.1016/j.jfca.2016.02.007

Kookana, R. S., Williams, M., Boxall, A. B., Larsson, D. J., Gaw, S., Choi, K., Yamamoto, H., Thatikonda, S., Zhu, Y. G., & Carriquiriborde, P. (2014). Potential ecological footprints of active pharmaceutical ingredients: An examination of risk factors in low-, middle-and high-income countries. *Philosophical Transactions of the Royal Society B: Biological Sciences, 369*(1656), 20130586. https://doi.org/10.1098/rstb.2013.0586

Koyande, A. K., Chew, K. W., Rambabu, K., Tao, Y., Chu, D. T., & Show, P. L. (2019). *Microalgae: A potential alternative to health supplementation for humans*. Elsevier B.V.

Kueh, B. L. (2007). *Pencirian sebatian bioaktif daripada Triphasia trifolia (Burm. F.) P. Wilson (Limau Kaya) sebagai bahan antibakteria dan bahan penyingkir radikal bebas* [Doctoral dissertation, Universiti Sains Malaysia, Pulau Pinang]. Universiti Sains Malaysia.

Kumar, V., Bansal, V., Madhavan, A., Kumar, M., Sindhu, R., Awasthi, M. K., Binod, P., & Saran, S. (2022). Active pharmaceutical ingredient (API) chemicals: A critical review of current biotechnological approaches. *Bioengineered, 13*(2), 4309–4327. https://doi.org/10.1080/21655979.2022.2045634.

Küster, A., & Adler, N. (2014). Pharmaceuticals in the environment: Scientific evidence of risks and its regulation. *Philosophical Transactions of the Royal Society B: Biological Sciences, 369*(1656), 20130587. https://doi.org/10.1098/rstb.2013.0587.

Laribi, B., Kouki, K., M'Hamdi, M., & Bettaieb, T. (2015). Coriander (*Coriandrum sativum* L.) and its bioactive constituents. *Fitoterapia, 103*, 9-26. https://doi.org/10.1016/j.fitote.2015.02.011.

Latiff, N. A., Ong, P. Y., Abd Rashid, S. N. A., Abdullah, L. C., Mohd Amin, N. A., & Fauzi, N. A. M. (2021). Enhancing recovery of bioactive compounds from *Cosmos caudatus* leaves via ultrasonic extraction. *Scientific Reports, 11*(1), 1-12. https://doi.org/10.1038/s41598-021-83457-3.

Lee, H. W., Bi, X., & Henry, C. J. (2022). Carotenoids, tocopherols and phylloquinone content of 26 green leafy vegetables commonly consumed in Southeast Asia. *Food Chemistry*, 132729. https://doi.org/10.1016/j.foodchem.2022.132729.

Lee, S. H., Jeong, Y. S., Song, J., Hwang, K. A., Noh, G. M., & Hwang, I. G. (2017). Phenolic acid, carotenoid composition, and antioxidant activity of bitter melon (*Momordica charantia* L.) at different maturation stages. *International Journal of Food Properties, 20*(sup3), S3078-S3087. https://doi.org/10.1080/10942912.2016.1237961

Meléndez-Martínez, A. J. (2019). An overview of carotenoids, apocarotenoids, and vitamin A in agro-food, nutrition, health, and disease. *Molecular Nutrition & Food Research, 63*(15), 1801045. https://doi.org/10.1002/mnfr.201801045

Nain, N. K., Nazefah, A. H., Ilina, C. C., Dzulkhairi, M. M., Syahrina, R. N., Fadzilah, Z. N., & Norhaizam, M. S. (2013). Towards a halal certification module and training programme development for the pharmaceutical industry. *Middle-East Journal of Scientific Research, 13*(13), 49-53.

National Toxic Network. (2015). Pharmaceutical pollution factsheet. https://ntn.org.au/wp-content/uploads/2015/05/Fact-Sheet-pharmaceutical-pollutants-1.pdf [10 Januari 2022].

Naz, T., Nazir, Y., Nosheen, S., Ullah, S., Halim, H., Fazili, A. B. A., Li, S., Mustafa, K., Mohamed, H., Yang, W., & Song, Y. (2020). Redirecting metabolic flux towards the mevalonate pathway for enhanced β-carotene production in *M. circinelloides* CBS 277.49. *BioMed Research International, 2020*, 1–12. https://doi.org/ 10.1155/2020/8890269.

Norhayati, Y., NorAini, M. F., Misri, K., Marziah, M., & Azman, J. (2011). α-Tocopherol, ascorbic acid and carotenoid content in *Centella asiatica* leaf tissues and callus cultures. *Pertanika Journal of Tropical Agricultural Science, 34*(2), 331-339.

OECD. (2019). Pharmaceutical residues in freshwater: Hazards and policy responses. *OECD Studies on Water, OECD Publishing*, Paris. www.oecd.org/water [22 Mac 2022].

Okmen, G., Kardas, S., Bayrak, D., Arslan, A., & Cakar, H. (2016). The antibacterial activities of *Crocus sativus* against mastitis pathogens and its antioxidant activities. *World Journal of Pharmacy and Pharmaceutical Sciences, 5*(3), 146-156.

Oladele, J. O., Ajayi, E. I., Oyeleke, O. M., Oladele, O. T., Olowookere, B. D., Adeniyi, B. M., & Oladiji, A. T. (2020). A systematic review on COVID-19 pandemic with special emphasis on curative potentials of Nigeria based medicinal plants. *Heliyon, e04897.*

Othman, A., Mariani, A., & Wan Lutfi, W. J. (2021). Active pharmaceutical ingredients (APIs) in Malaysian tap water: A way forward for upcoming legislative framework. *Environmental Policy and Law*, 1-8. https://doi.org/10.3233/EPL-201041.

Othman, R., Zaifuddin, F. A. M., & Hassan, N. M. (2014). Carotenoid biosynthesis regulatory mechanisms in plants. *Journal of Oleo Science, 63*(8), 753-760.

Othman, R., Mohd Zaifuddin, F. A., & Mohd Hassan, N. (2015). Potential sources for lipid soluble food colorants from selected Malaysian traditional vegetables. *Malaysian Journal of Analytical Sciences, 19*(1), 268–274.

Othman, R., Zaifuddin, F. A. M., & Hassan, N. M. (2017). Characterisation of carotenoid and total retinol equivalent content in ulam and medicinal species as alternative food intervention to combat vitamin A deficiency. *Journal of Pharmacy and Nutrition Sciences, 7*(3), 81-87.

Oyewole, O. I., Oladele, J. O., & Oladele, O. T. (2017). Methanolic leaf extract of *Ficus exasperata* attenuates arsenate-mediated hepatic and renal oxidative stress in rats. *Research Journal of Health Sciences, 5*(2), 115-123.

O'Flynn, D., Lawler, J., Yusuf, A., Parle-McDermott, A., Harold, D., Mc Cloughlin, T., & White, B. (2021). A review of pharmaceutical occurrence and pathways in the aquatic environment in the context of a changing climate and the COVID-19 pandemic. *Analytical Methods, 13*(5), 575-594. https://doi.org/10.1039/d0ay02098b.

Ramli, N., Amin, N., Zawawi, M., & Aziz, N. A. (2018). Healthcare service: Halal pharmaceutical in Malaysia, issues and challenges. *The Malaysian Journal of Consumer and Family Economics*, 101-111.

Rizk, E. M., El-Kady, A. T., & El-Bialy, A. R. (2014). Characterization of carotenoids (*lyco-red*) extracted from tomato peels and its uses as natural colorants and antioxidants of ice cream. *Annals of Agricultural Sciences, 59*(1), 53-61. https://doi.org/10.1016/j.aoas.2014.02.003.

Rodríguez-Rodríguez, E., Sánchez-Prieto, M., & Olmedilla-Alonso, B. (2020). Assessment of carotenoid concentrations in red peppers (*Capsicum annuum*) under domestic refrigeration for three weeks as determined by HPLC-DAD. *Food Chemistry: X, 6*, 100092. https://doi.org/10.1016/j.fochx.2020.100092.

Rodriguez-Concepcion, M., Avalos, J., Bonet, M. L., Boronat, A., Gomez-Gomez, L., Hornero-Mendez, D., Limon, M. C., & Zhu, C. (2018). A global perspective on carotenoids: Metabolism, biotechnology, and benefits for nutrition and health. *Progress in Lipid Research, 70*, 62-93. https://doi.org/10.1016/j.plipres.2018.01.002.

Said, M. M. T., Ismail, D., Abd Rahim, F., Rahman, A. A., & Khairuldin, W. M. K. F. W. (2018). Transformasi dan sumbangan Al-Razi (854-932M) dalam bidang perubatan. *Jurnal Islam Dan Masyarakat Kontemporari, 17*(1), 41-50.

Segneanu, A. E., Velciov, S. M., Olariu, S., Cziple, F., Damian, D., & Grozescu, I. (2017). Bioactive molecules profile from natural compounds. In *Asao T. & Md Asaduzzaman* (Eds.), *Amino Acid—New Insights and Roles in Plant and Animal* (pp. 209-228). InTech: Croatia.

Sidek, T., & Ahmad, R. (2017). Aplikasi al-istihalah dalam industri halal semasa menurut perspektif maqasid al-shariah. *International Journal of Humanities Technology and Civilization, 2*(1), 1-9.

Singh, G., Singh, A., Singh, P., Gupta, A., Shukla, R., & Mishra, V. K. (2021). Sources, fate, and impact of pharmaceutical and personal care products in the environment and their different treatment technologies. In *Kumar, A., Singh, V. K., Singh, P. & Mishra, V. K.* (Eds.), *Microbe Mediated Remediation of Environmental Contaminants* (pp. 391-407). Woodhead Publishing: Cambridge.

Sun, T., & Li, L. (2020). Toward the 'golden' era: The status in uncovering the regulatory control of carotenoid accumulation in plants. *Plant Science, 290*, 110331. https://doi.org/10.1016/j.plantsci.2019.110331

Suryanti, V., Marliyana, S. D., & Putri, H. E. (2016). Effect of germination on antioxidant activity, total phenolics, β-carotene, ascorbic acid and α-tocopherol contents of lead tree sprouts (*Leucaena leucocephala* (lmk.) de Wit). *International Food Research Journal, 23*(1), 167-172.

Tanvir, E. M., Hossen, M., Hossain, M., Afroz, R., Gan, S. H., Khalil, M., & Karim, N. (2017). Antioxidant properties of popular turmeric (*Curcuma longa*) varieties from Bangladesh. *Journal of Food Quality*, 8471785. https://doi.org/10.1155/2017/8471785.

Tuan, P. A., Kim, J. K., Park, N. I., Lee, S. Y., & Park, S. U. (2011). Carotenoid content and expression of phytoene synthase and phytoene desaturase genes in bitter melon (*Momordica charantia*). *Food Chemistry, 126*(4), 1686-1692. https://doi.org/10.1016/j.foodchem.2010.11.042.

UN Environment. (2019). Global Chemicals Outlook II: From legacies to innovative solutions. *United Nations Environment Programme*. https://wedocs.unep.org/bitstream/handle/20.500.11822/27651/GCOII_synth.pdf?sequence=1&isAllowed=y [22 Mac 2022].

USDA. (2021). Medicinal Botany. https://www.fs.usda.gov/wildflowers/ethnobotany/medicinal/index.shtml [28 April 2021].

Vikram, P., Chiruvella, K. K., Ripain, I. H. A., & Arifullah, M. (2014). A recent review on phytochemical constituents and medicinal properties of kesum (*Polygonum minus* Huds.). *Asian Pacific Journal of Tropical Biomedicine, 4*(6), 430-435. https://doi.org/10.1016/S2221-1691(14)60213-3.

Weber, F. A., aus der Beek, T., Bergmann, A., Carius, A., Grüttner, G., Hickmann, S., ... & Stolzenberg, H. C. (2016). Pharmaceuticals in the environment-the global perspective: Occurrence, effects, and potential cooperative action under SAICM. *German Environment Agency*.

Wurtzel, E. T. (2019). Changing form and function through carotenoids and synthetic biology. *Plant Physiology, 179*(3), 830-843. https://doi.org/10.1104/pp.18.01101.

Yabuzaki, J. (2017). Carotenoids Database: Structures, chemical fingerprints and distribution among organisms. *Database, 2017*, 1-11. https://doi.org/10.1093/database/bax029.

You, Y., Shahar, S., Haron, H., & Yahya, H. (2018). More ulam for your brain: A review on the potential role of ulam in protecting against cognitive decline. *Sains Malaysiana, 47*(11), 2713-2729.

You, Y. X., Shahar, S., Haron, H., Yahya, H. M., & Che Din, N. (2020a). High traditional Asian vegetables (ulam) intake relates to better nutritional status, cognition and mood among aging adults from low-income residential areas. *British Food Journal, 122*(10), 3179-3191. https://doi.org/10.1108/BFJ-04-2020-0302.

You, Y. X., Shahar, S., Mohamad, M., Yahya, H. M., Haron, H., & Abdul Hamid, H. (2020b). Does traditional Asian vegetables (ulam) consumption correlate with brain activity using fMRI? A study among aging adults from low-income households. *Journal of Magnetic Resonance Imaging, 51*(4), 1142-1153. https://doi.org/10.1002/jmri.26961.

Zarina, Z., Ruzaidi, C. M., Sam, S. T., Bakri, A. M. A., & Aminah, M. H. (2018). Bioactive compound analysis in seeds of *Leucaena leucocephala* (*Petai Belalang*). *AIP Conference Proceedings, 2030*(1), 020263. https://doi.org/10.1063/1.5065884.

Zheng, X., Giuliano, G., & Al-Babili, S. (2020). Carotenoid biofortification in crop plants: Citius, altius, fortius. *Biochimica et Biophysica Acta (BBA)-Molecular and Cell Biology of Lipids, 1865*(11), 158664. https://doi.org/10.1016/j.bbalip.2020.158664.

## Chapter 4

# Exploring Algae Pigments for Eco-Friendly Design and Energy Efficiency in Architecture

**Muhammad Hanafi Ismail[1], MSc,**
**Wan Syibrah Hanisah Wan Sulaiman[2,*], PhD**
**Rashidi Othman[3], PhD**
**Nursyafica Nadia Johari[4], PhD**
**and Aina Zulaikha Muzafar Shah[5]**
[2,5]International Institute for Halal Research and Training,
International Islamic University Malaysia, Kuala Lumpur, Malaysia
[1,3,4]Herbarium Unit, Department of Landscape Architecture, Kulliyyah of Architecture and Environmental Design, International Islamic University Malaysia,
Kuala Lumpur, Malaysia

## Abstract

The increasing demand for sustainable and energy-efficient solutions in the built environment has emphasized the innovative use of natural materials, focusing on algae pigments. This chapter explores the diverse applications of algae pigments in improving sustainability and energy efficiency within modern architecture. This chapter introduces algae and their diverse cell growth modes—autotrophic, mixotrophic, and heterotrophic—before exploring the potential of algae-based materials as sustainable alternatives in construction. The discussion examines current design trends incorporating algae into the built environment, highlighting their contribution to carbon sequestration, thermal insulation, and daylight optimization. It evaluates both the environmental benefits of

* Corresponding Author's Email: syibrahanis@iium.edu.my

In: Natural Pigments: Sources, Properties, and Health Benefits
Editors: Rashidi Othman, Qurratu Aini Mat Ali, Razanah Ramya et al.
ISBN: 979-8-90134-055-4

these applications and their significant potential to reduce energy consumption. Furthermore, this chapter provides a contextual analysis of algae utilization in Malaysia, showcasing regional practices and advancements. By integrating algae pigments into various aspects of architectural design and functionality, this chapter illustrates a comprehensive approach to achieving sustainable and energy-efficient built environments. This document synthesizes current research, practical applications, and future trends to inspire and guide architects, engineers, and sustainability practitioners about algae pigments in promoting sustainability within the built environment.

**Keywords**: algae pigments, sustainable design, energy efficiency

## Introduction

In 1890, the first microbiologist, Beijerinck, successfully cultivated pure cultures of Chlorella vulgaris, marking the beginning of the scientific investigation of microalgae near the close of the 19th century. This accomplishment facilitated the study into microalgae, including studies on photosynthesis. In 1919, Otto Warburg investigated Chlorella to deepen the understanding of oxygen produced during photosynthetic processes. Seaweed, commonly known as macroalgae, consists of enormous, multicellular algae that can be visible without magnification. On the other hand, microalgae are tiny, single-celled algae that can be classified as either eukaryotic, such as green algae (Chlorophyta), or prokaryotic, such as cyanobacteria (Chloroxybacteria) (Khan et al., 2018).

Algae are divided into two main groups based on their size and physical characteristics: microalgae and macroalgae. Microalgae comprise diverse photosynthetic microorganisms capable of converting carbon dioxide into valuable compounds, including bioactive substances used in animal feed, pharmaceuticals, biofuels, and food products (Elmeligy & Elhassan, 2019) In addition, the commonly utilized microalgae can be divided into four main groups: Chrysophyceae (which includes golden algae), Bacillariophyceae (which includes diatoms), and Cyanophyceae (which includes blue-green algae). The unique characteristics of the algae play a vital role in sustainability. Notably, one characteristic is their rapid growth rate with biomass doubling approximately every few hours (Elmeligy & Elhassan, 2019). Another significant characteristic is their photosynthetic capacity, which positions algae as essential contributors to sustainable biofuel production and $CO_2$

absorption boosting higher oil production and photosynthetic efficiency compared to other organisms. It is estimated that algae contribute around 50% of the Earth's total oxygen production, highlighting their critical importance in maintaining atmospheric balance and supporting life on the planet. Through photosynthesis, marine and freshwater algae transform carbon dioxide into oxygen, establishing themselves as crucial primary producers in aquatic ecosystems and key players in global oxygen cycles (Elmeligy & Elhassan, 2019). In addition, microalgae are recognized as promising natural sources of a variety of bioactive compounds, including polysaccharides such as paramylon, polyunsaturated fatty acids, and various pigments such as phycocyanin, phycoerythrin, and astaxanthin (Jareonsin & Pumas, 2021).

## Algae Cell Growth Mode (Autotrophic, Mixotrophic and Heterotrophic)

As shown in Figure 1, algae can grow under different nutritional modes, such as autotrophic, mixotrophic, and heterotrophic, which are characterized by the type of energy and carbon sources they use for growth and metabolism.

### *Autotrophic*

Autotrophic microalgae utilize light energy and carbon dioxide through photosynthesis, offering a sustainable approach to biomass production (Da Silva et al., 2019). The efficiency of such systems is significantly limited by several factors, including insufficient light penetration in densely populated cultures, heightened operational costs, and challenges associated with maintaining monocultures (Jareonsin & Pumas, 2021). The process is also inefficient in effectively reducing high levels of COD (Da Silva et al., 2021). Under photosynthetic autotrophic conditions, low microalgal biomass often results in reduced protein productivity, while the risk of microbial contamination further decreases production efficiency (Xie et al., 2023). While particular species exhibit adaptability to low-light conditions, the presence of incomplete metabolic pathways may necessitate supplementary carbon sources, thereby constraining their applicability in high-yield industrial contexts.

***Heterotrophic***

Heterotrophic microalgae can grow without light utilizing organic compounds as carbon and energy sources (Xie et al., 2023). Unlike autotrophic microalgae that depend on photosynthesis, heterotrophs metabolize organic substrates, such as glucose, facilitating rapid growth and biomass accumulation, thereby making them suitable for high-protein production. This cultivation mode has successfully produced high-value compounds in controlled dark conditions. Heterotrophic microalgae present several advantages over autotrophic systems, including higher cell density growth capacity, scalability in industrial fermenters, and compliance with established FDA standards (Jareonsin & Pumas, 2021). However, despite its advantages, heterotrophic cultivation presents certain limitations, including the high expense of organic carbon substrates, an elevated risk of contamination or competition from other microorganisms, and the lack of light-induced metabolites that are usually present in autotrophic systems (Jareonsin & Pumas, 2021).

***Mixotrophic***

Mixotrophic microalgae cultivation combines autotrophic and heterotrophic mechanisms, enabling cells to simultaneously utilize light energy and organic carbon sources such as glucose or acetate, for growth (Jareonsin & Pumas, 2021). This dual pathway allows the concurrent operation of photosynthetic and respiratory metabolisms, enhancing biomass productivity. In open-pond systems, small quantities of organic carbon are often added continuously to boost microalgal yields while minimizing the risk of bacterial contamination from excessive organic input (Perez-Garcia et al., 2011). Mixotrophic cultivation has proven effective in achieving high growth rates and significant biomass accumulation, while facilitating the production of valuable photosynthetic metabolites. For example, a cultivation of Chlorella vulgaris in a 10 L fermenter under mixotrophic conditions resulted in a protein content of 60.0% on a dry weight basis, demonstrating the effectiveness of this approach for generating nutrient-rich biomass (Xie et al., 2023).

## Natural Pigment from Algae

Algae are a diverse group of photosynthetic organisms recognized for their distinctive functional characteristics and a wide range of pigments, which present significant opportunities for sustainable applications. Chlorophyll plays a central role in the photosynthesis process in microalgae as the primary pigment. Phycobiliproteins and carotenoids (xanthophylls and carotenes) also serve as auxiliary or secondary pigments (Chen et al., 2017; Sijil et al., 2022). Chlorophylls are essential in converting solar energy into chemical energy during photosynthesis. Chlorophyll a is the most predominant type, optimally absorbing light in red and blue wavelengths, while chlorophyll b enhances absorption in the green wavelength (Paper et al., 2022).

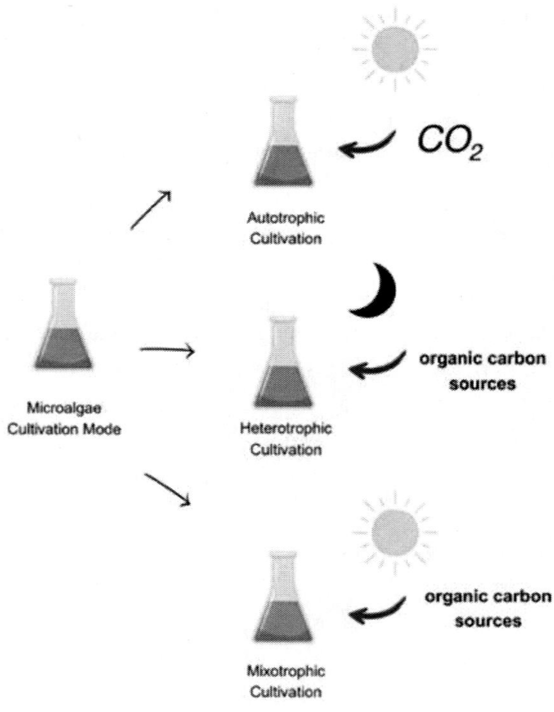


**Figure 1.** Algae Cell Growth Mode.

Furthermore, phycobiliproteins found predominantly in cyanobacteria, red algae, cryptomonads, and cyanelles, are essential for their physiological activities (Zuorro et al., 2021). It works synergistically with chorophyll to broaden the absorption spectrum during photosynthesis (Chen et al., 2023). Phycobiliproteins mostly consist of phycoerythrocyanin, phycoerythrin, and phycocyanin, depending on the specific phycobilin to which they are attached. Besides chlorophylls and phycobiliproteins, algae also possess a range of carotenoids that produce diverse red, orange, and yellow shades. These lipophilic molecules, including β-carotene, lutein, and zeaxanthin, enhance light absorption and protect cells from excessive light energy (Lim et al., 2022). Numerous types of pigments are abundant and widely distributed among algae, whereas some are exclusive to this group. As a result, algae constitute a perfect and indispensable source of pigments for modern industries.

## Algae Species Composition in Malaysia and Other Regions

The composition of algae species varies significantly across different geographical locations, influenced by environmental factors, such as salinity, temperature, light availability, and nutrient levels. In Malaysia, diverse habitats, such as coastal waters, river mouths, and estuarine zones, harbour a wide range of algal communities. Research conducted across multiple Malaysian states, specifically Kelantan, Negeri Sembilan, Selangor, Melaka, Perak, and Sarawak, has highlighted important patterns in the richness and distribution patterns of both marine and freshwater algae. Table 1 shows the comprehensive list of algae species identified in several localities across Malaysia, reflecting the ecological uniqueness of each area. These findings also provide comparative analyses of algae diversity with other regions, contributing to a broader understanding of algal biogeography.

**Table 1**. Algae species composition in Malaysia

| Species | Sources |
|---|---|
| Malaysia Region | |
| | |
| Kelantan, Negeri Sembilan, Selangor | Phang et al., 2015 |
| Chlorella vulgaris | |
| Chlorella volutis | |
| Chlorella vulgaris var. autotrophica | |
| Chlorella rotunda | |
| Chlorella vulgaris var. viridis | |
| Chlorella sorokiana | |
| Chlamydomonas | |
| Scenedesmus bijugatus | |
| Euglena | |
| Chlorococcum vitiosum | |
| Cyanosarcina chroococcoides | |
| Aphanocapsa planctonia | |
| | |
| Blue Lagoon, Melaka | Ismail et al., 2017 |

| Species | Sources |
|---|---|
| Glacilaria blodgettii | |
| Glacilaria vermiculophylla | |
| Glacilaria foliifera | |
| Glacilaria arcuata | |
| Analipus japonicus | |
| Caulerpa sertularioides | |
| Caulerpa lentillifera | |
| Caulerpa serrulata | |
| Caulerpa microphysa | |
| Caulerpa subserrata | |
| Caulerpa verticillate | |
| Chaetomorpha sp. | |
| Codium yezoense | |
| Udotea argentea | |
| Udotea javensis | |
| Ulva fasciata | |
| Ulva Lactuca | |
| Enteromorpha clathrate | |
| Enteromorpha intestinalis | |
| Dictyota dichotoma | |
| Dictyota ciliolata | |
| Dictyotaceae sp. | |
| Dictyopteris delicatula | |
| Dictyota dentata | |
| Lobophora variegata | |
| Padina japonica | |
| Padina minor | |
| Padina tetrastomatica | |
| Padina australis | |
| Sargassum polycystum | |
| Sargassum siliquosum | |
| Sargassum cristaefolium | |
| Sargaasum ilicifolium | |
| Sargassum asperifolium | |
| Sargassum oligocystum | |
| Turbinaria conoides | |
| Turbinaria ornata | |
| Turbinaria luzonensis | |
| Acanthophora spicifera | |
| Eucheuma alvarezii | |
| Jania decussata-dichotoma | |
| Lithophyllum pygmaeum | |
| Laurencia cartilaginea | |
| Perak | Yeong et al., 2012 |
| Ulva intestinalis | |
| Parvocaulis parvulus | |
| Sarawak | Wong et al., 2012 |
| Bryopsis indica | |
| Bryopsis pennata | |

**Table 1.** (Continued)

| Species | Sources |
|---|---|
| Caulerpa serrulate | |
| C. sertularioides | |
| C. taxifolia | |
| Codium papillaturn | |
| Tydernania expiditionis | |
| Avrainvillea erecta | |
| Udotea javensis | |
| Chaetomorpha minima | |
| Cladophora vagabunda | |
| Anadyomene stellata | |
| Bornetella oligospora | |
| Dasycladus sp | |
| Neomeris van-bosseae | |
| Acetabularia roxasii | |
| Enteromorpha flexuosa | |
| Scenedesmus sp | |

A more comprehensive perspective on global algae diversity emerges when comparing algal species in Malaysia with those from other regions. Table 2 presents the varieties of algae species identified in selected regions, such as Morocco, the Atlantic Ocean, the Baltic Sea, China, and Korea. These regions are marked by different climatic and oceanographic conditions that influence the presence and dominance of certain algal groups. Analyzing species composition in these areas allows for the identification of distribution patterns, endemic species, and ecological adaptations, which can inform future research on biodiversity conservation and climate-related changes in aquatic ecosystems.

**Table 2**. Algae species composition in other region

| Species | Sources |
|---|---|
| Other Region | |
| | |
| Morocco | Sabri et at., 2021 |
| Chlorococcum sp. (Morocco) | |
| Chlamydomonas debaryana (Morocco) | |
| Lomentaria articulata (Essaouira region, Morocco) | |
| Gracilaria cervicornis (Essaouira region, Morocco) | |
| Erythrotrichia welwitschia (Essaouira region, Morocco) | |
| Leptosiphonia schousboei (Essaouira region, Morocco) | |
| Polysiphonia atlantica (Essaouira region, Morocco) | |
| P. breviarticulata (Essaouira region, Morocco) | |

| Species | Sources |
|---|---|
| Champia parvula (Essaouira region, Morocco)<br>Cryptopleura ramose (Essaouira region, Morocco)<br>Phyllophora crispa (Essaouira region, Morocco)<br>Gymnogongrus griffithsiae (Essaouira region, Morocco)<br>Rhodymenia pseudopalmata (Essaouira region, Morocco)<br>Gelidium spinosum (Essaouira region, Morocco)<br>Mastocarpus stellatus (Essaouira region, Morocco)<br>Gracilaria bursa-pastoris (Essaouira region, Morocco)<br>Chondracanthus teedei (Essaouira region, Morocco)<br>Polysiphonia elongata (Essaouira region, Morocco) | |
| Atlantic<br>Palmaria palmata (North Atlantic)<br>Codium sp. (North Atlantic)<br>C. fragile (North Atlantic)<br>Ulva compressa (North Atlantic)<br>U. intestinalis (North Atlantic)<br>U. lactuca (North Atlantic)<br>U. rigida (North Atlantic)<br>Fucus spiralis (North Atlantic)<br>F. vesiculosus (North Atlantic)<br>Himanthalia elongata (North Atlantic)<br>Laminaria ochroleuca (North Atlantic)<br>Saccharina latissimi (North Atlantic)<br>Sargassum vulgare (North Atlantic)<br>Undaria pinnatifida (North Atlantic)<br>Chondrus crispus (North Atlantic)<br>Osmundea pinnatifida (North Atlantic)<br>Palmaria palmata (North Atlantic)<br>Porphyra sp. (North Atlantic)<br>Porphyra umbilicali (North Atlantic)<br>Pterocladiella capillacea (North Atlantic) | Pereira, 2023 |
| Baltic Sea<br>Hildenbrandia rubra (northern Baltic Sea)<br>Furcellaria lumbricalis (northern Baltic Sea)<br>Coccotylus truncatus (northern Baltic Sea)<br>Phyllophora pseudoceranoides (northern Baltic Sea)<br>Ceramium sp. (northern Baltic Sea)<br>Polysiphonia fibrillose (northern Baltic Sea)<br>Vertebrata fucoides (northern Baltic Sea)<br>Rhodomela confervoides (northern Baltic Sea) | Rinne & Kostamo, 2022 |
| China<br>Colpomenia peregrina (China)<br>Colpomenia claytoniae (China) | Xia et al., 2023 |

**Table 2.** (Continued)

| Species | Sources |
|---|---|
| Colpomenia sinuosa (China)<br>Ulva prolifera (Yellow Sea green, China)<br>Ulva linza (Yellow Sea green, China)<br>Ulva flexuosa (Yellow Sea green, China)<br>Ulva meridionalis (China<br>Ulva pertusa (China)<br>Stigeoclonium polyrhizum (China) | |
| Korea<br>Ceramium pacificum (Korea)<br>Cumathanmnion serrulatum (Korea)<br>Gayliella fimbriata (Korea)<br>Leptofauchea rhodymenioides (Korea)<br>Sorella pulchra (Korea)<br>Coelastrum astroideum<br>Kirchneriella lunaris<br>Fragilis sp.<br>Oocystis sp | Jeong et al., 2013 |

## Algae-Based Material and Design in Built Environment as Sustainable Solution

Algae offer a range of innovative solutions for improving environmental quality and supporting sustainable design initiatives. They play a crucial role in carbon-neutral energy production and contribute to the remediation of environmental pollutants, showcasing their potential to address pressing ecological challenges (Elmeligy & Elhassan, 2019). As a plant-based biomaterial, algae belong to a broader category of environmentally friendly resources that significantly reduce carbon emissions. Algae are notable among the diverse global plant resources for their abundance and increasing relevance in green innovation (Koç and Tuna, 2025). The concept of algae-based green communities is gaining global traction and has been implemented in several countries as part of eco-urban development strategies (Vassilev & Vassileva, 2016).

Microalgae mainly consist of proteins, lipids, carotenoids, and nucleic acids (Khoo et al., 2020). Microalgae convert energy from sunlight/artificial light, storing it as lipids or carbohydrates during photosynthesis process. The study by Caprariis et al. (2014) revealed that the green microalga Chlorella vulgaris acts as a biocatalyst in biophotovoltaic (BPV) systems, generating electricity upon exposure to light. This process occurred through an Ex electrogenic electron transfer mechanism, highlighting the potential of algae as a renewable and sustainable energy source. On top of that, algae are recognized as highly efficient photosynthetic organisms, harnessing solar energy into various products, including bioenergy (Ng et al., 2018).

The integration of microalgae with emerging bioreactive façade technologies, commonly referred to as algae-powered buildings, reflects a growing shift in sustainable architectural practices. These systems utilize the natural capabilities of microalgae for bioenergy production while also facilitating environmental functions, such as wastewater treatment and recycling. This dual-function approach underscores the potential of algae-based solutions to significantly contribute to green building strategies.

## Application of Algae as Carbon Sequestration in the Built Environment

A city can be considered as green or carbon-neutral if the GHG emissions remain below the critical threshold of a 2°C for global warming (Clark II and Cooke, 2016). The use of microalgae plays a multifaceted role in this context, facilitating $CO_2$ sequestration, wastewater treatment, nutrient absorption, and conservation of land and water resources. Microalgae also produce renewable energy sources such as bioethanol, biodiesel, and biogas (Ahmad et al., 2021). They demonstrate high efficiency in energy conversion via photosynthesis and can produce JP-8, jet fuel, and biodiesel, offering about 40% more energy per gallon than petrol (Khan et al., 2018). Research from the US Energy Department stated that over 3000 microalgal species might be used for biofuel production (Sheehan et al., 1998). However, despite extensive research, further advancements are required to achieve sustainable and economically viable biofuel production from microalgal biomass (Chen et al., 2017). The ability of algae to utilize photosynthesis for $CO_2$ absorption plays a critical role in diminishing the carbon footprint of buildings. Integrating algae cultivation systems can potentially reduce dependence on conventional carbon offsetting methods, thus enhancing carbon dioxide sequestration from urban environments. Algae are biological factories that convert solar energy into valuable metabolites (Chisti, 2006). The rising levels of pollution in metropolitan areas, as highlighted by recent crises in India's national capital, underscore the urgency for effective carbon sequestration strategies (Meeranayak, 2018). Various carbon sequestration methods exist, including terrestrial strategies (Singh & Ahluwalia, 2013) and the physical injection of $CO_2$ into the ocean. Algae possess a $CO_2$ fixation capacity up to 10–50 times higher than terrestrial plants, with species like *Chlorella* spp. showing particularly strong carbon sequestration potential (Ramanan et al., 2010). Open raceway ponds have extensive exposure for atmospheric $CO_2$ capture, while large-scale photobioreactors can enhance algal productivity by providing necessary elements. Effective models that improve air quality in highly polluted areas are essential. Closed raceway ponds have been developed and tested, achieving 95% $CO_2$ fixation efficiency during intermittent operation (Li et al., 2013).

Certain species of algae also demonstrate insulating properties, which can significantly impact building construction and energy efficiency. The implementation of microalgae-based materials enhances thermal insulation, leading to lower energy requirements for heating and cooling. The Microalgae-Based Bioreactive Façade (MA-BRF) serves multiple functions within buildings, acting as solar thermal collectors, photosynthetic biomass converters for biofuel production, thermal insulation materials, and adaptable shading devices (Casini, 2016). Research indicated that the energy exchange between Photo bioreactors (PBRs) and buildings can regulate thermal loads effectively. This symbiotic relationship prevents overheating in PBRs by filtering and shading thermal loads during hot weather while promoting microalgae biomass growth in cold weather (Pruvost et al., 2016). Additionally, the MA-BRF enhances the building's thermal insulation and supports improved comfort for occupants (Oncel et al., 2019). Research on various PBR prototypes integrated with different wall types have demonstrated improved insulation performance, reduced heat transfer, and promoting energy savings. The MA-BRF shows the potential for energy conservation through enhanced thermal insulation and acoustic performance provided by the water medium in microalgae. Additionally, temperature reduction resulted from shading, solar absorption, and the parasol effect generated during microalgal photosynthesis (Cervera Sardá & Vicente, 2016).

## Algae as Thermal Insulation in Buildings: Light Intensity in Heat Regulation

Algae-based thermal insulation for buildings presents a novel approach to energy efficiency by harnessing the natural properties of algae to regulate temperature through light intensity management. Algae panels, integrated into building facades, absorb sunlight and convert it into biomass while providing insulation. The effectiveness of these panels can be calculated by measuring the light intensity they receive and the corresponding heat absorption rate. Light intensity, often measured in lux or watts per square meter, directly impacts the rate of photosynthesis, influencing the thermal regulation capacity of the algae. By adjusting the algae layer thickness and optimizing the light intensity levels through shading or reflective materials, the system can effectively maintain indoor temperatures. Microalgae facades represent a cutting-edge innovation in architectural design. Research shows they can significantly lower a building's carbon footprint by reducing energy consumption and offer the added benefit of improving indoor air quality by helping to reduce air pollution (Warren, 2023).

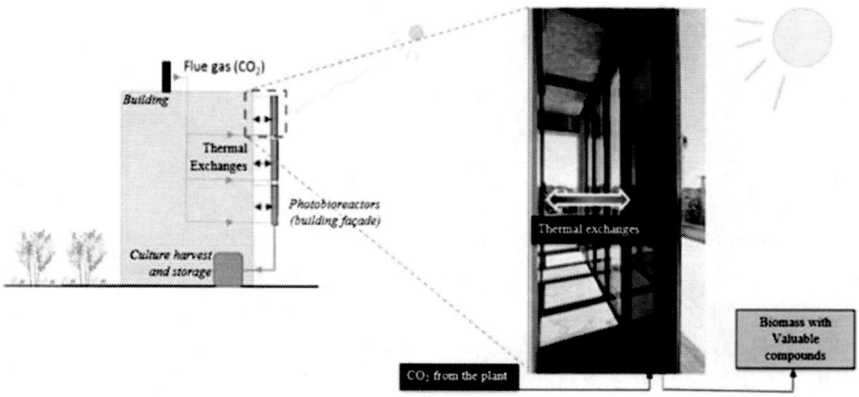


**Figure 2.** Microalgae culture in building-integrated Algae Panel (Pruvost et al., 2016).

Certain algae species possess insulating properties. Incorporating algae-based materials in building construction can enhance thermal insulation, thereby lowering the energy needed for heating and cooling. Algae panel can meet a building's thermal needs by functioning as solar thermal collectors, photosynthetic biomass converters for biofuel production, thermal insulation materials, and adaptable shading devices (Casini, 2016). The design of panels should be cultivated to make it properly function as Figure 2. Research indicates that energy exchange between algae panels and buildings aids in regulating thermal load. This symbiotic relationship prevents algae panels from overheating by filtering and shading thermal loads in hot weather, while also supporting microalgal biomass growth in cold weather (Pruvost et al.,

2016). Integrating the algae panel with solar collection and PV-integrated shading mechanisms allows the system to simultaneously generate electricity and provide shading (Al Dakheel and Tabet Aoul, 2017). Additionally, the algae panel enhances the building's thermal insulation, contributing to the occupants' thermal comfort (Öncel et al., 2016). Additionally, temperature reduction results from shading, solar absorption, and the parasol effect generated during microalgal photosynthesis (Cervera Sardá and Vicente, 2016). The best light intensity measurement for cooling a building focuses on reducing excess heat generated by sunlight, which can lead to increased cooling loads (Białek, 2022). To manage this effectively, architects and engineers often consider light intensity in terms of illuminance (measured in lux or lumens per square meter), and it is balanced with natural ventilation and shading strategies.

## Conclusion

Rapid urban growth is increasingly disrupting the balance of climate sustainability, heightening the risk of natural disasters as well as causing substantial social and economic impacts. Despite occupying only, a small fraction of the Earth's surface, cities contribute to the majority of carbon emissions and account for a significant share of global energy use. A large portion of greenhouse gas emissions also originates from highly concentrated urban areas. The current vulnerabilities of urban areas highlight the urgent need to plan, design, and construct green and resilient cities.

It is essential to plan and design future buildings and urban areas based on sustainable development principles, emphasizing self-sufficiency, zero waste and emissions, environmental preservation, and the use of clean energy and green fuels. Therefore, adopting comprehensive sustainable design solutions is crucial to enhance energy efficiency in the built environment. The success of these approaches should be tailored to the unique climatic conditions and economic capabilities of each country to ensure optimal performance and long-term sustainability. This approach can yield numerous benefits regarding sustainable energy, environment, and individual well-being. In this regard, researchers have been studying the potential of algae as a viable option for these objectives.

Integrating algae pigments into the built environment represents an innovative strategy in sustainable design, offering both aesthetic versatility and functional benefits such as solar shading, light filtering, and energy efficiency. Studies indicate that materials and colourants derived from algae can enhance sustainable architecture by reducing environmental impact, improving indoor comfort, and enabling carbon-neutral construction strategies.

## References

Ahmad, I., Yuzir, A., Mohamad, S. E., Iwamoto, K., & Abdullah, N. (2021, February). Role of microalgae in sustainable energy and environment. In *IOP Conference Series: Materials Science and Engineering* (Vol. 1051, No. 1, p. 012059). IOP Publishing. https://doi.org/10.1088/1757-899X/1051/1/012059.

Al Dakheel, J., & Tabet Aoul, K. (2017). Building applications, opportunities, and challenges of active shading systems: A state-of-the-art review. *Energies, 10*(10), 1672. https://doi.org/10.3390/en10101672.

Białek, A., & Dębska, L. (2022). Verification of the selected thermal comfort model in real and simulated conditions. In *E3S Web of Conferences* (Vol. 347, p. 04013). EDP Sciences. https://doi.org/10.1051/e3sconf/202234704013.

De Caprariis, B., De Filippis, P., Di Battista, A., Di, L., & Palma, M. S. (2014). Exoelectrogenic activity of a green microalgae, *Chlorella vulgaris*, in a bio-photovoltaic cells (BPVs). *Chemical Engineering Transactions, 38*, 553-558. https://doi.org/10.3303/CET1438083.

Casini, M. (2016). *Smart buildings: Advanced materials and nanotechnology to improve energy-efficiency and environmental performance*. Woodhead Publishing.

Cervera Sardá, R., & Vicente, C. A. (2016). Case studies on the architectural integration of photobioreactors in building facades. In *Nano and biotech based materials for energy building efficiency* (pp. 457-484). Elsevier.

Chen, K., Ríos, J. J., Pérez-Gálvez, A., & Roca, M. (2017). Comprehensive chlorophyll composition in the main edible seaweeds. *Food Chemistry, 228*, 625–633. https://doi.org/10.1016/j.foodchem.2017.02.054.

Chen, Z., Wu, W., Wen, Y., Zhang, L., Wu, Y., Farid, M. S., El-Seedi, H. R., Capanoglu, E., & Zhao, C. (2023). Recent advances of natural pigments from algae. *Food Production Processing and Nutrition, 5*(1). https://doi.org/10.1186/s43014-023-00055-w.

Chisti, Y. (2006). Microalgae as sustainable cell factories. *Environmental Engineering & Management Journal (EEMJ), 5*(3), 1-16.

Clark II, W., & Cooke, G. (2016). *Smart green cities: Toward a carbon neutral world.* Routledge.

Da Silva, T. L., Moniz, P., Silva, C., & Reis, A. (2019). The dark side of microalgae biotechnology: A heterotrophic biorefinery platform directed to ω-3 rich lipid production. *Microorganisms, 7*, 1–21. https://doi.org/10.3390/microorganisms7010009.

Da Silva, T. L., Moniz, P., Silva, C., & Reis, A. (2021). The role of heterotrophic microalgae in waste conversion to biofuels and bioproducts. *Processes, 9*, 1–24. https://doi.org/10.3390/pr9020029.

Elmeligy, D., & Elhassan, Z. (2019, November). The bio-adaptive algae contribution for sustainable architecture. In *IOP Conference Series: Earth and Environmental Science* (Vol. 397, No. 1, p. 012007). IOP Publishing. https://doi.org/10.1088/1755-1315/397/1/012007.

Ismail, A., Akmal, N., Ahmad, I., & Sarah Diyana, M. (2017). Biodiversity of macroalgae in Blue Lagoon, the Straits of Malacca, Malaysia and some aspects of changes in species composition. *Sains Malaysiana, 46*(1), 1-7.

Jareonsin, S., & Pumas, C. (2021). Advantages of heterotrophic microalgae as a host for phytochemicals production. *Frontiers in Bioengineering and Biotechnology, 9*(February). https://doi.org/10.3389/fbioe.2021.661973.

Jeong, S. Y., Won, B. Y., Kang, P. J., Kang, J. C., Kim, M. S., Nam, K. W., & Cho, T. O. (2013). New record of some red algal species (Rhodophyta) from Korea. *Journal of Ecology and Environment, 36*(4), 439–448. https://doi.org/10.5141/ecoenv.2013.439.

Khan, M. I., Shin, J. H., & Kim, J. D. (2018). The promising future of microalgae: Current status, challenges, and optimization of a sustainable and renewable industry for biofuels, feed, and other products. *Microbial Cell Factories, 17*(1), 1–21. https://doi.org/10.1186/s12934-018-0982-1.

Khoo, C. G., Lam, M. K., Mohamed, A. R., & Lee, K. T. (2020). Hydrochar production from high-ash low-lipid microalgal biomass via hydrothermal carbonization: Effects of operational parameters and products characterization. *Environmental Research, 188*, 109828. https://doi.org/10.1016/j.envres.2020.109828.

Li, S., Luo, S., & Guo, R. (2013). Efficiency of CO2 fixation by microalgae in a closed raceway pond. *Bioresource Technology, 136*, 267-272. https://doi.org/10.1016/j.biortech.2013.02.043.

Lim, K. C., Yusoff, F. M., Karim, M., & Natrah, F. M. I. (2022). Carotenoids modulate stress tolerance and immune responses in aquatic animals. *Reviews in Aquaculture, 15*(2), 872–894. https://doi.org/10.1111/raq.12486.

Meeranayak, U. F. J., & Shivasharana, C. T. (2018). Competitive and economically feasible cell wall disruption techniques for algal biofuel extraction. *International Journal of Science Research in Biological Sciences, 5*(6).

Ng, F. L., Phang, S. M., Periasamy, V., Beardall, J., Yunus, K., & Fisher, A. C. (2018). Algal biophotovoltaic (BPV) device for generation of bioelectricity using *Synechococcus elongatus* (Cyanophyta). *Journal of Applied Phycology, 30*(6), 2981-2988. https://doi.org/10.1007/s10811-018-1390-0.

Oncel, S. S., & Şenyay Öncel, D. (2019). Bioactive façade system symbiosis as a key for eco-beneficial building element. In *Environmentally-Benign Energy Solutions* (pp. 97-122). Springer International Publishing. https://doi.org/10.1007/978-3-030-15932-7_5.

Öncel, S. Ş., Köse, A., & Öncel, D. Ş. (2016). Façade integrated photobioreactors for building energy efficiency. In *Start-Up Creation* (pp. 237-299). Woodhead Publishing. https://doi.org/10.1016/B978-0-08-100401-1.00010-2.

Paper, M., Glemser, M., Haack, M., Lorenzen, J., Mehlmer, N., Fuchs, T., Schenk, G., Garbe, D., Weuster-Botz, D., Eisenreich, W., Lakatos, M., & Brück, T. B. (2022). Efficient green light acclimation of the green algae *Picochlorum sp.* triggering geranylgeranylated chlorophylls. *Frontiers in Bioengineering and Biotechnology, 10*. https://doi.org/10.3389/fbioe.2022.835213.

Pereira, L. (2023). Atlantic algae as food and their extracts. In *Exploration of Foods and Foodomics* (pp. 15–31). Springer.

Perez-Garcia, O., Escalante, F. M. E., de-Bashan, L. E., & Bashan, Y. (2011). Heterotrophic cultures of microalgae: Metabolism and potential products. *Water Research, 45*(1), 11–36. https://doi.org/10.1016/j.watres.2010.08.037.

Phang, S.-M., Mustafa, E. M., Ambati, R. R., Sulaiman, N. M. N., Lim, P.-E., Majid, N. A., Dommange, X., Schwob, C., & Liew, K.-E. (2015). Checklist of microalgae collected from different habitats in Peninsular Malaysia for selection of algal biofuel feed-stocks. *Malaysian Journal of Science, 34*(2), 148–174.

Pruvost, J., Le Gouic, B., Lepine, O., Legrand, J., & Le Borgne, F. (2016). Microalgae culture in building-integrated photobioreactors: Biomass production modelling and energetic analysis. *Chemical Engineering Journal, 284*, 850-861. https://doi.org/10.1016/j.cej.2015.08.091.

Ramanan, R., Kannan, K., Deshkar, A., Yadav, R., & Chakrabarti, T. (2010). Enhanced algal CO2 sequestration through calcite deposition by *Chlorella sp.* and *Spirulina platensis* in a mini-raceway pond. *Bioresource Technology, 101*(8), 2616-2622. https://doi.org/10.1016/j.biortech.2009.11.089.

Rinne, H., & Kostamo, K. (2022). Distribution and species composition of red algal communities in the northern Baltic Sea. *Estuarine, Coastal and Shelf Science, 269*, 107806. https://doi.org/10.1016/j.ecss.2022.107806.

Koç, Z. G., & Tuna, G. (2025). Design and manufacturing of building products based on biomaterials: A systematic literature review and a framework proposal based on the meta-synthesis method. *Megaron/Yildiz Technical University, Faculty Of Architecture E-Journal, 20*(2), 263-277. https://doi.org/10.48048/mtu.2025.67.

Sabri, H., Haroun, R., Bahammou, N., Hasni, M., Boundir, Y., Maarouf, A., & Cherifi, O. (2021). Intertidal benthic red algae (Rhodophyta) from Essaouira coastline (Morocco). *The Egyptian Journal of Aquatic Research, 47*(3), 277–282. https://doi.org/10.1016/j.ejar.2021.04.009.

Sheehan, J., Dunahay, T., Benemann, J., & Roessler, P. (1998). Look back at the US department of energy's aquatic species program: Biodiesel from algae; close-out report (No. NREL/TP-580-24190). National Renewable Energy Lab. (NREL), Golden, CO: United States.

Sijil, P. V., Cherita, C., Jethani, H., & Chauhan, V. S. (2022). Chapter 1. Microalgae as a renewable and sustainable source of high-value metabolites. In *Royal Society of Chemistry eBooks* (pp. 1–26). https://doi.org/10.1039/9781839164347-00001.

Singh, U. B., & Ahluwalia, A. S. (2013). Microalgae: A promising tool for carbon sequestration. *Mitigation and Adaptation Strategies for Global Change, 18*(1), 73-95. https://doi.org/10.1007/s11027-012-9386-7.

Vassilev, S. V., & Vassileva, C. G. (2016). Composition, properties and challenges of algae biomass for biofuel application: An overview. *Fuel, 181*, 1-33. https://doi.org/10.1016/j.fuel.2016.03.064.

Wong, S. C., Harah, Z. M., Sidik, B. J., & Arshad, A. (2012). Comparison of seaweed communities of the two rocky shores in Sarawak, Malaysia. *Coastal Marine Science, 35*(1), 78–84.

Xia, Z., Cao, X., Li, S., Cao, J., Tong, Y., Sun, Y., Liu, J., Zhao, S., Cui, Q., Zeng, Y., Chen, Z., He, P., & Zhang, J. (2023). Distribution of *Ulva prolifera*, the dominant species in green tides along the Jiangsu Province coast in the Southern Yellow Sea, China. *Journal of Sea Research, 196*, 102436. https://doi.org/10.1016/j.seares.2023.102436.

Xie, W., Li, X., Xu, H., Chen, F., Cheng, K. W., Liu, H., & Liu, B. (2023). Optimization of heterotrophic culture conditions for the microalgae *Euglena gracilis* to produce proteins. *Marine Drugs, 21*, 1–13. https://doi.org/10.3390/md21010001.

Yeong, H., Draisma, S. G. A., & Phang, S. (2012). Marine algae collected during the marine biodiversity mini expedition 2012 to Sembilan Group of Islands, Perak with one new record, *Parvocaulis parvulus* (Solms-Laubach) S. Berger et al. FOR MALAYSIA. *Malaysian Journal of Science, 31*(2), 98–110.

Zuorro, A., Leal-Jerez, A. G., Morales-Rivas, L. K., Mogollón-Londoño, S. O., Sanchez-Galvis, E. M., García-Martínez, J. B., & Barajas-Solano, A. F. (2021). Enhancement of phycobiliprotein accumulation in thermotolerant *Oscillatoria sp.* through media optimization. *ACS Omega, 6*(16), 10527–10536. https://doi.org/10.1021/acsomega.1c02013.

Warren, K., Milovanovic, J., & Kim, K. H. (2023). Effect of a microalgae facade on design behaviors: A pilot study with architecture students. *Buildings, 13*(3), 611. https://doi.org/10.3390/buildings13030611.

## Chapter 5

# Microalgae Pigments: Sustainable Solutions for a Greener Future

**Nursyafica Nadia Johari***, **PhD**
**and Rashidi Othman, PhD**
Department of Landscape Architecture,
Kulliyyah of Architecture and Environmental Design,
International Islamic University Malaysia, Kuala Lumpur, Malaysia

## Abstract

Microalgae pigments are extensively used in various industries such as the food, nutraceutical, pharmaceutical, aquaculture, and cosmetics industries. They have also been applied in clinical and research settings, where they function well as labels for receptors and antibodies. One of the most prominent characteristics of algae is their colour, which is determined by their pigments. These pigments are colourful chemical substances that are part of the photosynthetic system of microalgae. Due to their numerous potential industrial uses and health-promoting qualities, natural pigments derived from microalgae have been the subject of numerous studies. As consumers become more conscious of the link between nutrition, health, and illness prevention, microalgae represent a significant and exciting new field in biotechnology. The market is currently in high demand for microalgal pigments as naturally occurring, renewable colour enhancers for foods and feeds that also have health benefits. Moreover, these pigments, which exhibit remarkable dyeing properties even at extremely low concentrations (parts per million), find significant uses in the pharmaceutical and cosmetic industries respectively as biochemical tracers in immune assays and as

* Corresponding Author's Email: syaficanadia@gmail.com

In: Natural Pigments: Sources, Properties, and Health Benefits
Editors: Rashidi Othman, Qurratu Aini Mat Ali, Razanah Ramya et al.
ISBN: 979-8-90134-055-4

fluorescence-based indicators in skin creams that promote collagen synthesis. Natural pigments derived from microalgae not only provide colour, but they also have anti-inflammatory, anti-cancer, and antioxidant properties that make them a good substitute for artificial colorants. Therefore, the futures of all the microalgae and pigments derived from them that have been discussed are promising. Microalgae have been recognized by science as a natural source of nutraceuticals and colorants, and their potential has already been demonstrated in several industrial applications. Although there are still some issues to be resolved, the pigments derived from microalgae are likely to surpass synthetic forms due to their renewable and sustainable nature, as well as their additional health benefits.

**Keywords**: microalgae, pigments, chlorophyll, carotenoid, phycobiliproteins, astaxanthin

## Introduction

Microalgae pigments are extensively used in various industries such as food, nutraceutical, pharmaceutical, aquaculture, and cosmetics industry. It has also been applied in clinical and research settings, where it functions well as a label for receptors and antibodies (Begum et al., 2016). According to Reynolds (2006), microalgae are a diverse group of cryptogamic plants that include 13 major phyla and several smaller groups that have not yet received enough research. They come in a variety of forms, including filamentous, siphonaceous, colonial, and unicellular. Cyanobacteria are oxygenic photosynthetic prokaryotes that exhibit a great deal of diversity in their morphology, physiology, ecology, biochemistry, and other traits among the various phyla of microalgae. Cryptophytes are unicellular and widely distributed in both freshwater and marine environments, whereas Chlorophyta are unicellular, multicellular, filamentous, siphonous, and thallus algae that are primarily found in freshwater environments. Like monophytes, dinophytes have two different flagella and are mostly found in marine environments. (Reynolds, 2006). Different pigments are found in different phylum of microalgae. Cid et al., (2012) mentioned that microalgae is a primary producer and they are easily cultivated in photobioreactors under controlled conditions due to their small size. Microalgae are very sensitive, rapid, and cost-effective making them popular as tools for applications in creating a sustainable environment.

Both cyanobacteria and microalgae are photosynthetic organisms that make different types of pigments to capture light. Phycobiliproteins, carotenoids, and chlorophylls are the three main classes into which these pigments are typically divided. Depending on the phylum to which the organism belongs, microalgae and cyanobacteria have different types of pigments (Pagels et al., 2021). The most basic one is chlorophyll, which is found in all photosynthetic microalgae and cyanobacteria and is in charge of oxygenic photosynthetic activity.

The most well-known and commercially available carotenoids are lutein, Œ≤-carotene, and astaxanthin. Carotenoids can be both primary and secondary pigments, and their profiles vary depending on the species and growth conditions. The most well-known phycobiliproteins in cyanobacteria and red algae are phycocyanin and phycoerythrin. Phycobiliproteins are a unique class of pigments that are only found in these organisms and can function as the primary light absorber. When combined, the three pigment classes allow microalgae and cyanobacteria to harvest light as efficiently as possible across the entire light spectrum.

Microalgae and cyanobacteria pigments are highly desirable industrial ingredients for food, feed, pharmaceuticals, nutraceuticals, and cosmetics because of their colour and bioactive qualities, as well as being natural and environmentally friendly. The market's use of microalgae and cyanobacteria is still associated with high-added-value products, partly because their biomass production is smaller than that of synthetic alternatives. Nevertheless, the demand for natural colours (instead of synthetic ones) is driving up the need for a more sustainable source.

Microalgae are a broad class of unicellular, eukaryotic microorganisms that range in size and shape from 2 to 50 lm. The other main class of prokaryotes that are photosynthetic microorganisms are called cyanobacteria. Numerous phyla have been found to contain photosynthetic microorganisms, which can grow in fresh, brackish, or saltwater. Because they are photosynthetic organisms that account for 40% of global photosynthesis and produce nearly half of Earth's atmospheric oxygen at any given time, microalgae are essential to aquatic ecosystems. These organisms can grow photo autotrophically, using light and carbon dioxide as sources of energy and carbon, respectively, and converting them into biomass (Elisabeth et al., 2021). Apart from having a huge biodiversity. Merely thirty thousand of the estimated fifty thousand distinct types of microalgal species have been thoroughly examined (Richmond, 2004). There is still much untapped potential in this diversity. Microalgal biomass holds great promise as a source of alternative energy and as a rich source of various compounds of commercial interest. The application of microalgae in the industry is growing (Volk and Furkert, 2006).

In Malaysia, early studies of microalgae focused mainly on the taxonomy and ecological distribution of freshwater algae. The basic research in phycology has led to the publication of numerous checklists and monographs that have detailed Malaysia's microalgae diversity. The primary application of microalgae in the treatment of agro-industrial wastewater was the subject of early applied phytochemistry research. Such a system allowed for the effective removal of nitrogen from palm oil mill effluent (POME). Subsequently, it was demonstrated that the HRAP system could help treat agro-industrial wastewater, including landfill leachate, rubber effluent, and wastewater from sago starch factories and textile industries (Chu, 2012). Because of their high PUFA content, marine microalgae cultivated in POME were proposed as a potential source of feed for aquaculture in another study. The alga grows well on organic carbon and at high nitrogen levels, and it has potential uses in wastewater treatment and heavy metal removal. Despite being isolated from a freshwater environment, the strain can thrive in a variety of salinities.

PUFA and carotenoids, in particular, are among the targeted products from Malaysia's algal resources are microalgal lipids. For example, EPA and arachidonic acid are produced in significant quantities by the marine diatom Nitschschia inconspicua. It was discovered that this diatom's growth and fatty acid profiles were influenced by variables like carbon source, nitrate and silicate levels, and growth. The highest yields of biomass and EPA were found in cultures that were aerated with 5% CO2. The primary focus of current research on microalgal lipids is their potential application in the manufacturing of biodiesel. The identification of microalgal strains with high lipid production and the improvement of lipid synthesis in microalgae through the use of genomic, proteomic, and metabolomic techniques are among the priority areas (Chu et al., 1996).

## Microalgae Pigments

One of the most noticeable aspects of algae is their vibrant colour, which is primarily due to the presence of chemical pigments involved in their photosynthetic processes. These pigments are essential compounds in the photosynthetic network of microalgae and are typically categorized into three groups (Christaki et al., 2015):

1. Carotenoids – averaging 0.1–0.2% of dry weight, but can reach up to 14% in certain species, such as for β-carotene.
2. Chlorophylls - comprising approximately 0.5-1.0% of dry weight.
3. Phycobiliproteins - which can constitute up to 8% of dry weight.

To be considered suitable for commercial pigment production, microalgal strains must meet several criteria, including ease of cultivation, non-toxicity, high nutritional value, and the presence of digestible cell walls to ensure nutrient availability. As shown in Table 1, the most commonly used species include Dunaliella salina, Haematococcus pluvialis, Chlorella spp., Muriellopsis spp., Scenedesmus spp., Spirulina (Arthrospira) spp., and Porphyridium spp.

**Table 1.** Most frequent used species and their pigment (Christaki et al., 2015)

| Microalgae | Pigments |
|---|---|
| Dunaliela salina | b-carotene,<br>zeaxanthin,<br>chlorophylls a, b |
| Haematococcus pluvialis | astaxanthin,<br>canthaxanthin,<br>lutein,<br>chlorophylls a, b |
| Chlorella spp | astaxanthin,<br>canthaxanthin,<br>chlorophylls a, b |
| Scenedesmus spp. | lutein,<br>b-carotene |
| Muriellopsis spp | lutein |
| Spirulina spp. | b-carotene,<br>zeaxanthin,<br>phycocyanin,<br>allophycocyanin |
| Porphyridium spp. | phycoerythrin |

According to Masojídek et al. (2013), similar to higher plants, the photosynthetic apparatus of microalgae is organized within specialized organelles known as chloroplasts. These chloroplasts consist of several auxiliary light-harvesting pigments, an aqueous matrix called the stroma, and a system of lipoprotein membranes referred to as thylakoids. Certain pigments are responsible for the core processes of oxygenic photosynthesis, including light absorption, electron generation, and adenosine triphosphate (ATP) synthesis. During photosynthesis, electrons are transferred to the reaction center, where they are used to reduce the primary electron acceptor in the electron transport chain. This electron transport enables the final reduction of NAD(P)+ from ferredoxin and concurrently lowers the energy barrier for nutrient uptake. Through assimilatory reduction, inorganic compounds containing carbon, nitrogen, phosphorus, and sulfur (in the forms of $CO_2$, $NO_3^-$, $PO_4^{3-}$, and $SO_4^{2-}$) are converted into organic matter during photosynthesis and subsequently incorporated into the carbon skeletons of metabolic intermediates and cellular components. This process represents the primary mechanism for carbon fixation and oxygen release into the atmosphere. Therefore, photosynthesis serves as a fundamental source of energy and carbon for all living organisms on Earth—both directly and indirectly—supporting their metabolism and growth.

Microalgal pigments play a crucial role in photosynthetic processes such as light harvesting and energy transfer to the reaction centers. These pigments are organized into antenna complexes located within the thylakoid membranes of the chloroplasts in eukaryotic microalgae (Masojídek et al., 2013). Pigments are generally classified into three main groups (Masojídek et al., 2013):

1. Chlorophylls (chl),
2. Carotenoids
3. Phycobiliproteins

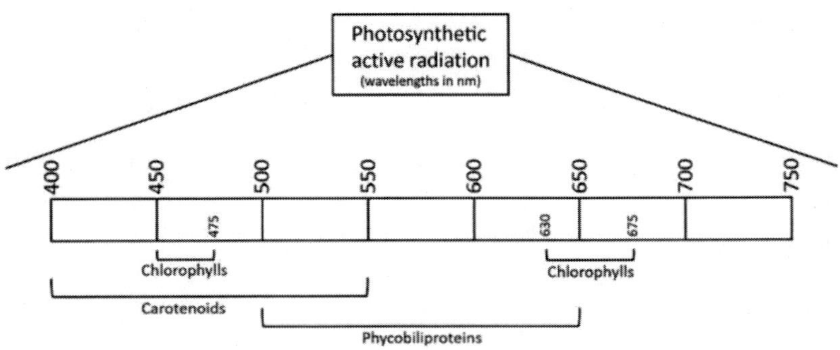


**Figure 1.** Schematic representation of the absorption range of three main classes of pigments (chlorophylls, carotenoids, and phycobiliproteins) in the visible light spectrum (Pagels et al., 2020).

The pigment composition within the photosynthetic machinery often varies between microalgal groups. All photosynthetic organisms possess chlorophyll a as a core component of the reaction center. Additionally, chlorophyll b, c, or d may be present as accessory pigments in the antenna complexes, extending the range of light absorption. In some specific microalgal groups, chlorophylls exhibit two primary absorption bands: 450–475 nm and 630–675 nm (see Figure 1).

Phycobiliproteins, a major class of light-harvesting pigment-protein complexes, are typically found in cyanobacteria and red algae. These pigments are organized into complexes known as phycobilisomes, which are attached to the cytoplasmic face of the thylakoid membranes (Pagels et al., 2020). As hydrophilic pigments, phycobiliproteins absorb light in the visible spectrum, specifically between 500 and 650 nm (see Figure 1). Working synergistically with chlorophyll a (Chl a), they enhance light-harvesting efficiency. Phycobiliproteins are generally classified into four main types (Dumay & Morançais, 2016; Pagels et al., 2020):

1. Phycocyanin,
2. Phycoeryhrin,
3. Phycoerythrocyanin
4. Allophycocyanin

Carotenoids, another essential pigment group, are widely recognized as terpenoid compounds that are critical for the survival of photosynthetic organisms (Guedes et al., 2011). Carotenoids can be classified based on their function within the cell. Primary carotenoids—such as α-carotene, β-carotene, and lutein—are vital for cell survival and directly participate in photosynthesis. In contrast, secondary carotenoids—including astaxanthin and canthaxanthin—are synthesized in response to specific environmental stimuli via a process known as *carotenogenesis* (Guedes et al., 2011; Zhang et al., 2014).

Carotenoids serve two primary functions in photosynthesis. First, within the absorption range of 400–550 nm (refer to Figure 1), they act as accessory pigments by transferring energy to Chl a, thereby broadening the spectrum of light usable for photosynthetic activity. Second, they protect the photosynthetic apparatus by quenching reactive oxygen species formed under high light stress. Beyond their biological functions, microalgal pigments are highly valued in various industrial applications due to their desirable properties. These include use in the food, feed, pharmaceutical, nutraceutical, and cosmetic industries. For example, the antioxidant capacities of β-carotene and phycocyanin have been extensively documented (Abed et al., 2009; Guedes et al., 2011).

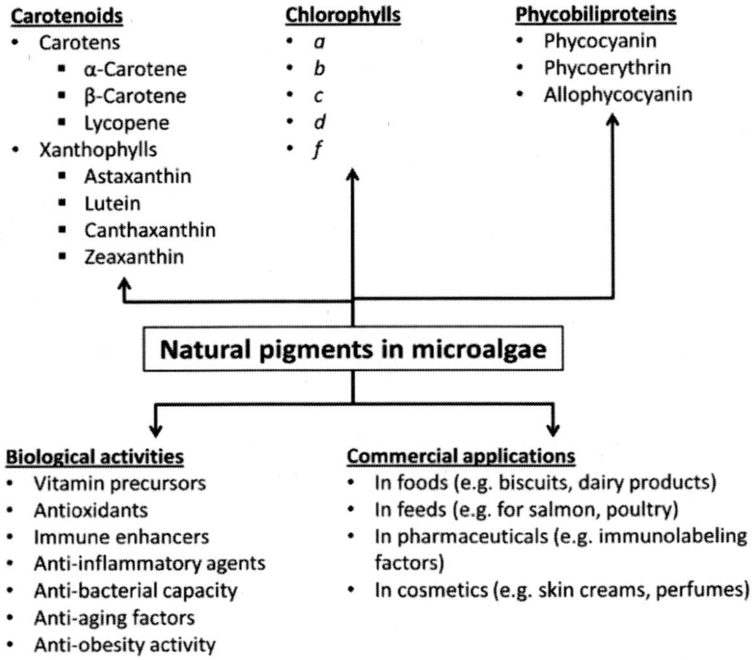


**Figure 2.** Distribution, biological activities, and commercial applications of natural pigments derived from microalgae (Christaki et al. 2015).

Microalgae are considered promising industrial pigment sources not only for their pigment production capacity but also due to several advantages over traditional plant sources. These include their vast natural diversity and distribution, their ability to grow on non-arable land, and in some cases, their capability to fix atmospheric nitrogen—eliminating the need for nitrogen supplementation. Additionally, unlike heterotrophic organisms, many microalgae species do not require carbon supplementation, making production potentially more cost-effective (Panda et al., 2006). Due to these advantages, microalgae are already being used industrially to produce pigments and other valuable bioproducts. Although the market prices of pigments are generally higher than those of other microalgal components, pigments remain the main source of revenue for industries, particularly in the food, cosmetic, and healthcare sectors. The increasing awareness among consumers about the relationship between diet, health, and disease prevention has fueled interest in natural pigments. As a result, microalgae have emerged as a promising and dynamic field in biotechnology (see Figure 2).

Among the wide range of compounds derived from microalgae, natural pigments stand out as one of the most significant categories for further investigation. These pigments not only serve as natural colorants but also exhibit anti-inflammatory, antioxidant, and anti-cancer properties, making them a promising alternative to synthetic colorants.

### *Carotenoids*

Carotenoids are among the most commercially valuable groups of algal pigments. These isoprenoid molecules, ranging in colour from yellow to red, are among the oldest pigments found in nature. Due to their lipid solubility, carotenoids are absorbed alongside dietary fats and enter the bloodstream bound to various lipoproteins (Lordan et al., 2011). As humans and animals (with the exception of the aphid genus) cannot synthesise carotenoids, they must obtain them through their diet. Carotenoids are generally classified into two groups: xanthophylls (oxygenated molecules such as astaxanthin, lutein, and canthaxanthin) and carotenes (hydrocarbons such as α-carotene, β-carotene, and lycopene).

Additionally, carotenoids are categorised as either primary or secondary. Primary carotenoids, such as lutein and β-carotene, are directly involved in photosynthesis and form structural and functional components of the photosynthetic apparatus. Besides their photoprotective roles—such as filtering, quenching, or scavenging—they also contribute to light harvesting (Christaki et al., 2015). In contrast, secondary carotenoids like astaxanthin and canthaxanthin are not bound to the photosynthetic machinery. Instead, they accumulate in the cytoplasm after being synthesised in the chloroplast (Mulders et al., 2014). The overproduction of secondary carotenoids is induced exclusively by certain environmental and cultural stressors, including high light intensity, nutrient depletion, temperature fluctuations, pH shifts, high salinity, and oxidative stress. These conditions trigger the process of carotenogenesis in microalgae, resulting in the formation of carotenoids that serve solely as filters and photoprotective pigments, enabling algae to better cope with stress.

### *Chlorophylls*

According to Ferruzzi et al. (2002), green pigments known as chlorophylls are found in bacteria, higher plants, and algae. Chlorophyll is a fat-soluble substance with bioactive properties such as antimutagenic and antioxidant effects. The extractable amount of this pigment from microalgae biomass ranges from 0.5% to 1.5% on a dry weight basis. Due to their vivid green colour and biological activities, chlorophylls tetrapyrroles with a centrally bound magnesium ion have attracted considerable interest from commercial industries. This natural green pigment is used in the food, cosmetic, and pharmaceutical sectors. Despite being recognized as natural pigments, chlorophylls present certain limitations in practical use. They are sensitive to heat and light and are chemically unstable in acidic environments. To overcome these issues, chlorophylls are often converted into more stable forms, such as sodium copper chlorophyllin, by replacing the magnesium ion with sodium or copper. Chlorophyllin, a water-soluble derivative of chlorophyll, is widely used as a food additive and colourant in the food industry.

There are several structural types of chlorophyll, including chlorophyll a (blue-green), b (brilliant green), c (yellow-green), d (forest green), and f (emerald green). Chlorophyll a and b are the most common and are found in most photosynthetic organisms, while chlorophylls c, d, and f are found only in certain species of microalgae, algae, and photosynthetic bacteria (Christaki et al., 2015). Chlorella, a genus of unicellular green eukaryotic microalgae that thrives in freshwater, is valued for its high nutritional content and is used as aquaculture feed and as a food supplement in various forms (e.g., tablets, powders, capsules). One of the most well-known species, *Chlorella vulgaris*, was first described in 1890. It is capable of converting solar energy into chemical energy through photosynthesis and requires nutrients such as carbon and nitrogen for optimal growth. *Chlorella* grows rapidly and can withstand various environmental stresses (Safi et al., 2014).

Under standard growth conditions, *Chlorella* cells are rich in lipids (including essential fatty acids), vitamins, minerals, and proteins (comprising 40–60% of the biomass). The cells also contain significant amounts of chlorophylls a and b, which can reach up to 4.5% of dry weight. Due to its high chlorophyll content, *Chlorella* is often referred to as "emerald food." In addition to chlorophylls, it also contains other pigments such as lutein, β-carotene, c-astaxanthin, and astaxanthin, though in smaller quantities.

Although chlorophyll is a natural pigment, its practical application faces certain challenges. For example, it becomes unstable in food products with fluctuating pH levels (Hosikian et al., 2010), and the pigment turns dull or pale when the magnesium ion is lost (Humphrey, 2004). Furthermore, excessive production of chlorophylls may not be economically viable for microalgae-based pigment production systems (Mulders et al., 2014). However, under optimal conditions, fast-growing species like *Chlorella* with their high chlorophyll content remain promising candidates for commercial pigment production.

### *Phycocyanin*

Phycobiliproteins are protein pigments that absorb light. They constitute 40–50% of all soluble proteins and are located in the supramolecular phycobilisomes on the exterior of the thylakoid membrane. These molecules are water-soluble and can be divided into three primary groups: phycoerythrin (red pigment), allophycocyanin (light-blue pigment), and phycocyanin (blue pigment). Each group differs in absorption spectra, colour, and chemical structure. Phycobiliproteins are chromophore-bearing polypeptides composed of $\alpha$ and $\beta$ subunits, each weighing about 20 kDa. They possess functional properties and are used as fluorescent markers (Pan-Utai et al., 2018).

According to Pulz and Gross (2004), *Spirulina (Arthrospira)* sp., which produces phycocyanin, and *Porphyridium* sp., which produces phycoerythrin, are two of the most well-known cyanobacteria species used in the commercial production of phycobiliproteins. *Spirulina* biomass is highly valued for its nutritional content and potential for the extraction of bioactive compounds with added value. It is known to contain minerals (Fe, Mg, Ca, P, Cr, Cu, Na, and Zn), pigments (phycocyanin, chlorophylls, and carotenoids), essential fatty acids ($\gamma$-linolenic acid), phenolic compounds, biopeptides, and enzymes. These compounds exhibit a variety of functional properties, including immunomodulatory, antioxidant, anti-inflammatory, anticancer, antimicrobial, and prebiotic effects.

Phycocyanin, the primary phycobiliprotein present in this microalga, is a naturally occurring blue pigment and one of the value-added products derived from *Spirulina*. In 2013, the Food and Drug Administration approved the use of this pigment as a colour additive in food matrices. It is currently used as a natural colourant in foods such as beverages and candies. Additionally, due to its therapeutic properties (antioxidant, anti-inflammatory, and anticancer), it is also applied in the pharmaceutical industry, cosmetics (e.g., lipsticks and eyeliners), and as a fluorescent marker.

*Spirulina platensis*, which accounts for up to 25% of the dry biomass (w/w), is the most relevant source for commercial phycocyanin production.

Photoautotrophic cultures of *S. platensis* cultivated in open raceway ponds are commonly used for this purpose. Light which is specifically its wavelength, intensity, and source is a crucial factor influencing phototrophic production. Additionally, the culture medium should have a substantial quantity of salts (such as carbonate and bicarbonate) and high alkalinity (pH 9.5–9.8). According to Chen et al. (2013), both biomass and phycocyanin productivity increased significantly when light intensity was raised from 100 to 700 μmol/m²/s. *Spirulina* has also been cultivated in photobioreactors, which are preferred over open raceways due to lower contamination risk and higher biomass productivity (5–20 times more than open raceways).

### *β-Carotene*

Carotenoids are a class of molecules that display hues ranging from orange and yellow to red and brown. They are broadly categorized into two groups: carotenes and xanthophylls, both of which share a common structure composed of isoprenoid units. The primary distinction between them lies in their chemical structures and functional groups. Carotenes, such as β-carotene ($C_{40}H_{56}$), are pure hydrocarbon compounds, whereas xanthophylls contain oxygenated derivatives, such as hydroxyl and ketone groups at the end rings for example, astaxanthin ($C_{40}H_{52}O_4$). Due to the presence of these oxygenated groups (–OH and –CO), xanthophylls are relatively hydrophilic (Gong and Bassi, 2016). Carotenoids have a wide range of applications across various industries. They are used as active ingredients in food colouring additives (e.g., β-carotene (E160), nutraceuticals, cosmetics, pharmaceuticals, and animal feed, such as in salmon aquaculture. Moreover, carotenoids are recognized as potent antioxidants with numerous health benefits.

According to Ye et al. (2008), β-carotene is the most common carotenoid. It is stored in lipid globules located in the interthylakoid spaces of chloroplasts and is classified as a primary carotenoid, meaning it is directly involved in photosynthesis. Due to its structural role in the photosynthetic apparatus, β-carotene also provides photoprotection and aids in light absorption. In its natural state, β-carotene exists as a mixture of two isomers: all-trans and 9-cis-trans, which are challenging to synthesize. Natural β-carotene has demonstrated antioxidant and anticancer properties, offering protection against photoaging of the skin in both humans and animals, and effectively reducing cholesterol levels, thus lowering the risk of cardiovascular diseases. Additionally, the natural form of β-carotene is more readily absorbed by the human body compared to its synthetic counterpart. β-carotene, a lipid-soluble pigment, also serves as a vitamin A precursor. It undergoes enzymatic conversion into retinal and subsequently into retinol (vitamin A). Its high provitamin A activity is attributed to its ability to produce two molecules of retinol from a single β-carotene molecule. Today, β-carotene is widely used as an active ingredient in antioxidant supplements and as a natural colouring agent in products such as baked goods, soft drinks, and margarine.

Among microalgae, *Dunaliella salina*, a halotolerant green microalga, is the most abundant source of β-carotene (Xu and Harvey, 2019). Under stressful conditions, such as high salinity, intense light, extreme temperatures, or nutrient deficiency, D. salina can accumulate β-carotene up to 10% of its dry biomass. A two-stage cultivation strategy is commonly employed for β-carotene production. The first stage, known as the "greening stage," involves growing D. salina under optimal conditions. Once the cell concentration reaches a certain level, stress conditions are introduced, initiating the "reddening phase," during which the cells transition from green to orange as they accumulate carotenoids. The European Commission has approved β-carotene derived from *Dunaliella salina* as a food coloring agent (E160a(iv) Algal Carotenes).

### *Astaxanthin*

Astaxanthin is an orange-red, fat-soluble pigment that belongs to the carotenoid group, just like β-carotene. It is synthesized by certain plants, algae, and bacteria, and accumulates in some fish, crustaceans, and birds through natural food chains. Astaxanthin is considered the most powerful antioxidant found in nature, exhibiting antioxidant activity that is ten times stronger than that of other carotenoids such as lutein, zeaxanthin, β-carotene, and canthaxanthin. Furthermore, research has shown that this pigment has neuroprotective properties, provides protection against ultraviolet (UV) light, and prevents the oxidation of essential polyunsaturated fatty acids (Panis and Carreon, 2016). According to Shah et al. (2016), astaxanthin is a secondary carotenoid, while β-carotene is classified as a primary carotenoid. Astaxanthin is synthesized in the chloroplast but accumulates in the cytoplasm. Its production can be induced by environmental stressors such as oxidative stress, high salinity, intense light, nutrient starvation, and temperature fluctuations. Under such stress, secondary carotenoids are produced in excess and function as photoprotective pigments to aid in cellular adaptation.

One of the earliest uses of astaxanthin was in aquaculture, where it was incorporated into fish feed to impart a red coloration to the flesh and shells of salmon, trout, prawns, and langoustines. Today, astaxanthin is also widely used in dietary supplements due to its anti-aging, anti-inflammatory, and anticancer properties. It is available from both natural sources (extracted from crustaceans, yeast, and microalgae) and chemical synthesis. However, approximately 95% of the astaxanthin on the market is synthetic, which has an antioxidant capacity up to 20 times lower than that of the natural form. Notably, the FDA has only approved natural astaxanthin for human consumption. The green freshwater unicellular microalga *Haematococcus pluvialis* can accumulate high amounts of astaxanthin—up to 3.8–5% of its dry biomass (Fang et al., 2019). This microalga is widely used in food products, cosmetics, pharmaceuticals, nutraceuticals, and aquaculture. The requirements for astaxanthin production in *Haematococcus pluvialis* are similar to those for β-carotene production in *Dunaliella salina.* For both microalgae, stressful environmental conditions are essential for triggering pigment accumulation. In response to stress, astaxanthin accumulates in lipid globules within the cytoplasm of *Haematococcus pluvialis*. Stress conditions include salt stress, high solar irradiance, elevated temperatures, and nitrogen and phosphorus deprivation.

*Haematococcus pluvialis* displays four different cell morphologies capable of producing astaxanthin: haematocyst (aplanospores), microzooid (zoospores), palmella, and macrozooid (zoospores). In the first stage, referred to as the green motile stage, the cells exist as palmella, microzooids, and macrozooids. In the second stage, known as the red non-motile stage, the cells become aplanospores. During this transition, the cells undergo a colour change from green to reddish orange as astaxanthin accumulates.

## Factors Affecting the Microalgae Pigment Production

The production of microalgal pigments is influenced by various environmental parameters (Begum et al., 2016), including:

1. Temperature
2. Irradiances
3. Coloured wavelength
4. Photoperiods
5. pH
6. Nutrient limitation
7. Nitrogen supplements
8. Salinity

9. Pesticides
10. Heavy metals

Both environmental and nutritional factors can significantly affect the productivity and cellular composition of microalgae cultures. Prasanna et al. (2004) reported that light intensity, light wavelength, temperature, and nutrient availability regulate the tetrapyrrole content and composition in cyanobacteria, depending on the surrounding environmental parameters. According to Dipak and Lele (2005), the carotenoid content in microalgae can be enhanced through cell division inhibition, nitrogen starvation, high salinity, and exposure to high radiation and elevated temperatures. Table 2 below presents a summary of factors affecting microalgal pigment production as reported by various authors.

***Application of Microalgae Pigments***

Microalgae, which include both cyanobacteria and microalgae, are considered a potential source of high-value nutrients such as lipids, proteins, carbohydrates, and pigments. The consumption of microalgae by humans as dietary supplements has led to the development of industrial-scale microalgae production worldwide. Historically, cyanobacterial species such as *Spirulina* and green algal species such as *Chlorella* have been successfully cultivated at an industrial scale (Khanra et al., 2018). Seo et al. (2013) also noted that, apart from biomass, microalgae produce a variety of pigment molecules such as β-carotene, chlorophyll, and carotenoids which are used as colorants in the food and cosmetics industries. Algal strains such as *Chlorella sp., Dunaliella sp.,* and *Scenedesmus sp.,* along with cyanobacterial strains like *Spirulina sp. and Nostoc sp.,* are sources of fine chemicals and nutrient-rich food supplements.

Additionally, pigments derived from microalgae are used in the cosmetics industry in anti-aging creams and revitalizing or regenerating care products to nourish and heal damaged skin (Tominaga et al., 2012; Zhu, 2015). For example, *Haematococcus pluvialis*, a microalga, is recognized as a natural source of astaxanthin, a keto-carotenoid. In the poultry and aquaculture industries, astaxanthin a red pigment and precursor of vitamin A is essential for embryo development and cell reproduction (Goto et al., 2015). Furthermore, astaxanthin is gaining popularity as a human dietary supplement due to its superior antioxidant properties compared to lutein, β-carotene, α-carotene, lycopene, canthaxanthin, and vitamin E.

**Table 2.** Factors that affect microalgal pigment production

| Factors | Discussion(s) | Author(s) |
|---|---|---|
| Light and temperature | 1. Most important factors that affect the overall biomass production especially in phototrophic microalgal cultures.<br>2. Different spectral proportion of light (red: far red, blue: red, green:red, and blue: green) has effect on the relative pigment composition and likely act as photomorphogenic signals in microalgae. Individual light regimen could affect the productivity of microalgae.<br>3. An adaptive response is triggered by variations in the levels of these proteins and chlorophyll, enabling effective light harvesting. Cyanobacteria prefer low light levels and increase the synthesis of phycobiliproteins.<br>4. Because of their low specific maintenance energy rate and pigment composition, cyanobacteria prefer low light intensities. Several reports have suggested that more accessory pigment is synthesised at low irradiance.<br>5.. Cell density affects the amount of light available, and the result of cells shading one another over time affects the pigment content of microalgal cells.<br>6. The study examined the impact of varying experimental irradiance on pigment production. The results indicated that the optimal intensity for producing phycobiliproteins in blue-green microalgae (namely, S. maxima, Anabaena NCCU-9, and Synechococcus NKBG) was 25 µmolphotons/m2 /sec. Conversely, Nostoc UAM 206 exhibited better pigment production at 60 µmolphotons/m2 /sec. Higher light intensity (135 µmolphotons/m2/sec) increased Spirulina's production of phycocyanin and phycoerythrin.<br>7. The amount of light present has a significant impact on how much pigment accumulates in microalgal cells. Chlorophyll content per cell was found to be inversely correlated, meaning that green algae (Dunaliella salina, 66.9 mg/g) produced more chlorophyll at low intensity and less at high irradiance (2300 lux).<br>8. It is found that the chlorophyll content was 11.6 mg/g at 54 µmolphoton/m2/sec, while it was 14.7 mg/g at low light intensity (27 µmolphoton/m2/sec). The cellular age and composition of the cultivation medium have an impact on the amount of chlorophyll in the biomass of S. platensis. Research has demonstrated that when microalgae are cultivated in high light conditions, their biomass chlorophyll content is lower, while when they are cultivated in low light conditions, their biomass chlorophyll content is higher. As a result, the amount of chlorophyll and light intensity are inversely correlated.<br>9. Strong light caused oxidative stress, which increased the synthesis of astaxanthin. Consequently, excessive photo oxidation produces molecules of active oxygen.<br>. | Carvalho and Malcata (2003)<br>Kagawa and Suetsugu (2007)<br>Grossman et al. (1993)<br>Begum et al. (2016)<br>Begum et al. (2016)<br>Madhyastha and Vatsala (2007)<br>Chauhan and Pathak (2010)<br>Begum et al. (2016)<br>) |

| Factors | Discussion(s) | Author(s) |
|---|---|---|
| | 10. When compared to cultures conducted under continuous light sources, flashing light increased the rate of astaxanthin production in H. pluvialis per photon by at least four times. This indicates that light quality is more important than quantity.<br>11. The culture density, cell maturity, medium nutrient profile, and light path were all impacted by irradiance.<br>12. The three primary photoreceptor systems that are involved in the diurnal variations of the photosynthetic pigment content are the B/UV light photoreceptor (BLP), the green light photoreceptor (GLP), and the red/far red photoreceptors.<br>13. Light particles are absorbed by phycobiliproteins, which then transmit the energy to the active chlorophyll molecules in the photosystem. The production of phycobiliproteins in blue-green microalgae (Anabaena NCCU-9) is not stimulated by coloured light. Nonetheless, a number of studies have shown that blue light increases the production of chlorophyll in Spirulina sp. and stimulates the production of phycobiliproteins in Anacystis nidulans, Synechococcus sp., Calothrix 7601, Nostoc UAM 206, and N. muscorum, among other blue-green microalgae.<br>14. Temperature is the most fundamental element affecting all living things because it has an impact on cell biochemistry and metabolic processes. Furthermore, the ideal temperature for growth and the threshold for extreme values vary widely amongst microalgae strains. Microalgae, however, are stressed by abrupt temperature changes. Elevated temperatures promote the build-up of carotenoids, such as β-carotene, in blue-green microalgae.<br>15.The ideal temperature for Anabaena sp. to produce phycobiliproteins was found to be 30°C. Other researchers have found that the ideal temperature for S. platensis, Anaebaena sp., Nostoc sp., and Spirulina sp. was 25°C, 35°C, and 36°C.<br>16. The ideal temperature for Synechococcus sp. to produce phycobiliproteins was 36°C. A higher concentration of chlorophyll was obtained at 28°C because higher temperatures may have caused osmotic pressure changes that would have harmed the cells. Similarly, green microalgae (H. pluvialis) produced astaxanthin best at a temperature of 28°C | Salguero et al. (2003)<br>Kim et al. (2006)<br>Wang et al. (2003)<br>Begum et al. (2016)<br>Hemlata and Fatma (2009)<br>Garcia-Gonzalez et al. (2005)<br><br>Chaneva et al. (2007)<br><br>Chauhan and Pathak (2010) |
| pH | 1. The interaction of pH, irradiance, and available inorganic carbon on the development and makeup of pigments in the cyanobacterium Nostoc sp. strain UAM206, which was isolated from rice fields. The increase in pH had no effect on the amount of chlorophyll a, but it did increase the amount of phycocyanin, phycoerythrin, and allophycocyanin.<br>2. For blue-green microalgae like S. platensis, the maximum production of phycobiliproteins was achieved at pH 8.<br>3. pH 9 was found to be the ideal value for Nostoc sp. phycobiliprotein production. Similarly, it was discovered that pH 9 was ideal for S. platensis to produce chlorophyll.<br>4. The optimal pH for β-carotene production in the blue-green microalgae D. salina was found to be 7.5.<br>5. For green microalgae Scenedesmus almeriensis, pH 8 was ideal for carotenoid production; for chlorococcum citriforme and Neospongiococcus gelatinosum, pH 7 was ideal. | Hemlata and Fatma (2009)<br>Hemlata and Fatma (2009)<br>Chauhan and Pathak (2010)<br>Garcia-Gonzalez et al. (2005)<br>Del Campo et al. (2000) |

**Table 2.** (Continued)

| Factors | Discussion(s) | Author(s) |
| --- | --- | --- |
| Salinity | 1. Microalgae's ability to produce pigment is significantly impacted by salinity. According to reports, phycobilisomes detaching from the thylakoid membrane allows sodium ions to enter quickly, which reduces photosynthesis.<br>2. Energy transfer from phycobiliproteins to the photosynthetic reaction centre and the absorption of additional mineral nutrients can also cause the detachment.<br>3. Microalgae are grown in a saline environment therefore, osmosis is essential to pigment content.<br>4. The increased sodium chloride (NaCl) concentration creates a hypertonic solution in the cell's surrounding environment. This condition results in a net flow of water molecules out of the cell, shrinking the cell and causing damage to both the cell and its constituent parts.<br>5. The production of β-carotene and chlorophyll is greatly impacted by salinity. The authors observed that while β-carotene production in blue green microalgae increased in response to salinity, the chlorophyll content decreased. | Rafiqul et al. (2003)<br>Hasegawa et al. (2005)<br>Dipak and Lele (2005)<br>Hemlata and Fatma (2000)<br>Begum et al. (2016) |
| Other factors (pesticides, heavy metals, nutrient limitation) | 1. Under thiobencarb pesticide stress, blue-green microalgae like Nostoc sphaeroids and Anabena variabilis produced less phycobiliproteins.<br>2. More pesticide exposure on the phycobiliproteins' intracellular thylakoid membrane causes more damage to the proteins, which leads to their detachment.<br>3. Nitrogen is one of the principal nutrient requirements of growth media for any cell. The culture organisms may be in stress due to the absence of nitrogen or starvation. During nitrogen starvation, the microalgae cease to divide since nitrogen is the primary requirement for all the metabolic activities of the cell. Nitrogen starvation conditions can cause an excessive formation of free radicals on nitrogen starved cells and consequently β-carotene content increase markedly to 7.05 pg/cell from 1.65 pg/cell.<br>4. Cyanobacteria need a specific source of nitrogen. The highest amount of phycobiliprotein was produced by blue-green microalgae (Anabaena NCCU-9) in a nitrogen-free environment.<br>5. While Fischerella sp. produced more phycobiliproteins under nitrate grown cells than in nitrogen-free media, Anabaena 7120 produced more phycobiliprotein in nitrogen-free media than in nitrate grown cultures.<br>6. It has been reported that adding iron to media increases astaxanthin synthesis. The process of iron electro valencies and counterions affects astaxanthin accumulation and cell growth.<br>7. Astaxanthin synthesis was boosted by photosynthetic carbon fixation, and when precise nitrate dosage is provided, it is also produced more efficiently in outdoor conditions. | Xia (2005)<br>Prasad et al. (2005)<br>Moller et al. (2000)<br>Hemlata and Fatma (2009)<br>Soltani et al. (2007)<br>Cai et al. (2009)<br>Garcia-Malea et al. (2009) |

**Table 3.** Commercial use of microalgal pigments

| Microalgal pigments | Discussion (s) | Author(s) |
|---|---|---|
| Phycobiliproteins | 1. Used in the commercial sector as natural dyes. Thus, phycocyanin is widely used as food pigment to replace the current synthetic pigments.<br>2. Because of their spectral characteristics, phycobiliproteins are important components of fluorescent-based detection systems, particularly for flow cytometry.<br>3. Phycocyanin's anti-inflammatory, neuroprotective, hepatoprotective, and antioxidant qualities make it a useful pharmaceutical ingredient. | Santiago-Santos et al. (2004)<br>Begum et al. (2016)<br>Sekar et al. (2008) |
| Chlorophylls | 1. Chlorophyll are present in microalgae. Rhodophyta and cyanobacteria are rich in chlorophyll a, the primary photosynthetic pigment. While chlorophylls c, d, and e are found in freshwater diatoms, chlorophyta and euglenophyta, which resemble higher plants and marine microalga, contain chlorophyll b. Chlorophyll is a vital substance that is utilised in cosmetic and pharmaceutical products as an additive.<br>2. Because of its stability, chlorophyll a has been used widely as a colouring agent. This material is typically derived from higher plants that also synthesise other types of chlorophyll. However, S. platensis only possesses chlorophyll a. The largest source of chlorophyll, which is used in colourants as an alternative to artificial colouring, is found in Spirulina sp.<br>3. Chlorophyll compounds' ability to heal wounds and reduce inflammation makes them useful in medicine. and there is a reduction in the risk of colorectal cancer as a result of consuming chlorophyll. | Begum et al. (2016)<br>Begum et al. (2016)<br>Balder et al. (2006) |
| β-carotene | 1. Out of all the carotenoids found in nature, β-carotene is arguably the most significant. Provitamin A is one of these molecules, and it has a variety of biological actions and functions, especially with regard to human health.<br>2. β-carotene is converted by the body through body tissue to vitamin A, it has many positive effects on the human body. According to reports, vitamin A is essential for human health because it boosts immunity, guards against skin conditions, cataracts, and night blindness. β-carotene is utilised in the preparation of healthy foods and as a pro-vitamin A (retinol) in multivitamin preparations.<br>3. Research has linked β-carotene to a lower risk of cancer and other degenerative diseases. It also possesses immune-modulating, anti-aging, and anticancer qualities. β-carotene from Dunaliella sp. contains 40% 9-cis and 50% all-trans stereoisomers, which is important for reducing the incidence of various cancer types and degenerative diseases, according to a few epidemiological researchers. | Dipak and Lele (2005)<br>Spolaore et al. (2006)<br>Begum et al. (2016) |

**Table 3.** (Continued)

| Microalgal pigments | Discussion (s) | Author(s) |
|---|---|---|
| Astaxanthin | 1. Unlike β-carotene, astaxanthin has no provitamin activity. Researchers have discovered that astaxanthin is useful for preventing low density protein oxidation in vitro, which can be used to stop the development of arteriosclerosis, coronary heart disease, and ischemic brain damage. | Begum et al. (2016) |
| | 2. Due to its stronger antioxidant activity, which is more than 500 times more effective than α-tocoferol and 10 times higher than β-carotene, astaxanthin functions as a super vitamin E. It has a protective effect against aflatoxin carcinogenicity and an inhibitory effect on lipid peroxidation mediated by active form of oxygen because of its stronger antioxidant activities. | Begum et al. (2016) |
| | 3. Astaxanthin's potent bioactive antioxidant qualities make it a potential treatment for the neuronal damage linked to age-related macular degeneration. Because astaxanthin can cross the blood-brain barrier and does not form crystals in the eyes, it can be used to treat a variety of central nervous system injuries, including Parkinson's disease, Alzheimer's disease, ischemic reperfusion injury, spinal cord injuries, and other conditions. | Begum et al. (2016) |

Microalgae are also regarded as a reliable and abundant source of plant-based protein (Becker, 2007). Nutritional studies on various microalgae have shown that they produce both high quantities and high-quality proteins containing essential amino acids (Hosseini et al., 2022). Recently, microalgae have shown promise as an alternative expression host for the synthesisof recombinant proteins (Specht et al., 2010). Phycobiliproteins are a class of auxiliary photosynthetic pigment-protein complexes produced by some algae species, such as cyanobacteria and red algae (*Rhodophyta*), to harvest light (Khanra et al., 2018). These proteins are in high demand in the pharmaceutical industry and are specifically used as fluorophores in biological research.

Due to their anti-inflammatory, neuroprotective, hepatoprotective, and antioxidant properties, microalgal pigments have long been recognized for their potential in industrial applications (Spolaore et al., 2006). They are widely used across various industries, including (Dufossé et al., 2005):

1. Food
2. Nutraceuticals
3. Pharmaceuticals
4. Aquaculture
5. Phycocyanin

Additionally, microalgal pigments have been utilized in clinical and research laboratories as effective labels for antibodies and receptors. Begum et al. (2016) also noted that the eco-friendly nature of microalgae production has increased interest in using microalgal pigments in aquaculture for feed formulation and supplementation. Phycobiliproteins, chlorophylls, β-carotene, and astaxanthin are already being commercially used in various fields, as outlined in Table 3.

Primary pigments such as astaxanthin, β-carotene, chlorophyll a, b, and c, xanthophylls, and phycobiliproteins have numerous potential applications in medicine, biomedical research, diagnostics, and as food colorants in dairy, cosmetics, and other products. Due to their non-toxic and non-carcinogenic nature, these natural pigments are becoming increasingly favoured over synthetic alternatives. The pigment content of microalgae is influenced by both the type of microalgae and their cultivation environment. Several factors affect microalgal pigment production, including temperature, salinity, irradiance, light wavelength, photoperiods, pH, nutrient limitation, nitrogen supplementation, pesticides, and heavy metals. Therefore, these environmental and nutritional parameters should be carefully considered during pigment production, given the broad range of potential applications. In addition, microalgae are widely recognized as a promising source of various high-value products for biotechnological exploitation. Table 4 presents selected microalgal pigment products and their respective producers

**Table 4.** Biotechnological applications of products (Chu, 2012)

| Product | Applications | Microalgal Producers |
|---|---|---|
| Phycobiliproteins<br>Phycocyanin | Natural dye for health food and cosmetics (lipsticks and eyeliners) antioxidants | Spirulina platensis |
| Phycoerythrin | Flurescent agent, tool for bimedical research, diagnostic tool | Red algae (e.g. Porphyridium cruentum) |
| Carotenoids<br>β-carotene | Food colourant; antioxidant; cancer-preventive properties | Dunaliella salina |
| Astaxanthin | Pigmentary for salmon, antioxidant | Haematococcus pluvialis |

Phycobiliproteins are water-soluble pigments that function as auxiliary photosynthetic pigments in *Rhodophyta* (red algae), glaucophytes, cryptophytes, and cyanobacteria. The three major types phycocyanin, allophycocyanin, and phycoerythrin exhibit distinct spectral characteristics. The phycobiliprotein content in microalgae is influenced by environmental factors, particularly light intensity and spectral quality. For example, in Spirulina platensis, phycocyanin content can range from 0.11% to 12.7% of dry weight depending on light intensity. *Spirulina* and *Porphyridium* are two microalgae species that have been commercially exploited for phycobiliprotein production. These pigments are primarily used as fluorescent markers in diagnostic tools for biomedical research and as natural colorants in food and cosmetic products. For instance, Dainippon Ink and Chemicals (Sakura) markets phycocyanin under the brand name Lina Blue, which is used in chewing gum, ice sherbets, popsicles, candies, soft drinks, dairy products, and wasabi. A different formulation of this pigment is also used in natural cosmetics such as lipsticks and eyeliners. In immunology laboratories, phycobiliproteins serve as fluorescent markers in techniques like fluorescence microscopy and fluorescence-activated cell sorting (FACS), where they label biological molecules such as antibodies and receptors (Chu, 2012).

Recent studies suggest that phycocyanin offers a broad range of therapeutic applications in addition to its health-promoting properties. Its antioxidant capacity, which protects human erythrocytes from lysis caused by peroxyl radicals, has been shown to be 16 times greater than Trolox (a vitamin

E analog) and 20 times greater than vitamin C. Chu (2012) also reported that phycocyanin reduces oxidative stress and has demonstrated protective effects on human pancreatic cells and in experimental arthritis in rats. Additionally, selenium-enriched phycocyanin from *Spirulina platensis* has been shown to prevent atherosclerosis in hamsters by inhibiting oxidative stress-inducing factors. According to Chu (2012), carotenoids are derived from 5-carbon isoprene units polymerized enzymatically into highly conjugated 40-carbon structures. In algae, they function as accessory pigments during photosynthesis. Carotenoids containing only hydrocarbons are classified as carotenes, while those with hydroxyl, xylene, or oxo groups are classified as xanthophylls. Among the 400 known carotenoids, β-carotene and astaxanthin are the most commercially utilized, with others like lutein, zeaxanthin, and lycopene used less frequently. Carotenoids are applied in cosmetics, as animal feed additives, and as natural food colorants. Nutritionally, some carotenoids especially β-carotene serve as provitamin A. Their therapeutic value lies primarily in their antioxidant, anti-inflammatory, and anticancer properties.

*Dunaliella salina*, a green alga, is mass-cultivated in open ponds to produce β-carotene. When grown under high-salinity conditions, this alga can accumulate β-carotene up to 14% of its dry weight. Currently, approximately 25 industrial facilities in Australia, the United States, and China are dedicated to its cultivation. Unlike synthetic β-carotene, which contains only trans-isomers, *Dunaliell*a produces both cis- and trans-isomers, resulting in higher antioxidant activity. Commercial products include dried and powdered *Dunaliella* for animal and human consumption, as well as β-carotene extracts. Studies have shown that *Dunaliella*-derived β-carotene can reduce atherosclerosis risk by inhibiting low-density lipoprotein (LDL) oxidation and positively influencing blood lipid profiles such as triglycerides, cholesterol, and HDL levels. Astaxanthin, another high-value carotenoid, is predominantly produced by the green microalga *Haematococcus pluvialis*. Under stress conditions, the alga accumulates astaxanthin up to 4–5% of its dry weight as it transitions from the motile flabellate stage to a red, thick-walled resting stage. Astaxanthin is widely used as a natural feed additive in aquaculture, particularly for enhancing the pigmentation of salmon, trout, and prawns, and as a poultry feed additive for yolk coloration. It has also entered the nutraceutical market in encapsulated form and is sold as a dietary supplement due to its exceptional antioxidant potency reported to be 100 times greater than α-tocopherol.

Astaxanthin's health benefits include protection against various conditions such as cancer, inflammatory diseases, metabolic syndrome, diabetes, diabetic nephropathy, neurological disorders, and eye-related conditions. These protective effects are primarily attributed to its ability to prevent oxidative damage and inhibit apoptosis or necrosis caused by oxidative stress (Chu, 2012).

## Conclusion

Therefore, microalgae have great potential across various markets and are excellent sources of high-value compounds with significant health benefits. These include lipids, proteins, pigments, vitamins, minerals, polysaccharides, and polyunsaturated fatty acids. The abundance of microalgae in aquatic systems and the richness of their bioactive compounds have sparked global interest in developing microalgae-based products. Among the wide array of compounds derived from microalgae, natural pigments represent one of the most promising categories for further investigation. These pigments not only provide natural coloration but also possess anti-inflammatory, anticancer, and antioxidant properties, making them excellent alternatives to synthetic colourants.

The microalgae species and their derived pigments discussed herein have a promising future. The scientific community has recognized microalgae as a natural source of colourants and nutraceuticals, with their potential already demonstrated across several industrial sectors. Although certain challenges remain, the renewable, sustainable nature of microalgae and the additional health benefits of their pigments suggest that these natural compounds will soon surpass synthetic alternatives.

## References

Abed, R. M., Dobretsov, S., & Sudesh, K. (2009). Applications of cyanobacteria in biotechnology. *Journal of Applied Microbiology, 106*(1), 1-12. https://doi.org/10.1111/j.1365-2672.2008.04016.x.

Becker, E. W. (2007). Micro-algae as a source of protein. *Biotechnology Advances, 25*(2), 207-210. https://doi.org/10.1016/j.biotechadv.2007.01.002.

Begum, H., Yusoff, F. M., Banerjee, S., Khatoon, H., & Shariff, M. (2016). Availability and utilization of pigments from microalgae. *Critical Reviews in Food Science and Nutrition, 56*(13), 2209-2222. https://doi.org/10.1080/10408398.2014.972050.

Cai, M., Li, Z., & Qi, A. (2009). Effects of iron electrovalence and species on growth and astaxanthin production of *Haematococcus pluvialis*. *Chinese Journal of Oceanology and Limnology, 27*, 370-375. https://doi.org/10.1007/s00343-009-0111-3.

Carvalho, A. P., & Malcata, F. X. (2003). Kinetic modeling of the autotrophic growth of *Pavlova lutheri*: Study of the combined influence of light and temperature. *Biotechnology Progress, 19*(4), 1128-1135. https://doi.org/10.1021/bp034058g.

Chaneva, G., Furnadzhieva, S., Minkova, K., & Lukavsky, J. (2007). Effect of light and temperature on the cyanobacterium *Arthronema africanum*, a prospective phycobiliprotein producing strain. *Journal of Applied Phycology, 19*, 537-544. https://doi.org/10.1007/s10811-007-9224-2.

Chauhan, U. K., & Pathak, N. (2010). Effect of different conditions on the production of chlorophyll by *Spirulina platensis*. *Journal of Algal Biomass Utilization, 1*(4), 89-99..

Chen, C. Y., Kao, P. C., Tsai, C. J., Lee, D. J., & Chang, J. S. (2013). Engineering strategies for simultaneous enhancement of C-phycocyanin production and CO2 fixation with *Spirulina platensis*. *Bioresource Technology, 145*, 307-312. https://doi.org/10.1016/j.biortech.2013.02.092.

Christaki, E., Bonos, E., & Florou-Paneri, P. (2015). Innovative microalgae pigments as functional ingredients in nutrition. In *Handbook of Marine Microalgae* (pp. 233-243). Academic Press. https://doi.org/10.1016/B978-0-12-417012-9.00021-4.

Chu, W. L. (2012). Biotechnological applications of microalgae. *IEJSME, 6*(1), S24-S37.

Chu, W. L., Phang, S. M., & Goh, S. H. (1996). Environmental effects on growth and biochemical composition of *Nitzschia inconspicua* Grunow. *Journal of Applied Phycology, 8*, 389-396. https://doi.org/10.1007/BF02186343.

Cid, A., Prado, R., Rioboo, C., Suarez-Bregua, P., & Herrero, C. (2012). Use of microalgae as biological indicators of pollution: Looking for new relevant cytotoxicity endpoints. In M. N. Johnsen (Ed.), *Microalgae: Biotechnology, Microbiology and Energy* (pp. 311-323). Nova Science Publishers.

Del Campo, J. A., Moreno, J., Rodríguez, H., Vargas, M. A., Rivas, J., & Guerrero, M. G. (2000). Carotenoid content of chlorophycean microalgae: Factors determining lutein accumulation in *Muriellopsis sp.* (Chlorophyta). *Journal of Biotechnology, 76*(1), 51-59. https://doi.org/10.1016/S0168-1656(99)00257-X.

Dipak, P., & Lele, S. (2005). Carotenoid production from microalga, *Dunaliella salina*. *Indian Journal of Biotechnology, 4*(8), 476-483.

Dufossé, L., Galaup, P., Yaron, A., Arad, S. M., Blanc, P., Murthy, K. N. C., & Ravishankar, G. A. (2005). Microorganisms and microalgae as sources of pigments for food use: A scientific oddity or an industrial reality?. *Trends in Food Science & Technology, 16*(9), 389-406. https://doi.org/10.1016/j.tifs.2005.07.003.

Dumay, J., & Morançais, M. (2016). Proteins and pigments. In *Seaweed in Health and Disease Prevention* (pp. 275-318). Academic Press.

Elisabeth, B., Rayen, F., & Behnam, T. (2021). Microalgae culture quality indicators: A review. *Critical Reviews in Biotechnology, 41*(4), 457-473. https://doi.org/10.1080/07388551.2020.1744109.

Fang, N., Wang, C., Liu, X., Zhao, X., Liu, Y., Liu, X., ... & Zhang, H. (2019). De novo synthesis of astaxanthin: From organisms to genes. *Trends in Food Science & Technology, 92*, 162-171. https://doi.org/10.1016/j.tifs.2019.08.007.

Ferruzzi, M. G., Böhm, V., Courtney, P. D., & Schwartz, S. J. (2002). Antioxidant and antimutagenic activity of dietary chlorophyll derivatives determined by radical scavenging and bacterial reverse mutagenesis assays. *Journal of Food Science, 67*(7), 2589-2595. https://doi.org/10.1111/j.1365-2621.2002.tb08708.x.

García-González, M., Moreno, J., Manzano, J. C., Florencio, F. J., & Guerrero, M. G. (2005). Production of *Dunaliella salina* biomass rich in 9-cis-β-carotene and lutein in a closed tubular photobioreactor. *Journal of Biotechnology, 115*(1), 81-90. https://doi.org/10.1016/j.jbiotec.2004.07.001.

García-Malea, M. C., Acién, F. G., Del Río, E., Fernández, J. M., Cerón, M. C., Guerrero, M. G., & Molina-Grima, E. (2009). Production of astaxanthin by *Haematococcus pluvialis*: Taking the one-step system outdoors. *Biotechnology and Bioengineering, 102*(2), 651-657. https://doi.org/10.1002/bit.22087.

Gong, M., & Bassi, A. (2016). Carotenoids from microalgae: A review of recent developments. *Biotechnology Advances, 34*(8), 1396-1412. https://doi.org/10.1016/j.biotechadv.2016.08.005.

Goto, M., Kanda, H., & Machmudah, S. (2015). Extraction of carotenoids and lipids from algae by supercritical CO2 and subcritical dimethyl ether. *The Journal of Supercritical Fluids, 96*, 245-251. https://doi.org/10.1016/j.supflu.2014.10.003.

Grossman, A. R., Schaefer, M. R., Chiang, G. G., & Collier, J. L. (1993). Environmental effects on the light-harvesting complex of cyanobacteria. *Journal of Bacteriology, 175*(3), 575-582. https://doi.org/10.1128/jb.175.3.575-582.1993.

Guedes, A. C., Amaro, H. M., & Malcata, F. X. (2011). Microalgae as sources of carotenoids. *Marine Drugs, 9*(4), 625-644. https://doi.org/10.3390/md9040625.

Hasegawa, P. M., Bressan, R. A., Zhu, J. K., & Bohnert, H. J. (2000). Plant cellular and molecular responses to high salinity. *Annual Review of Plant Biology, 51*(1), 463-499. https://doi.org/10.1146/annurev.arplant.51.1.463.

Hemlata, & Fatma, T. (2009). Screening of cyanobacteria for phycobiliproteins and effect of different environmental stress on its yield. *Bulletin of Environmental Contamination and Toxicology, 83*, 509-515. https://doi.org/10.1007/s00128-009-9762-x.

Hoseini, S. M., Khosravi-Darani, K., & Mozafari, M. R. (2013). Nutritional and medical applications of spirulina microalgae. *Mini Reviews in Medicinal Chemistry, 13*(8), 1231-1237. https://doi.org/10.2174/1389557511313080008.

Hosikian, A., Lim, S., Halim, R., & Danquah, M. K. (2010). Chlorophyll extraction from microalgae: A review on the process engineering aspects. *International Journal of Chemical Engineering, 2010*. https://doi.org/10.1155/2010/381258.

Hosseini, S. F., Rezaei, M., & McClements, D. J. (2022). Bioactive functional ingredients from aquatic origin: A review of recent progress in marine-derived nutraceuticals. *Critical Reviews in Food Science and Nutrition, 62*(5), 1242-1269. https://doi.org/10.1080/10408398.2020.1787489.

Humphrey, A. M. (2004). Chlorophyll as a color and functional ingredient. *Journal of Food Science, 69*(5), C422-C425. https://doi.org/10.1111/j.1365-2621.2004.tb11867.x.

Kagawa, T., & Suetsugu, N. (2007). Photometrical analysis with photosensory domains of photoreceptors in green algae. *FEBS Letters, 581*(3), 368-374. https://doi.org/10.1016/j.febslet.2006.12.019.

Khanra, S., Mondal, M., Halder, G., Tiwari, O. N., Gayen, K., & Bhowmick, T. K. (2018). Downstream processing of microalgae for pigments, protein and carbohydrate in industrial application: A review. *Food and Bioproducts Processing, 110*, 60-84. https://doi.org/10.1016/j.fbp.2017.12.004.

Kim, Z. H., Kim, S. H., Lee, H. S., & Lee, C. G. (2006). Enhanced production of astaxanthin by flashing light using *Haematococcus pluvialis*. *Enzyme and Microbial Technology, 39*(3), 414-419. https://doi.org/10.1016/j.enzmictec.2005.10.023.

Lordan, S., Ross, R. P., & Stanton, C. (2011). Marine bioactives as functional food ingredients: Potential to reduce the incidence of chronic diseases. *Marine Drugs, 9*(6), 1056-1100. https://doi.org/10.3390/md9061056.

Madhyastha, H. K., & Vatsala, T. M. (2007). Pigment production in *Spirulina fussiformis* in different photophysical conditions. *Biomolecular Engineering, 24*(3), 301-305. https://doi.org/10.1016/j.bioeng.2007.02.003.

Masojídek, J., Torzillo, G., & Koblížek, M. (2013). Photosynthesis in microalgae. In *Handbook of Microalgal Culture: Applied Phycology and Biotechnology* (pp. 21-36). Wiley-Blackwell. https://doi.org/10.1002/9781118556707.ch3.

Moller, A. P., Biard, C., Blount, J. D., Houston, D. C., Ninni, P., Saino, N., & Surai, P. F. (2000). Carotenoid-dependent signals: Indicators of foraging efficiency, immunocompetence or detoxification ability. *Poultry and Avian Biology Reviews, 11*(3), 137-160.

Mulders, K. J., Lamers, P. P., Martens, D. E., & Wijffels, R. H. (2014). Phototrophic pigment production with microalgae: Biological constraints and opportunities. *Journal of Phycology, 50*(2), 229-242. https://doi.org/10.1111/jpy.12154.

Pagels, F., Pereira, R. N., Vicente, A. A., & Guedes, A. C. (2021). Extraction of pigments from microalgae and cyanobacteria—A review on current methodologies. *Applied Sciences, 11*(11), 5187. https://doi.org/10.3390/app11115187.

Pagels, F., Salvaterra, D., Amaro, H. M., & Guedes, A. C. (2020). Pigments from microalgae. In *Handbook of Microalgae-Based Processes and Products* (pp. 465-492). Academic Press. https://doi.org/10.1016/B978-0-12-818820-1.00021-7.

Panda, B., Jain, P., Sharma, L., & Mallick, N. (2006). Optimization of cultural and nutritional conditions for accumulation of poly-β-hydroxybutyrate in *Synechocystis sp.* PCC 6803. *Bioresource Technology, 97*(11), 1296-1301. https://doi.org/10.1016/j.biortech.2005.08.023.

Panis, G., & Carreon, J. R. (2016). Commercial astaxanthin production derived by green alga *Haematococcus pluvialis*: A microalgae process model and a techno-economic assessment all through production line. *Algal Research, 18*, 175-190. https://doi.org/10.1016/j.algal.2016.06.015.

Pan-Utai, W., Kahapana, W., & Iamtham, S. (2018). Extraction of C-phycocyanin from *Arthrospira* (*Spirulina*) and its thermal stability with citric acid. *Journal of Applied Phycology, 30*, 231-242. https://doi.org/10.1007/s10811-017-1297-2.

Prasad, S. M., Kumar, D., & Zeeshan, M. (2005). Growth, photosynthesis, active oxygen species, and antioxidants responses of paddy field cyanobacterium *Plectonema boryanum* to endosulfan stress. *The Journal of General and Applied Microbiology, 51*(2), 115-123. https://doi.org/10.2323/jgam.51.115.

Prasanna, R., Pabby, A., Saxena, S., & Singh, P. K. (2004). Modulation of pigment profiles of *Calothrix elenkenii* in response to environmental changes. *Journal of Plant Physiology, 161*(10), 1125-1132. https://doi.org/10.1078/0176-1617-01293.

Pulz, O., & Gross, W. (2004). Valuable products from biotechnology of microalgae. *Applied Microbiology and Biotechnology, 65*, 635-648. https://doi.org/10.1007/s00253-004-1613-6.

Rafiqul, I. M., Hassan, A., Sulebele, G., Orosco, C. A., Roustaian, P., & Jalal, K. C. A. (2003). Salt stress culture of blue-green algae *Spirulina fusiformis*. *Pakistan Journal of Biological Sciences, 6*(7), 1420-1423.

Reynolds, C. S. (2006). *The Ecology of Phytoplankton*. Cambridge University Press.

Richmond, A. (Ed.). (2004). *Handbook of Microalgal Culture: Biotechnology and Applied Phycology* (Vol. 577). Blackwell Science.

Safi, C., Zebib, B., Merah, O., Pontalier, P. Y., & Vaca-Garcia, C. (2014). Morphology, composition, production, processing and applications of *Chlorella vulgaris*: A review. *Renewable and Sustainable Energy Reviews, 35*, 265-278. https://doi.org/10.1016/j.rser.2014.04.017.

Salguero, A., de la Morena, B., Vigara, J., Vega, J. M., Vilchez, C., & León, R. (2003). Carotenoids as protective response against oxidative damage in *Dunaliella bardawil*. *Biomolecular Engineering, 20*(4-6), 249-253. https://doi.org/10.1016/j.biomoleng.2003.08.005.

Santiago-Santos, M. C., Ponce-Noyola, T., Olvera-Ramírez, R., Ortega-López, J., & Cañizares-Villanueva, R. O. (2004). Extraction and purification of phycocyanin from *Calothrix* sp. *Process Biochemistry, 39*(12), 2047-2052. https://doi.org/10.1016/j.procbio.2003.10.001.

Sekar, S., & Chandramohan, M. (2008). Phycobiliproteins as a commodity: Trends in applied research, patents, and commercialization. *Journal of Applied Phycology, 20*(2), 113-136. https://doi.org/10.1007/s10811-007-9236-y.

Seo, Y. C., Choi, W. S., Park, J. H., Park, J. O., Jung, K. H., & Lee, H. Y. (2013). Stable isolation of phycocyanin from *Spirulina platensis* associated with high-pressure extraction process. *International Journal of Molecular Sciences, 14*(1), 1778-1787. https://doi.org/10.3390/ijms14011778.

Shah, M. M. R., Liang, Y., Cheng, J. J., & Daroch, M. (2016). Astaxanthin-producing green microalga *Haematococcus pluvialis*: From single cell to high value commercial products. *Frontiers in Plant Science, 7*, 531. https://doi.org/10.3389/fpls.2016.00531.

Soltani, N., Khavari-Nejad, R. A., Yazdi, M. T., & Shokravi, S. (2007). Growth and some metabolic features of cyanobacterium *Fischerella sp.* FS18 in different combined nitrogen sources. *J. Sci. Repub. Iran, 18*, 123–128.

Specht, E., Miyake-Stoner, S., & Mayfield, S. (2010). Micro-algae come of age as a platform for recombinant protein production. *Biotechnology Letters, 32*, 1373-1383. https://doi.org/10.1007/s10529-010-0330-7.

Spolaore, P., Joannis-Cassan, C., Duran, E., & Isambert, A. (2006). Commercial applications of microalgae. *Journal of Bioscience and Bioengineering, 101*(2), 87-96. https://doi.org/10.1263/jbb.101.87.

Tominaga, K., Hongo, N., Karato, M., & Yamashita, E. (2012). Cosmetic benefits of astaxanthin on human subjects. *Acta Biochimica Polonica, 59*(1), 93-96.

Volk, R. B., & Furkert, F. H. (2006). Antialgal, antibacterial, and antifungal activity of two metabolites produced and excreted by cyanobacteria during growth. *Microbiological Research, 161*(2), 180-186. https://doi.org/10.1016/j.micres.2005.10.002.

Wang, B., Zarka, A., Trebst, A., & Boussiba, S. (2003). Astaxanthin accumulation in *Haematococcus Pluvialis* (Chlorophyceae) as an active photoprotective process under high irradiance. *Journal of Phycology, 39*(6), 1116-1124. https://doi.org/10.1046/j.1529-8817.2003.02018.x.

Xia, J. (2005). Response of growth, photosynthesis, and photoinhibition of the edible cyanobacterium *Nostoc sphaeroides* colonies to thiobencarb herbicide. *Chemosphere, 59*(4), 561-566. https://doi.org/10.1016/j.chemosphere.2004.11.002.

Xu, Y., & Harvey, P. J. (2019). Carotenoid production by *Dunaliella salina* under red light. *Antioxidants, 8*(5), 123. https://doi.org/10.3390/antiox8050123.

Ye, Z. W., Jiang, J. G., & Wu, G. H. (2008). Biosynthesis and regulation of carotenoids in *Dunaliella*: Progresses and prospects. *Biotechnology Advances, 26*(4), 352-360. https://doi.org/10.1016/j.biotechadv.2008.03.001.

Zhang, J., Sun, Z., Sun, P., Chen, T., & Chen, F. (2014). Microalgal carotenoids: Beneficial effects and potential in human health. *Food & Function, 5*(3), 413-425. https://doi.org/10.1039/c3fo60228a.

Zhu, L. (2015). Biorefinery as a promising approach to promote the microalgae industry: An innovative framework. *Renewable and Sustainable Energy Reviews, 41*, 1376-1384. https://doi.org/10.1016/j.rser.2014.08.042.

# Chapter 6

# The Spectrum of Natural Pigments: Diversity, Industrial Applications, and Future Prospects

**Nur Hanie Mohd Latiff[1,*], PhD**
**and Rashidi Othman[2], PhD**
[1]International Institute for Halal Research and Training,
International Islamic University Malaysia, Kuala Lumpur, Malaysia
[2]Herbarium Unit, Department of Landscape Architecture,
Kulliyyah of Architecture and Environmental Design,
International Islamic University Malaysia, Kuala Lumpur, Malaysia

## Abstract

Natural pigments are vast and multifaceted, promising significant advancements in various industries, including food, pharmaceuticals, textiles, and environmental sustainability. These pigments encompass a wide array of compounds derived from botanical, animal, and mineral sources. They offer a safer and more sustainable option, imbuing artworks, textiles, cosmetics, and other goods with a unique character and depth originating from nature's palette. Plant-based pigments, such as chlorophylls, carotenoids, and anthocyanins, showcase colours ranging from green to red. In contrast, animal-derived pigments like cochineal extract yield unique red and purple shades. Mineral pigments like iron oxides contribute earthy tones suitable for various applications. Continuous research and innovation suggest that natural pigments are poised to play a significant role in the future of colouration and material science, offering a renewable and environmentally friendly solution to meet evolving consumer needs. There is a noticeable shift toward natural additives over synthetic alternatives in consumer choices, prompting the

* Corresponding Author's Email: nurhanielatiff@iium.edu.my

In: Natural Pigments: Sources, Properties, and Health Benefits
Editors: Rashidi Othman, Qurratu Aini Mat Ali, Razanah Ramya et al.
ISBN: 979-8-90134-055-4

food industry to anticipate a considerable move towards plant-based pigments. This transition is bolstered by continued research to enhance the stability and intensity of these pigments. Additionally, marine natural pigments and bacterial pigments show promising therapeutic applications, offering biocompatibility, bioavailability, and environmentally friendly solutions. Thus, the future potential of natural pigments lies in their ability to meet the demands of various industries for sustainable, safe, and bioactive colourants. Continuous research and technological advancements pave the way for their expanded use, enhancing stability, colour range, and application across industries.

**Keywords**: natural pigments, bioactive colourant, enhancing stability, colour range

## Introduction

Introducing the essence of nature's hues: natural pigments. These organic colourants derived from nature possess a rich historical significance in artistic and practical application. From the vibrant reds of ochre to the deep blues of indigo, natural pigments offer a spectrum of hues that captivate the senses and evoke a connection to the natural world. Delving into their origins, properties, and diverse uses reveals a fascinating journey through culture, artistry, and sustainability.

Current trends in natural pigment research are showcasing a diverse approach that spans various fields, reflecting a growing interest in sustainable and health-conscious alternatives to synthetic dyes. A notable focus is the exploration and industrial application of pigment-producing microorganisms, such as cyanobacteria, macroalgae, and microalgae. Research has emphasized the production of carotenoids and phycocyanin production on an industrial level, including developing toxin-free strains for broader consumer acceptance (Dufossé, 2023). In the textile industry, there is a shift towards natural dyes and pigments. Researchers are investigating methods to enhance these colourants' fastness and exploring environmentally sustainable extraction techniques. The increasing importance of ultraviolet protection and the antibacterial properties of natural dyes in textile applications is becoming more prominent (Mamun et al., 2023). In the food industry, the application of natural pigments is expanding due to their bioactive properties, offering health benefits beyond their colouring function. This has led to a demand for non-toxic, eco-sustainable, and biodegradable pigments (Di Salvo et al., 2023).

The scarcity of natural edible blue pigments has prompted significant research into microbial fermentation metabolites, focusing on improving extraction processes and exploring their bioactive properties (Huang & Zhu, 2022). The marine environment, rich in biodiversity, is increasingly recognised for its potential to provide pigments with bioactive properties relevant to the food, pharmaceutical, cosmetic, and textile industries. This highlights the importance of sustainability and environmental safety (Muñoz-Miranda & Iñiguez-Moreno, 2023). Concerns over the safety of synthetic dyes have driven the food, pharmaceutical, and cosmetic industries to search for natural colourings with additional health benefits, such as antibacterial and antioxidant properties. This has led to the exploration of novel microbial strains for pigment production (Aman Mohammadi et al., 2022). Recent advancements in synthetic biology and optimization of microbial cultivation represent significant trends in scaling up the production of natural pigments (Lyu et al., 2022). The therapeutic potential of natural pigments is also being explored, focusing on their bioactive properties and the development of advanced extraction and identification techniques (Pasdaran et al., 2023). The search for natural pigments with high stability and productivity, such as prodigiosin from microbial sources, which is a significant trend due to its potential applications in food and pharmaceutical industries (Paul et al., 2020). Lastly, innovative methods like freeze-drying are being investigated to extract high-quality natural food colourants, with an interest in using industrial waste and natural by-products (Różyło, 2020).

Natural pigments not only offer a safer and more sustainable option but also imbue artworks, textiles, cosmetics, and other goods with a unique character and depth that reflects nature's palette. Furthermore, the use of natural pigments in contemporary design and architecture enhances authenticity and timelessness to spaces, creating environments that resonate harmony and a connection to the natural world. Whether used in traditional crafts or modern innovations, natural pigments inspire creativity and deepen the appreciation for the beauty and diversity of Earth's resources.

## The Natural Pigments Types

Natural pigments, which are essential for imparting colour and enhancing the appeal of various products. They can be broadly categorized into several types based on their chemical structure, source, and application. These pigments are derived from a wide range of natural sources, including plants, fungi, bacteria, insects, algae, and marine organisms. Due to their safety and environmental benefits over synthetic alternatives, they are widely utilized in the food, cosmetics, and textile industries. Other notable pigments include porphyrins, found in plants and plant parts, which significantly contribute to the natural colour palette, along with carotenoids, anthocyanins, and chlorophylls (Singh et al., 2023). Microbial fermentation has emerged as a promising method for producing natural pigments, including flavonoids, isoprenoids, porphyrins, and others, offering a sustainable alternative that is not affected by seasonal or regional limitations (Derksen1 & Bechtold2, 2023). The diversity of natural pigments, such as naphthoquinones, xanthones, benzotropolones, alkaloids, and terpenoids, underscores their importance in various applications, from food colouring to therapeutic agents (Rodriguez-Amaya & Carle, 2021). In summary, natural pigments encompass many compounds, including carotenoids, anthocyanins, flavonoids, chlorophylls, betalains, phycobiliproteins, and porphyrins, all derived from numerous natural sources. These pigments are integral for imparting colour and value to a multitude of products, highlighting the shift towards natural alternatives driven by safety and environmental reasons (Swami et al., 2020). Figure 1 illustrates various types of natural pigments and their compounds.

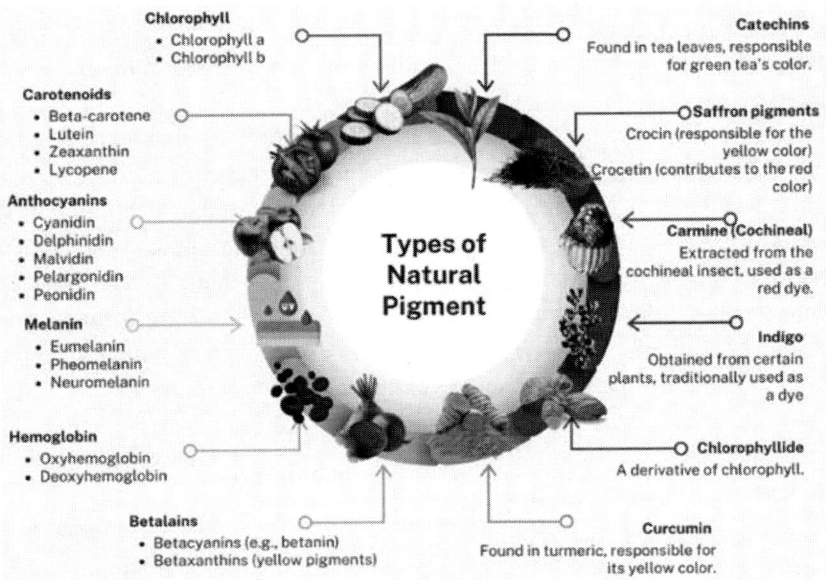


**Figure 1.** Basic diagram for types of natural pigments.

## Carotenoids

A significant group of natural pigments, characterized by orange, yellow, and red hues, is widely distributed among bacteria, fungi, and plants (Mapelli-Brahm et al., 2023; Martínez-Espinosa, 2022). This group includes pigments, such as lutein, zeaxanthin, and β-carotene (Nabi et al., 2023; Pasdaran et al., 2023). They are derived from a variety of sources, including plants, microbes, and marine organisms, are widely utilized in extensive applications across multiple fields, owing to their bioactive properties and a growing consumer preference for natural over synthetic additives. Additionally, they are recognized for their antioxidant properties, making them valuable in food, pharmaceuticals, and cosmetics (Tiwari et al., 2022). Carotenoids are naturally synthesized by plants and microorganisms, serving as colourants and offer potential health benefits, such as a decreased risk of certain diseases like cancer, cataracts, and macular degeneration (Joshi et al., 2023). since humans cannot synthesize carotenoids de novo, these compounds must be obtained through dietary sources, primarily from fruits and vegetables, which are considered significant sources (Cheng et al., 2020). The extraction and production of carotenoids have evolved to include sustainable and green technologies, moving away from conventional methods that often use hazardous solvents. Recent advancements highlight the use of greener solvents and bioprocessing technologies to improve the yield and efficiency of carotenoid production, thereby enhancing safety for consumers and the environment (Cheng et al., 2020; Mendes-Silva et al., 2020). Additionally, microbial synthesis of carotenoids presents an alternative to plant-based sources, offering advantages, such as controlled cultivation conditions, reduced production times, and the ability to utilize lignocellulosic waste as a substrate, which contributes to sustainability (Martínez-Espinosa, 2022; Singh & Mukherjee, 2022). Carotenoids are widely used in the food, pharmaceutical, and cosmetic industries due to their colourant properties, antioxidant activity, and health-promoting effects. They are found in various products, including food additives to skin care formulations, where they help prevent premature skin aging induced by oxidative stress (Mapelli-Brahm et al., 2023; Mussagy et al., 2019). The exploration of novel sources, including marine organisms and extremophiles from environments like the Caatinga domain, has enabled sustainable carotenoid productions, supporting global efforts towards environmental conservation and the development of green technologies (Srivastava et al., 2022).

## Anthocyanins and Flavonoids

Anthocyanins and flavonoids are crucial polyphenolic compounds that enhance plant colouration and have important implications for human health. Anthocyanins, a subclass of flavonoids, are water-soluble pigments responsible for the vivid orange-red, purple, and blue hues observed in various fruits, vegetables, and flowers (Giusti et al., 2023). They share a common diphenylpropane skeleton (C6C3C6) and are synthesized through the phenylpropanoid pathway, contributing plant aesthetics and serving protective mechanisms against various environmental stresses (Câmara et al., 2022; Guo et al., 2022). Fruits and vegetables, particularly berries, grapes, red cabbage, apples, and coloured grains, are the primary sources of anthocyanins, with berries identified as the most significant contributors to anthocyanin content among plant sources (Posadino et al., 2023). These natural pigments are not only pivotal for plant life but also hold extensive applications in human health and industry due to their potent antioxidant, anti-inflammatory, and anticancer properties (Mohammed & Khan, 2022; Sangeeta et al., 2023; Vaishali et al., 2023). In the food and pharmaceutical industries, anthocyanins are valued for their use as natural colourants, offering a safer alternative to synthetic dyes. The stability of these compounds, which is influenced by factors, such as pH, temperature, and light, is a critical for their application in food products, textiles, and cosmetics (Qi et al., 2023). Moreover, the health benefits associated with these factors, particularly in preventing and treating cardiovascular diseases, neurodegenerative disorders, cancer, and diabetes, have been extensively documented. These observed effects are attributed to their ability to modulate oxidative stress and inflammation, as well as their potential to improve visual health (Mattioli et al., 2020; Zhang, 2022). Recent research has also explored the potential of anthocyanins to improve the efficacy of cancer therapies when combined with conventional drugs, highlighting their role in developing treatments that demonstrate fewer side effects (Roszkowski, 2023). The ongoing study of anthocyanins and flavonoids continues to reveal various benefits, underscoring their importance in both plant biology and human health.

## Chlorophylls and Chlorophyllins

Chlorophylls are naturally occurring pigments that responsible for the green colour in plants and are crucial in converting light energy into chemical energy, a process fundamental to photosynthesis (Dufossé, 2023). These pigments are abundant in all forms of algae and cyanobacteria, playing a primary role in light harvesting for photosynthesis (Roca et al., 2024). The mechanism of chlorophylls involves the capture of sunlight, with the absorbed energy being rapidly and efficiently transferred among pigments within the light-harvesting antenna systems to the reaction centres, enabling charge separation, an essential step in the functionality of the photosynthetic apparatus (Ebrahimi et al., 2023). As chlorophyll derivatives, chlorophyllins are widely utilized for their stability and safety. However, they are susceptible to modification and affected by the activity of endogenous plant enzymes, which limits their direct application as colourants (Herrera & Roca, 2023). Chlorophyllins are water-soluble derivatives obtained from chlorophyll, offering an attractive green colour for various applications, including food colourants identified as E140 and E141 in the European Union (Tamiaki, 2022). Chlorophylls and chlorophyllins originate from various sources, including terrestrial plants and microalgae. Microalgae are particularly valuable due to their adaptability, flexibility of cultivation, and the feasibility of the extraction process, making them highly productive for the food and feed industries (Clark et al., 2022). The recovery of chlorophylls from food by-products is also an interesting approach to enhance the sustainability of food production, addressing the standard disposal of these by-products, leading to environmental pollution (Willows, 2020). Beyond their pigment properties, chlorophylls and chlorophyllins are recognized for their health benefits, including antioxidant, anti-inflammatory, anticancer, and anti-obesity effects (Fernandes et al., 2020). Moreover, chlorophyllins have been proposed to improve the safety and effectiveness of antiviral treatments, such as Molnupiravir for COVID-19, by potentially reducing adverse genetic changes and enhancing antiviral efficacy (Solymosi & Mysliwa-Kurdziel, 2017). Additionally, chlorophyll-based drugs have been identified as effective radioprotectors, inhibiting radical reactions involved in lipid oxidation, suggesting their therapeutic potential in managing conditions related to oxidative stress. Furthermore, chlorophyll has been utilized in organic syntheses as an organo-photocatalyst, offering a green alternative to traditional photocatalysts (Banu & Yadav, 2022).

## Betalains

Betalains are a class of water-soluble, nitrogen-containing vacuolar pigments divided into two subclasses: yellow-orange betaxanthins and red-violet betacyanins (Meena et al., 2022). These pigments are synthesized in the cells of Caryophyllales, such as red beets and opuntia, and are also found in certain fungi (Zannou et al., 2023). Betalains are responsible for the vibrant colouration of flowers, bracts, stems, and leaves in the genus Amaranthus, showcasing their wide chemical diversity and bioactive properties (Diaz et al., 2023). They are highly valued for their potent antioxidant activity, which can neutralize reactive oxygen species and free radicals, and are linked to various health benefits. These include correcting metabolic disorders, reducing cardiovascular disease risk, anticancer effects, and protecting against neurodegenerative disorders (Araujo-León et al., 2023). Betalains are increasingly extracted from natural sources as safe alternatives to synthetic dyes in the food industry due to their attractive colours and health-promoting properties (Popescu et al., 2023). Their applications extend beyond food colourants to the pharmaceutical and cosmetic sectors, where their antimicrobial properties are effective against a wide range of microbes, including multidrug-resistant bacteria (Wijesinghe & Choo, 2022). Additionally, betalains have been explored for their use in the textile industry, particularly in sustainable dyeing processes (Deng et al., 2023). While betalains present several advantages, their instability during processing and storage poses challenges for broader application (Pereira et al., 2022). Recent advancements in extraction methods and encapsulation techniques aim to enhance their stability and bioavailability, expanding their use across various industrial applications. Genetic engineering has also been employed to develop new sources of betalains, such as carrots, which could be used as functional vegetables or in breeding for health-promoting properties (Pereira et al., 2022). This multifaceted approach to improving the stability and applicability of betalains underscores their potential as natural colourants and bioactive compounds in a range of industries.

## Phycobiliproteins

Phycobiliproteins are a group of coloured, water-soluble proteins found in cyanobacteria, algae, and certain other microorganisms, known for their role in photosynthesis as light-harvesting antenna pigments (Carabantes &

Dufossé, 2023; Chini Zittelli et al., 2023; Shrivastava, 2023). These proteins are categorized into three main types based on their absorption spectra: phycoerythrin, phycocyanin, and allophycocyanin. Each type absorbs light at different wavelengths, facilitating photosynthesis (Ma et al., 2024). The primary sources of phycobiliproteins include cyanobacteria such as Arthrospira platensis, as well as from both micro and macro algae within the phylum Rhodophyta (Carmona et al., 2022; Kovaleski et al., 2022). Phycobiliproteins have garnered attention for their potential biotechnological applications, particularly as natural pigments. Their vibrant colours and fluorescent properties (exhibit red, blue, and purple shades), make them viable alternatives to natural colourants in food, beverages, and cosmetics (Chen et al., 2022; Dagnino-Leone et al., 2022; Qin et al., 2023). Beyond colouration, phycobiliproteins are also utilized in clinical immunological analysis as fluorescent markers due to their brilliant colours and specific light absorption properties (García et al., 2021; Mahanil et al., 2021). Furthermore, these proteins have shown promise in various pharmaceutical applications, including antioxidant, anti-tumor, anti-inflammatory, and antidiabetic effects, highlighting their potential beyond merely serving as colourants. Research into the extraction and purification of phycobiliproteins from natural sources has been a pivotal area of study, aiming to improve yield, stability, and functionality for industrial use. Despite challenges related to their chemical instability and the high purity requirements for certain applications, advancements in biotechnology and the development of novel extraction methods pave the way for more efficient and cost-effective production processes. The exploration of extremophile microorganisms for developing more stable phycobiliprotein variants further underscores the ongoing efforts to enhance the applicability of these natural pigments in various industries.

## Porphyrins

Porphyrins are a class of naturally occurring macrocyclic compounds characterized by a large ring composed of four connected pyrrole subunits connected by methine bridges, forming a highly conjugated system with 26 electrons. This unique structure facilitates light absorption in the visible range, contributing to their distinctive colours (Brothers & Senge, 2022; Liu et al., 2022). Porphyrins play crucial roles in various biological processes as they act as ligands, binding iron in heme, constituting the active site of hemoglobin in red blood cells. Additionally, plants form magnesium complexes that are essential for photosynthesis (Landrou, 2023). Their widespread distribution in nature and involvement in critical life-sustaining functions highlight their importance (Tahoun et al., 2021). The sources of porphyrins are diverse, for instance, the red-coloured heme in hemoglobin facilitates oxygen transport in bloodstream, whereas the green chlorophylls in plants and some bacteria are utilized for photosynthesis (Mathius, 2022). These compounds are prevalent in living organisms and have also been discovered in the fossil record, indicating their high chemical stability and ability to survive for hundreds of millions of years (Li et al., 2023). Porphyrins have found extensive applications beyond their biological roles. They are used as catalysts in the industrial synthesis of drug molecules, due to their unique chemical properties (Di Salvo et al., 2023). In the field of photocatalysis and tumor therapy, porphyrin nanostructures and photosensitizers are employed for their excellent light absorption properties and strong phototoxicity with minimal side effects (Park et al., 2021). Additionally, porphyrins are prevalent in the food industry, pharmacology, and the textile and printing sectors, where natural pigments are acknowledged for their bioactive compounds, which may confer health benefits (Estrada-Montaño et al., 2022). This wide range of applications underscores the versatility and significance of porphyrins as natural pigments and functional materials in both biological and industrial applications (Chandra et al., 2000).

## The Natural Pigments Derivation

The increasing consumer demand for natural additives in food, pharmaceuticals, and cosmetics has fostered significant research into the extraction of natural pigments from various sources. This area of study has led to the development and optimization of numerous innovative and environmentally friendly extraction techniques to enhance the efficiency, sustainability, and safety of the extraction process. Ultrasonication, or ultrasound-assisted extraction (UAE), is highlighted for its effectiveness in extracting plant pigments. The process utilizes cavitation to improve extraction efficiency and is characterized by low installation and maintenance costs and reduced energy consumption. This method is particularly beneficial for extracting anthocyanins, betalains, carotenoids, and chlorophyll from plant sources (Kumar et al., 2023). Supercritical fluid extraction (SFE) is another advanced technique, which is notable for its waste-free production, shortened extraction times, automation, and lower solvent consumption. SFE is particularly efficient in extracting pure antimicrobial pigments, showcasing high potential for application in various industries due to the biological and antimicrobial properties of these pigments (Abou Elmaaty et al., 2022). Microwave-assisted extraction (MAE) and pressurized liquid extraction (PLE) are also employed for their effective pigment extraction from natural sources, offering advantages related to speed and minimized solvent use (Lu et al., 2023). Enzyme-assisted extraction (EAE) enhances the extraction yield and quality of pigments by breaking down cell walls, thus facilitating their release (Molina et al., 2023). Other notable techniques include pulsed electric field extraction, which uses short bursts of high voltage to permeabilize cell membranes, and subcritical water extraction, applying water at high temperatures and pressures to solubilize pigments (Carreira-Casais et al., 2021; Fatima et al., 2023). Additionally, methods like ohmic heating and electric pulse technologies have been explored for their potential to disrupt cell structures, thereby improving pigment extraction from microalgae and cyanobacteria (Pagels et al., 2021). Each extraction technique offers unique advantages regarding efficiency, environmental impact, and applicability across different pigment sources and industries, reflecting the diverse and innovative approaches being developed to meet the growing demand for natural pigments (Miranda et al., 2021; Ohtsu et al., 2023; Rizkiyah et al., 2023).

## Future Insight of Natural Pigments

The future of natural pigments appears promising and multifaceted, driven by a resurgence in their application across various industries, including food, textiles, pharmaceuticals, and cosmetics, owing to their eco-friendly and health-promoting properties. The shift towards natural colourants is largely motivated by the environmental and health hazards associated with synthetic dyes, alongside consumer demand for sustainable and non-toxic alternatives (Di Salvo et al., 2023; Sarkar et al., 2022). Natural pigments from plants, microbes, and marine organisms offer vibrant hues and additional benefits, such as antioxidant, anticancer, and anti-inflammatory properties. These attributes render them highly desirable in the food and pharmaceutical industries (Mamun et al., 2023; Yadav et al., 2023). The textile sector is increasingly focusing on replacing synthetic dyes with natural ones, spurred by concerns over water pollution and the non-biodegradability of synthetic dyes. The industry is exploring eco-friendly extraction techniques and the use of mordants to improve the fastness of natural dyes, alongside investigating their UV protection and antimicrobial properties (Huang & Zhu, 2022). Similarly, the food industry is increasingly incorporating natural pigments due to their bioactive compounds, with ongoing research focused on enhancing their stability and intensity for broader application (Dufossé, 2023; Fernández-López et al., 2020). Marine and microbial sources of natural pigments are significant due to their biocompatibility and stability, offering a rich palette of colours absent in terrestrial plants and presenting novel applications in therapeutics and food colouration (Karuppan et al., 2014; Manivasagan et al., 2018; Zaki & Jai, 2020). However, challenges remain regarding these pigments' large-scale production and regulatory approval, indicating a need for additional research and development. Overall, the future of natural pigments is contingent upon advancements in extraction and stabilisation techniques, the expansion of their colour range, and the utilisation of their health benefits, all while maintaining environmental sustainability and safety.

## Conclusion

The exploration and application of natural pigments across various fields underscore their versatile functionality and potential as sustainable alternatives to synthetic counterparts. Natural pigments offer a sustainable, eco-friendly, and versatile alternative for various applications, including healthcare, food, renewable energy, and cosmetics. Advancements in extraction methods and integration into emerging technologies, position natural pigments as versatile and effective solutions. Their diverse biological activities, consumer preferences, and potential industrial applications make them a valuable focus for future research and development. Thus, they are poised to influence the fields of colouration and material science significance.

## References

Abou Elmaaty, T., Sayed-Ahmed, K., Elsisi, H., & Magdi, M. (2022). Optimization of extraction of natural antimicrobial pigments using supercritical fluids: A review. *Processes, 10*(10), 2111. https://doi.org/10.3390/pr10102111.

Aman Mohammadi, M., Ahangari, H., Mousazadeh, S., Hosseini, S. M., & Dufossé, L. (2022). Microbial pigments as an alternative to synthetic dyes and food additives: A brief review of recent studies. *Bioprocess and Biosystems Engineering, 45*(1), 1-12. https://doi.org/10.1007/s00449-021-02672-1.

Araujo-León, J. A., Aguilar-Hernández, V., Sánchez-delPino, I., Brito-Argáez, L., Peraza-Sánchez, S. R., Xingú-López, A., & Ortiz-Andrade, R. (2023). Analysis of Red Amaranth (*Amaranthus cruentus* L.) betalains by LC-MS. *Journal of the Mexican Chemical Society, 67*(3), 227-239. https://doi.org/10.29356/jmcs.v67i3.1517.

Banu, S., & Yadav, P. P. (2022). Chlorophyll: The ubiquitous photocatalyst of nature and its potential as an organo-photocatalyst in organic syntheses. *Organic & Biomolecular Chemistry, 20*(44), 8584-8598. https://doi.org/10.1039/d2ob01948d.

Boutchkova, M., Doshi, H., Durnev, A., & Molchanov, A. (2012). Precarious politics and return volatility. *The Review of Financial Studies*, 1111-1154. https://doi.org/10.1093/rfs/hhs065.

Brothers, P. J., & Senge, M. O. (2022). An introduction to porphyrins for the twenty-first century. *Fundamentals of Porphyrin Chemistry: A 21st Century Approach, 1*, 1-8. https://doi.org/10.1002/9781119672590.ch1.

Câmara, J. S., Locatelli, M., Pereira, J. A., Oliveira, H., Arlorio, M., Fernandes, I., Perestrelo, R., Freitas, V., & Bordiga, M. (2022). Behind the scenes of anthocyanins—from the health benefits to potential applications in food, pharmaceutical, and cosmetic fields. *Nutrients, 14*(23), 5133. https://doi.org/10.3390/nu14235133.

Carabantes, A. G., & Dufossé, L. (2023). Microalgal phycobiliproteins for food/feed applications. In *Handbook of Food and Feed from Microalgae* (pp. 161-169). Elsevier. https://doi.org/10.1016/B978-0-12-819989-4.00014-7.

Carmona, R., Murillo, M. C., Lafarga, T., & Bermejo, R. (2022). Assessment of the potential of microalgae-derived phycoerythrin as a natural colourant in beverages. *Journal of Applied Phycology, 34*(6), 3025-3034. https://doi.org/10.1007/s10811-022-02562-0.

Carreira-Casais, A., Lourenço-Lopes, C., Otero, P., Rodriguez, M. C., Pereira, A. G., Echave, J., Soria-Lopez, A., Chamorro, F., Prieto, M. A., & Simal-Gandara, J. (2021). Application of green extraction techniques for natural additives production. In *Natural Food Additives* (pp. 1-16). IntechOpen. https://doi.org/10.5772/intechopen.98984.

Chandra, R., Tiwari, M., Kaur, P., Sharma, M., Jain, R., & Dass, S. (2000). Metalloporphyrins—Applications and clinical significance. *Indian Journal of Clinical Biochemistry, 15*, 183-199. https://doi.org/10.1007/BF02918359.

Chen, H., Qi, H., & Xiong, P. (2022). Phycobiliproteins—a family of algae-derived biliproteins: Productions, characterization and pharmaceutical potentials. *Marine Drugs, 20*(7), 450. https://doi.org/10.3390/md20070450.

Chen, J., Hong, H., & Stein, J. (2001). Forecasting crashes: Trading volume, past returns, and conditional skewness in stock prices. *Journal of Financial Economics, 59*(3), 345-381. https://doi.org/10.1016/S0304-405X(00)00079-3.

Cheng, S.-H., Khoo, H. E., Kong, K. W., Prasad, K. N., & Galanakis, C. M. (2020). Extraction of carotenoids and applications. In *Carotenoids: Properties, Processing and Applications* (pp. 259-288). Elsevier. https://doi.org/10.1016/B978-0-12-817732-1.00013-9.

Chini Zittelli, G., Lauceri, R., Faraloni, C., Silva Benavides, A. M., & Torzillo, G. (2023). Valuable pigments from microalgae: Phycobiliproteins, primary carotenoids, and fucoxanthin. *Photochemical & Photobiological Sciences, 22*(8), 1733-1789. https://doi.org/10.1039/D3PP00088H.

Clark, N. F., Taylor-Robinson, A. W., & Heimann, K. (2022). Could chlorophyllins improve the safety profile of beta-d-N4-hydroxycytidine versus N-hydroxycytidine, the active ingredient of the SARS-CoV-2 antiviral molnupiravir? *Therapeutic Advances in Drug Safety, 13*, 20420986221107753. https://doi.org/10.1177/20420986221107753.

Dagnino-Leone, J., Figueroa, C. P., Castañeda, M. L., Youlton, A. D., Vallejos-Almirall, A., Agurto-Muñoz, A., Pérez, J. P., & Agurto-Muñoz, C. (2022). Phycobiliproteins: Structural aspects, functional characteristics, and biotechnological perspectives. *Computational and Structural Biotechnology Journal, 20*, 1506-1527. https://doi.org/10.1016/j.csbj.2022.05.010.

Deng, Y.-J., Duan, A.-Q., Liu, H., Wang, Y.-H., Zhang, R.-R., Xu, Z.-S., & Xiong, A.-S. (2023). Generating colourful carrot germplasm through metabolic engineering of betalains pigments. *Horticulture Research, 10*(4), uhad024. https://doi.org/10.1038/s41535-023-00678-3.

Derksen1, G. C., & Bechtold2, T. (2023). Natural colourants—Quinoid, naphthoquinoid, and anthraquinoid dyes. In *Handbook of Natural Colourants* (pp. 271-315). Wiley.

Di Salvo, E., Lo Vecchio, G., De Pasquale, R., De Maria, L., Tardugno, R., Vadalà, R., & Cicero, N. (2023). Natural pigments production and their application in food, health, and other industries. *Nutrients, 15*(8), 1923. https://doi.org/10.3390/nu15081923.

Diaz, Y. A., Qi, Z., Yang, B., López, N. M. M., Urbano, M. B., & Cianciosi, D. (2023). Betalains: The main bioactive compounds of *Opuntia* spp and their possible health benefits in the Mediterranean diet. *Mediterranean Journal of Nutrition and Metabolism, 1-10*. https://doi.org/10.1007/s12349-023-00223-6.

Dufossé, L. (2023). Natural colourants from cyanobacteria and algae. In *Handbook of Natural Colourants* (pp. 417-437). Wiley. https://doi.org/10.1002/9781119682804.ch22.

Ebrahimi, P., Shokramraji, Z., Tavakkoli, S., Mihaylova, D., & Lante, A. (2023). Chlorophylls as natural bioactive compounds existing in food by-products: A critical review. *Plants, 12*(7), 1533. https://doi.org/10.3390/plants12071533.

Estrada-Montaño, A. S., Gómez-Benítez, V., Camacho-Dávila, A., Rivera, E., Morales-Morales, D., & Zaragoza-Galán, G. (2022). Metalloporphyrins: Ideal catalysts for olefin epoxidations. *Journal of Porphyrins and Phthalocyanines, 26*(12), 821-836. https://doi.org/10.1142/S1088424622500350.

Fatima, I., Munir, M., Qureshi, R., Hanif, U., Gulzar, N., & Sheikh, A. A. (2023). Advanced methods of algal pigments extraction: A review. *Critical Reviews in Food Science and Nutrition*, 1-18. https://doi.org/10.1080/10408398.2023.2204346.

Fernandes, A. S., Nass, P. P., Oliveira, Á., & Zepka, L. Q. (2020). Chlorophylls as food additives. In *Pigments from Microalgae Handbook* (pp. 391-420). Springer. https://doi.org/10.1007/978-3-030-42057-4_15.

Fernández-López, J. A., Fernández-Lledó, V., & Angosto, J. M. (2020). New insights into red plant pigments: More than just natural colourants. *RSC Advances, 10*(41), 24669-24682. https://doi.org/10.1039/D0RA07210A.

García, A. B., Longo, E., & Bermejo, R. (2021). The application of a phycocyanin extract obtained from *Arthrospira platensis* as a blue natural colourant in beverages. *Journal of Applied Phycology, 33*(5), 3059-3070. https://doi.org/10.1007/s10811-021-02324-3.

Giusti, M. M., Miyagusuku-Cruzado, G., & Wallace, T. C. (2023). Flavonoids as natural pigments. In *Handbook of Natural Colourants* (pp. 371-390). Wiley. https://doi.org/10.1002/9781119682804.ch19.

Goodell, J. W., & Vahamaa, S. (2013). US presidential elections and implied volatility: The role of political uncertainty. *Journal of Banking and Finance, 37*(7), 1108-1117. https://doi.org/10.1016/j.jbankfin.2013.01.003.

Guo, Y., Zhang, H., Shao, S., Sun, S., Yang, D., & Lv, S. (2022). Anthocyanin: A review of plant sources, extraction, stability, content determination and modifications. *International Journal of Food Science & Technology, 57*(12), 7573-7591. https://doi.org/10.1111/ijfs.15914.

Herrera, M., & Roca, M. (2023). Microalgal chlorophylls for food/feed applications. In *Handbook of Food and Feed from Microalgae* (pp. 147-160). Elsevier. https://doi.org/10.1016/B978-0-12-819989-4.00013-6.

Hong, H., & Stein, J. (2003). Differences of opinion, short-sales constraints, and market crashes. *Review of Financial Studies, 16*(3), 485-525. https://doi.org/10.1093/rfs/hhg012.

Huang, L., & Zhu, L. (2022). Research progress of natural edible blue pigment. *Academic Journal of Science and Technology, 2*(3), 20-25..

Hutton, A. P., Marcus, A. J., & Tehranian, H. (2009). Opaque financial reports, R2, and crash risk. *Journal of Financial Economics, 92*(1), 67-86. https://doi.org/10.1016/j.jfineco.2008.07.003.

Joshi, K., Kumar, P., & Kataria, R. (2023). Microbial carotenoid production and their potential applications as antioxidants: A current update. *Process Biochemistry, 128*, 190-205. https://doi.org/10.1016/j.procbio.2023.01.019.

Karuppan Perumal, K. P., Sekarenthiran, S. C., Velusamy Karuppuraj, V. K., Sadasivam Mona, S. M., Mahalingam Ponsugumari, M. P., Ramayanam, B. M., & Devi Sagarika, D. S. (2014). Prospective aspects of myco-chrome as promising future textiles. In *Proceedings of the 8th International Conference on Mushroom Biology and Mushroom Products (ICMBMP8)* (pp. 412-416).

Kim, J.-B., Li, Y., & Zhang, L. (2011a). CFOs versus CEOs: Equity incentives and crashes. *Journal of Financial Economics, 101*(3), 713-730. https://doi.org/10.1016/j.jfineco.2011.03.003.

Kim, J.-B., Li, Y., & Zhang, L. (2011b). Corporate tax avoidance and stock crash risk: Firm-level analysis. *Journal of Financial Economics, 101*(3), 639-662. https://doi.org/10.1016/j.jfineco.2011.03.004.

Kovaleski, G., Kholany, M., Dias, L. M., Correia, S. F., Ferreira, R. A., Coutinho, J. A., & Ventura, S. P. (2022). Extraction and purification of phycobiliproteins from algae and their applications. *Frontiers in Chemistry, 10*, 1065355. https://doi.org/10.3389/fchem.2022.1065355.

Kumar, G., Upadhyay, S., Yadav, D. K., Malakar, S., Dhurve, P., & Suri, S. (2023). Application of ultrasound technology for extraction of colour pigments from plant sources and their potential bio-functional properties: A review. *Journal of Food Process Engineering, 46*(6), e14238. https://doi.org/10.1111/jfpe.14238.

Landrou, G. (2023). Porphyrins for medical and photocatalytic applications (Doctoral dissertation, Πανεπιστήμιο Κρήτης. Σχολή Θετικών και Τεχνολογικών Επιστημών. Τμήμα Χημείας. Τομέας Ανόργανης Χημείας. Εργαστήριο Ανόργανης Χημείας).

Li, Q., Li, S., & Xu, L. (2018). National elections and tail risk: International evidence. *Journal of Banking and Finance, 92*, 113-128. https://doi.org/10.1016/j.jbankfin.2018.04.004.

Li, X., Tang, C., Zhang, L., Song, M., Zhang, Y., & Wang, S. (2023). Porphyrin-based covalent organic frameworks: Design, synthesis, photoelectric conversion mechanism, and applications. *Biomimetics, 8*(2), 171. https://doi.org/10.3390/biomimetics8020171.

Liu, S., Li, Z., Tong, H., Zhong, Y., & Bai, F. (2022). Porphyrin self-assembled nanostructures and applications. In *Self-Assembly of Materials and Their Applications* (pp. 149-182). IntechOpen. https://doi.org/10.5772/intechopen.98178.

Lu, X., Li, W., Wang, Q., Wang, J., & Qin, S. (2023). Progress on the extraction, separation, biological activity, and delivery of natural plant pigments. *Molecules, 28*(14), 5364. https://doi.org/10.3390/molecules28145364.

Lyu, X., Lyu, Y., Yu, H., Chen, W., Ye, L., & Yang, R. (2022). Biotechnological advances for improving natural pigment production: A state-of-the-art review. *Bioresources and Bioprocessing, 9*(1), 8. https://doi.org/10.1186/s40643-022-00486-7.

Ma, J., Hu, J., Sha, X., Meng, D., & Yang, R. (2024). Phycobiliproteins, the pigment-protein complex form of natural food colourants and bioactive ingredients. *Critical Reviews in Food Science and Nutrition, 64*(10), 2999-3017. https://doi.org/10.1080/10408398.2023.2220617.

Mahanil, K., Sensupa, A., Pekkoh, J., Tragoolpua, Y., & Pumas, C. (2021). Application of phycobiliproteins from *Leptolyngbya* sp. KC45 for natural illuminated colourant beverages. *Journal of Applied Phycology, 33*(6), 3747-3760. https://doi.org/10.1007/s10811-021-02441-2.

Mamun, M. A. A., Haji, A., Mahmud, M. H., Repon, M. R., & Islam, M. T. (2023). Bibliometric evidence on the trend and future direction of the research on textile colouration with natural sources. *Coatings, 13*(2), 413. https://doi.org/10.3390/coatings13020413.

Manivasagan, P., Bharathiraja, S., Santha Moorthy, M., Mondal, S., Seo, H., Dae Lee, K., & Oh, J. (2018). Marine natural pigments as potential sources for therapeutic applications. *Critical Reviews in Biotechnology, 38*(5), 745-761. https://doi.org/10.1080/07388551.2017.1412779.

Mapelli-Brahm, P., Gómez-Villegas, P., Gonda, M. L., León-Vaz, A., León, R., Mildenberger, J., Rebours, C., Saravia, V., Vero, S., & Vila, E. (2023). Microalgae, seaweeds, and aquatic bacteria, archaea, and yeasts: Sources of carotenoids with potential antioxidant and anti-inflammatory health-promoting actions in the sustainability era. *Marine Drugs, 21*(6), 340. https://doi.org/10.3390/md21060340.

Martínez-Espinosa, R. M. (2022). Introductory chapter: Overview of new perspectives and applications of carotenoids. In *IntechOpen*.

Mathius, M. (2022). Synthesis of highly conjugated porphyrinoid systems. *Illinois State University*.

Mattioli, R., Francioso, A., Mosca, L., & Silva, P. (2020). Anthocyanins: A comprehensive review of their chemical properties and health effects on cardiovascular and neurodegenerative diseases. *Molecules, 25*(17), 3809. https://doi.org/10.3390/molecules25173809.

Meena, N. K., Meena, V. S., Verma, M., & Mishra, S. (2022). Plant extracts as colouring agents. In *Plant Extracts: Applications in the Food Industry* (pp. 187-207). Elsevier. https://doi.org/10.1016/B978-0-12-821602-4.00013-9.

Mendes-Silva, T. d. C. D., da Silva Andrade, R. F., Ootani, M. A., Mendes, P. V. D., da Silva, M. R. F., Souza, K. S., dos Santos Correia, M. T., da Silva, M. V., & de Oliveira, M. B. M. (2020). Biotechnological potential of carotenoids produced by extremophilic microorganisms and application prospects for the cosmetics industry. *Advances in Microbiology, 10*(8), 397-410. https://doi.org/10.4236/aim.2020.108028.

Miranda, P. H. S., dos Santos, A. C., de Freitas, B. C. B., de Souza Martins, G. A., Boas, E. V. d. B. V., & Damiani, C. (2021). A scientific approach to extraction methods and stability of pigments from Amazonian fruits. *Trends in Food Science & Technology, 113*, 335-345. https://doi.org/10.1016/j.tifs.2021.07.003.

Mohammed, H. A., & Khan, R. A. (2022). Anthocyanins: Traditional uses, structural and functional variations, approaches to increase yields and products' quality, hepatoprotection, liver longevity, and commercial products. *International Journal of Molecular Sciences, 23*(4), 2149. https://doi.org/10.3390/ijms23042149.

Molina, A. K., Corrêa, R. C., Prieto, M. A., Pereira, C., & Barros, L. (2023). Bioactive natural pigments' extraction, isolation, and stability in food applications. *Molecules, 28*(3), 1200. https://doi.org/10.3390/molecules28031200.

Muñoz-Miranda, L. A., & Iñiguez-Moreno, M. (2023). An extensive review of marine pigments: Sources, biotechnological applications, and sustainability. *Aquatic Sciences, 85*(3), 68. https://doi.org/10.1007/s00027-023-00960-0.

Mussagy, C. U., Winterburn, J., Santos-Ebinuma, V. C., & Pereira, J. F. B. (2019). Production and extraction of carotenoids produced by microorganisms. *Applied Microbiology and Biotechnology, 103*, 1095-1114. https://doi.org/10.1007/s00253-018-9747-4.

Nabi, B. G., Mukhtar, K., Ahmed, W., Manzoor, M. F., Ranjha, M. M. A. N., Kieliszek, M., Bhat, Z. F., & Aadil, R. M. (2023). Natural pigments: Anthocyanins, carotenoids, chlorophylls, and betalains as colourants in food products. *Food Bioscience, 52*, 102403. https://doi.org/10.1016/j.fbio.2023.102403.

Ohtsu, T., Ono, R., Kinou, Y., Matsukura, S., & Ohtani, N. (2023). Extraction of natural blue emissive pigments from *Fraxinus lanuginosa* by column chromatography. *Molecular Crystals and Liquid Crystals, 762*(1), 49-62. https://doi.org/10.1080/15421406.2023.2178759.

Pagels, F., Pereira, R. N., Vicente, A. A., & Guedes, A. C. (2021). Extraction of pigments from microalgae and cyanobacteria—A review on current methodologies. *Applied Sciences, 11*(11), 5187. https://doi.org/10.3390/app11115187.

Park, J. M., Hong, K.-I., Lee, H., & Jang, W.-D. (2021). Bioinspired applications of porphyrin derivatives. *Accounts of Chemical Research, 54*(9), 2249-2260. https://doi.org/10.1021/acs.accounts.1c00327.

Pasdaran, A., Zare, M., Hamedi, A., & Hamedi, A. (2023). A review of the chemistry and biological activities of natural colourants, dyes, and pigments: Challenges, and opportunities for food, cosmetics, and pharmaceutical application. *Chemistry & Biodiversity, 20*(8), e202300561. https://doi.org/10.1002/cbdv.202300561.

Paul, T., Bandyopadhyay, T. K., Mondal, A., Tiwari, O. N., Muthuraj, M., & Bhunia, B. (2020). A comprehensive review on recent trends in production, purification, and applications of prodigiosin. *Biomass Conversion and Biorefinery*, 1-23. https://doi.org/10.1007/s13399-020-01056-5.

Pereira, C., Dias, M. I., Pinela, J., Roriz, C. L., Ferreira, I. C. F. R., & Barros, L. (2022). Chapter 11 - Betalains. In C. M. Galanakis (Ed.), *Nutraceutical and Functional Food Components* (Second Edition) (pp. 461-507). Academic Press. https://doi.org/10.1016/B978-0-12-821016-3.00011-4.

Popescu, V., Blaga, A. C., Cașcaval, D., & Popescu, A. (2023). *Beta vulgaris* L.—A source with a great potential in the extraction of natural dyes intended for the sustainable dyeing of wool. *Plants, 12*(10), 1933. https://doi.org/10.3390/plants12101933.

Posadino, A. M., Giordo, R., Ramli, I., Zayed, H., Nasrallah, G. K., Wehbe, Z., Eid, A. H., Gürer, E. S., Kennedy, J. F., & Aldahish, A. A. (2023). An updated overview of cyanidins for chemoprevention and cancer therapy. *Biomedicine & Pharmacotherapy, 163*, 114783. https://doi.org/10.1016/j.biopha.2023.114783.

Qi, Q., Cai, D., Yu, X., Shi, J., Bai, W., & Yan, N. (2023). Anthocyanins in Subtropical Fruits. *Anthocyanins in Subtropical Fruits*, 1-31.

Qin, Z., Wang, X., Gao, S., Li, D., & Zhou, J. (2023). Production of natural pigments using microorganisms. *Journal of Agricultural and Food Chemistry, 71*(24), 9243-9254. https://doi.org/10.1021/acs.jafc.3c01474.

Rizkiyah, D. N., Putra, N. R., Yunus, M. A. C., Veza, I., Irianto, I., Aziz, A. H. A., Rahayuningsih, S., Yuniarti, E., & Ikhwani, I. (2023). Insight into green extraction for Roselle as a source of natural red pigments: A review. *Molecules, 28*(3), 1336. https://doi.org/10.3390/molecules28031336.

Roca, M., Chen, K., & Pérez-Gálvez, A. (2024). Chlorophylls. In *Handbook on Natural Pigments in Food and Beverages* (pp. 193-226). Elsevier. https://doi.org/10.1016/B978-0-12-819989-4.00009-5.

Rodriguez-Amaya, D. B., & Carle, R. (2021). Alterations of natural pigments. In *Chemical Changes During Processing and Storage of Foods* (pp. 265-327). Elsevier. https://doi.org/10.1016/B978-0-12-821015-6.00012-5.

Roszkowski, S. (2023). Application of polyphenols and flavonoids in oncological therapy. *Molecules, 28*(10), 4080. https://doi.org/10.3390/molecules28104080.

Różyło, R. (2020). Recent trends in methods used to obtain natural food colourants by freeze-drying. *Trends in Food Science & Technology, 102*, 39-50. https://doi.org/10.1016/j.tifs.2020.06.009.

Sangeeta, S., Rai, S., Bisht, S., & Nazim, M. (2023). Chemical properties of anthocyanins sourced from different subtropical fruits. In *Anthocyanins in Subtropical Fruits* (pp. 33-58). CRC Press. https://doi.org/10.1201/9781003242200-5.

Santa-Clara, P., & Valkanov, R. (2003). The presidential puzzle: Political cycles and the stock market. *The Journal of Finance, 58*, 1841-1872. https://doi.org/10.1111/1540-6261.00588.

Sarkar, S., Khatun, D., Dutta, B., & Bandopadhyay, R. (2022). Journey of natural pigments from ancient antiquity to present: Insights on sustainable development. *Indian Journal of History of Science, 57*(4), 330-334. https://doi.org/10.16943/ijhs/2022/58569.

Shrivastava, A. (2023). Extraction and native PAGE separation of phycobiliproteins from some cyanobacteria collected from their natural habitats. *International Journal of Current Science Research and Review, 6*, 4682-4687.

Singh, A., & Mukherjee, T. (2022). Application of carotenoids in sustainable energy and green electronics. *Materials Advances, 3*(3), 1341-1358. https://doi.org/10.1039/d2ma00383b.

Singh, T., Pandey, V. K., Dash, K. K., Zanwar, S., & Singh, R. (2023). Natural bio-colourant and pigments: Sources and applications in food processing. *Journal of Agriculture and Food Research, 100628*. https://doi.org/10.1016/j.jafr.2023.100628.

Solymosi, K., & Mysliwa-Kurdziel, B. (2017). Chlorophylls and their derivatives used in food industry and medicine. *Mini Reviews in Medicinal Chemistry, 17*(13), 1194-1222. https://doi.org/10.2174/1389557516666170804105817.

Srivastava, N., Srivastava, R., & Gaurav, K. (2022). Biotechnological production of carotenoid from oleaginous red yeast and its applications. *Asian Journal of Biotechnology and Bioresource Technology, 8*(2), 15-29. https://doi.org/10.9734/ajbt/2022/v8i21443.

Swami, S. B., Ghgare, S. N., Swami, S. S., Shinde, K. J., Kalse, S. B., & Pardeshi, I. L. (2020). Natural pigments from plant sources: A review. *The Pharma Innovation Journal, 9*(10), 566.

Tahoun, M., Gee, C. T., McCoy, V. E., Sander, P. M., & Müller, C. E. (2021). Chemistry of porphyrins in fossil plants and animals. *RSC Advances, 11*(13), 7552-7563. https://doi.org/10.1039/d1ra08791a.

Tamiaki, H. (2022). Chlorophylls. In *Fundamentals of Porphyrin Chemistry: A 21st Century Approach* (pp. 743-776). Wiley.

Tiwari, S., Yawale, P., & Upadhyay, N. (2022). Carotenoids: Extraction strategies and potential applications for valorization of under-utilized waste biomass. *Food Bioscience, 48*, 101812. https://doi.org/10.1016/j.fbio.2022.101812.

Vaishali, N., Shahana, A., Asfiya, S., & Naziya, M. (2023). Anthocyanins as natural plant-derived antifungal agents. *The Journal of Plant Science Research, 39*(1), 29-38.

Wijesinghe, V. N., & Choo, W. S. (2022). Antimicrobial betalains. *Journal of Applied Microbiology, 133*(6), 3347-3367. https://doi.org/10.1111/jam.15524.

Willows, R. D. (2020). Biosynthesis of chlorophyll and bilins in algae. In *Photosynthesis in Algae: Biochemical and Physiological Mechanisms* (pp. 83-103). Elsevier. https://doi.org/10.1016/B978-0-12-817538-9.00006-6.

Yadav, S., Tiwari, K. S., Gupta, C., Tiwari, M. K., Khan, A., & Sonkar, S. P. (2023). A brief review on natural dyes, pigments: Recent advances and future perspectives. *Results in Chemistry, 5*, 100733. https://doi.org/10.1016/j.rechem.2023.100733.

Zaki, N. A. M., & Jai, J. (2020). Plant-based pigments: Challenges and future perspective for natural food colourants. *Malaysian Journal of Chemical Engineering and Technology (MJCET), 3*(1), 44-49.

Zannou, O., Oussou, K. F., Chabi, I. B., Odouaro, O. B., Deli, M. G., Goksen, G., Vahid, A. M., Kayodé, A. P., Kelebek, H., & Selli, S. (2023). A comprehensive review of recent development in extraction and encapsulation techniques of betalains. *Critical Reviews in Food Science and Nutrition, 1-18.* https://doi.org/10.1080/10408398.2023.2210475.

Zhang, Y. (2022). Study on the dietary intake of anthocyanins in Chinese people and effect of anthocyanins on cardiovascular disease. *Highlights in Science, Engineering and Technology, 19*, 58-65.

## Chapter 7

# From Colour to Cure: Unveiling the Antimicrobial Power of Plant Pigments

**Qurratu Aini Mat Ali[1,*], PhD**
**Farah Ayuni Mohd Hatta[2], PhD**
**Rashidi Othman[3], PhD**
**and Mastura Mohtar[4], PhD**

[1,2]Institute of Islam Hadhari, Universiti Kebangsaan Malaysia, Universiti Kebangsaan Malaysia Bangi, Selangor, Malaysia
[3]Department of Landscape Architecture, Kulliyyah of Architecture and Environmental Design, International Islamic University Malaysia, Kuala Lumpur, Malaysia
[4]Biology Laboratory of Bioprospecting Program, Natural Products Division, Forest Research Institute Malaysia, Kuala Lumpur, Malaysia

## Abstract

Plant pigments, including carotenoids, chlorophyll, and betalains, are a diverse group of compounds with unique chemical structures that contribute to their vibrant colours and potential antimicrobial properties. These properties have found applications in the food industry for extending the shelf life of perishable goods and enhancing food safety. Moreover, the incorporation of plant pigment extracts or derivatives into pharmaceutical formulations holds promise for combating drug-resistant microbes. This study investigated the antimicrobial efficacy of selected plant pigments using minimum inhibitory concentration (MIC) and minimum bactericidal concentration (MBC) assays. Chlorophyll,

---

* Corresponding Author's Email: qurratu.mat@gmail.com

In: Natural Pigments: Sources, Properties, and Health Benefits
Editors: Rashidi Othman, Qurratu Aini Mat Ali, Razanah Ramya et al.
ISBN: 979-8-90134-055-4

extracted from green algae (*Chlorella vulgaris; C. vulgaris*), exhibited a potent bactericidal activity against both *Staphylococcus aureus* (*S. aureus*) and *Escherichia coli* (*E. coli*) at a concentration of 2.5 μg/μL. Similarly, bixin, extracted from annatto (*Bixa orellana* L.) seeds, also exhibited bactericidal activity by effectively inhibiting *S. aureus* and *E. coli* at concentrations of 5 μg/μL and 2.5 μg/μL, respectively. In contrast, the remaining four plant extracts required concentrations of ≥ 5 μg/μL to inhibit the growth of both microbes. These findings highlight the potential of natural pigments as sustainable alternatives to synthetic antibiotics, offering a multitude of benefits beyond their colour. Further research is crucial to unravel the complex relationship between the structural diversity of plant pigments and their antimicrobial activity.

**Keywords**: plant pigment, carotenoid, chlorophyll, betalains, colour, antimicrobial properties

## Introduction

Colour is the main criterion that is visually detected, and it is usually one of the main factors that determines the selection of foods, textiles, cosmetics, and so on. Colour can improve the appearance and quality of products to attract buyers and can be used to cover or replace the original colour of products that have faded or been lost during the production process (Ghosh et al., 2022). Nowadays, natural colourants are regaining attention and demand due to their health benefits and advantages (Benucci et al., 2022; Griffiths, 1990). Plants are known for their vibrant colours, which are primarily attributed to pigments such as anthocyanins, carotenoids, chlorophyll, and betalains. These pigments not only contribute to the aesthetic appeal of plants but also play an essential role in plant growth. Moreover, plant pigments exhibit antimicrobial properties that can inhibit the growth of bacteria and fungi. These antimicrobial properties make plant pigments valuable components, especially in food preservation and pharmaceutical applications (Molina et al., 2023; Venil et al., 2020).

## Pigment Classification and Source

Heritage Pigments, which are defined as a group of compounds with concentrated colours, are used to colour other materials. They are divided into two main groups, namely, water-soluble pigments (hydrophilic molecules), and lipid-soluble pigments (hydrophobic molecules). These pigments are produced from various sources, including plants and microorganisms, that contain bioactive compounds of a unique biological nature (Chaudhary, 2024; Muthusamy et al., 2024). Natural colourants based on these pigments are one of the potential colourants to replace synthetic colourants. Natural pigments from microbial sources consist of bacterial and algal species. Betalains (red-yellow), and anthocyanins (red-blue-violet) are natural pigments from plant sources that belong to the hydrophilic group, while chlorophyll (green), and carotenoids (yellow-orange-red) come under the hydrophobic group (Juri'c et al., 2022; Sharma et al., 2021b; Rodriguez-Amaya, 2016).

These main pigments from natural plant-based sources can produce various colours. For example, red can be obtained from tomatoes and berries, yellow from saffron flowers, orange from saffron, blue from butterfly pea flowers, and green from leafy vegetables. These pigments not only provide colour variations, but also possess functional properties, acting as antioxidants and antimicrobials (Sharma et al., 2021a; de Andrade Lima et al., 2019). In addition, they contribute to the production of natural colourants in the food industry (Ghosh et al., 2022). Based on previous research, anthocyanins are the most interesting colourants as they have been studied extensively since the 1930s compared to carotenoids, chlorophyll, and betalains. Anthocyanins are used extensively because they can produce the widest range of colours, namely red, blue, magenta, purple, and pink (Zhao et al., 2022; de Oliveira et al., 2021; Enaru et al., 2021; Fu et al., 2021; Roy & Rhim, 2021; Alappat & Alappat, 2020; Mekapogu et al., 2020; Nassour et al., 2020; Swami et al., 2020; Khoo et al., 2017). Therefore, only carotenoid pigments, chlorophyll, and betalains were chosen for this study to examine the antimicrobial role of these pigments.

Carotenoids are fat-soluble tetraterpene pigments that produce yellow, red and orange colours. These pigments are found in fungi, photosynthetic bacteria, algae, plants, and animals (Ghosh et al., 2022). In addition, carotenoids are also important pigments in photosynthetic organs that enable the photosynthesis process to be carried out (Maoka, 2020). Focusing on plants, carotenoids are found in yellow, orange and red fruits and green leafy vegetables (Bohn, 2019). Carotenoid pigments are closely related to chlorophyll, as every green plant contains these two pigments. In simple terms, these two pigments can be seen when certain factors occur to change the green colour of the plant to yellow, orange, red, or brown (Lev-Yadun, 2022; Maoka, 2020). For example, chillies or tomatoes, which are originally green, will degrade and turn orange or red when they are ripe (Naeem et al., 2023).

Overall, carotenoids are produced by living organisms that undergo photosynthesis, especially plants, and some non-photosynthetic organisms, such as fungi, which are classified in the subclass of isoprenoid metabolites (Rodríguez-Concepción et al., 2018).

Chlorophyll is a green photosynthetic pigment belonging to the main class of tetrapyrroles that act as photoreceptors. Chlorophyll only absorbs light energy in red and blue wavelengths, which makes plants green. The cellulose cell wall of green plants is the main component that diffuses and reflects visible light in green plants (Virtanen et al., 2022). Chlorophyll is the most abundant pigment in nature. It works to give a green colour to all green plants on earth and can be found in green plants, algae, and cyanobacteria (Cai et al., 2021; da Silva & Lombardi, 2020). This green pigment is one of the natural pigments produced in green vegetables and fruits because it is the main component in the photosynthetic process, which is responsible for absorbing energy from sunlight and changing carbon dioxide and water into carbohydrates (Juri'c et al., 2022; Ghosh et al., 2022). There are six main classes of chlorophyll, namely chlorophyll *a* (found in all green plants, algae, and cyanobacteria), chlorophyll *b* (found in green plants and green algae), chlorophyll *c* (found in diatoms, dinoflagellates, and brown algae (seaweed)), chlorophyll *d* (only found in red algae that live at the bottom of water bodies), chlorophyll *e* (found in some golden algae), and chlorophyll *f*, which has been newly discovered (Ghosh et al., 2022; Mega, 2022).

Betalains are water-soluble nitrogen pigments that are divided into two main structural groups, namely betacyanins, which have a red-purple colour range, and betaxanthins, which are yellow (Polturak & Aharoni, 2018). Previous studies have reported that betalains in the environment originate from approximately 50 sources of betacyanin (purple) and 20 sources of betaxanthin (yellow) (Coy-Barrera, 2020; Cai et al., 2005). In addition to producing attractive colours in plants, betalains have also been shown to possess high antioxidant properties. Studies were previously conducted extensively on the potential health benefits of betalain pigments, such as their anti-cancer, anti-inflammatory, anti-diabetic, and other properties (Polturak & Aharoni, 2018; Khan, 2016; Clifford et al., 2015). Furthermore, their stability varies according to their pH, making these pigments suitable for use in the food industry as a natural colourant for food products (Molina et al., 2023). Betalain pigments are unique in that they can change to pink, purple, brown, and liquid yellow colours under certain conditions, especially pH (Sadowska-Bartosz & Bartosz, 2021; Sigwela, 2020). In addition, betalain pigments have an exclusive relationship with anthocyanins, where, even though they share the same or almost the same function and colour, and produce almost the same colour (namely, pink, purple, warp, and melt yellow) under certain conditions, they never exist together in the same plant (Sadowska-Bartosz & Bartosz, 2021; Sturm & Hensel, 2017; Gandía-Herrero et al., 2016).

## Pigment Bioactivity

From a scientific point of view, the term "bioactive" is an alternative term for "biologically active". In other words, a bioactive compound or substance is defined as a substance that has the property of biological activity, also known as bioactivity. In the medical field, bioactive substances tend to have an effect or trigger a reaction in the tissues of living things. In short, every bioactive substance will affect living things, either negatively or positively, depending on certain factors (Zhang et al., 2020; Guaadaoui et al., 2014). In this study, plants were used as the basic source for the production of compounds or bioactive substances to ensure that the results obtained would have a positive effect and benefit the whole of nature, including humans as consumers.

Apart from their antioxidant activity, plant-based natural colourants also have antimicrobial capabilities, which are an added value for any products that utilise this source. Usually, antibiotics are used to prevent bacterial growth or infection. However, the use of antibiotics in excess or over a long period of time can lead to dangerous side effects because the level of bacterial resistance will keep increasing, and a higher dose will be required to inhibit them (Mancuso et al., 2021; Breijyeh et al., 2020; Serwecińska, 2020). Therefore, there is a need to enhance natural colourants from plants that contain a high rate of antimicrobial activity as an added value to improve the quality of products and further promote the health of consumers (De Mejia et al., 2020; Favela-González et al., 2020; Jafarzadeh et al., 2020).

## Antimicrobial Potential for the Food Industry

Based on previous studies, plant pigments contain potential bioactive and antimicrobial compounds. It has been proven that these functional properties are present in natural colourants. Thus, they might be alternatives to replace synthetic bioactive compounds, which are known to have long-term side effects on human health and the environment (Nieto et al., 2023; Câmara et al., 2020; De Mejia et al., 2020). Several factors can cause the human body to be infected by viruses, parasites, fungi, or bacteria. However, the main factor is hygiene, since infection often occurs when a person's hygiene is neglected. For example, food contamination can occur if there is hygienic deterioration throughout the process between the time of preparation and the time when the food is consumed (Gallo et al., 2020; Hirose et al., 2024). Foodborne diseases are usually caused by viruses, parasites, bacteria, or toxins from contaminated raw materials or water sources. Some cases of foodborne diseases can cause illnesses of the digestive system, such as nausea, diarrhoea, vomiting, and can even result in death (Pattabhiramaiah & Mallikarjunaiah, 2023; Pan et al., 2020; Jarriyawattanachaikul et al., 2016). In fact, more than 50,000 mortalities due to infectious diseases are reported annually worldwide (Ahmad & Wajid, 2013).

Furthermore, some of the issues that are gaining attention are related to foodborne microbial outbreaks involving prepared and packaged food. The original concept of packed food was to enable consumers to enjoy food easily and quickly without having to go through a complicated preparation process (Sen et al., 2021; Majid et al., 2018). However, this issue of microbial contamination has urged the food packaging industry to produce food packaging innovations that can survive the attack and development of microbes, or antimicrobial packaging that can maintain the quality and safety of food (Fadiji et al., 2023; Manzoor et al., 2023; Ahari & Soufiani, 2021). The common microbial species involved in foodborne disease are *Staphylococcus aureus* (*S. aureus*) and *Escherichia coli* (*E. coli*). *S. aureus* is often found and spread from skin to skin. Normally, it might be transferred to food during the preparation process. Meanwhile, *E. coli* can spread through raw or undercooked food and contaminated water. Thus, both microbes have the potential to contaminate food products, especially packed food.

Plant extracts containing antimicrobial compounds have the potential to prevent and control foodborne diseases and further protect food from viruses, parasites, fungi, or bacteria (Sar et al., 2023; Pan et al., 2020; Ullah et al., 2020). As reported by Jarriyawattanachaikul et al. (2016), plants have the ability to treat and reduce contamination by foodborne pathogens, such as *E. coli*. Previous scientific studies have also proven that herbs and spices possess antimicrobial properties (Okmen et al., 2016). Plants also contain various natural chemical elements that detect microbial activity as well as preservative properties that are low in side effects and safe for humans and animals (Nguyen et al., 2020; Zhang et al., 2019; Priyadarshi et al., 2018).

### Antimicrobial Potential for the Pharmaceutical Industry

The use of plants to treat illnesses has been practised since ancient times. Thus, the use of plant sources as medicine to heal infections is not a not idea. It has been proven by the WHO that people living in developing countries still rely on plants as their main source of health care, such as self-medication, health foods, functional foods, and home care remedies. There is still a wide range of benefits to be explored within plants since it is estimated that only around 10% of all the plants in the plant kingdom have been isolated for medicinal purposes, and among the 300,000 plants species worldwide, only 2% of them have been examined for their antimicrobial potential (Ahmad & Wajid, 2013).

Among the challenges to accepting plants as antimicrobial agents is their inconsistent quality. One way to maintain the quality and confirm the efficacy and safety of plants as antimicrobial agents is to conduct further and thorough research to ensure that the results are compatible with existing antibiotics and offer better advantages (Ahmad & Wajid, 2013). This is because with technological advancements, many synthetic drugs have been invented, including antibiotics for the treatment of several illness and diseases. However, there is still uncertainty over the use of these synthetic antibiotics as, over time, microbials are becoming more resistant towards existing antibiotics (Viswanad et al., 2017). Moreover, the irrational use of antibiotics has contributed to the multi-drug resistance of bacterial strains (Ahmad & Wajid, 2013; Khan et al., 2014). On encountering this issue, researchers have started to focus on the use of plant extracts or crude plant products as potential antimicrobial agents, and even the developed countries are turning their attention to the health benefits of plants.

## Method

### Plant Selection

Six plants, namely, annatto (*Bixa orellana*), to represent a carotenoid pigment, *cekur manis* (*Sauropus androgynus*; *S. androgynus*), *kelor* (*Moringa oleifera*; *M. oleifera*), *selasih* (*Ocimum basilicum*; *O. basilicum*), and green algae (*Chlorella vulgaris; C. vulgaris*) to represent chlorophyll pigments, and dragon fruit (*Hylocereus polyrhizus*; *H. polyrhizus*) to represent a betalain pigment, were tested. They were selected for their high antioxidant content, which suggests that they may also possess significant antimicrobial properties.

## Plant Extraction

The plant pigments were extracted by means of a conventional method, which consisted of simple and cheap procedures that succeeded in enhancing the quality and quantity of the extracts (Zin et al., 2020; Ngamwonglumlert et al., 2017). The procedures were carried out with reference to the methodology explained by Othman (2009), with several modifications. A methanol solvent in the ratio of 1 g of sample to 10 ml of methanol was used to extract the carotenoid and betalain pigments. The samples in methanol were then kept below -20°C for 24 hours for optimum extraction. Meanwhile, for the extraction of chlorophyll, 1 g of the sample was hydrated with 5 ml of distilled water before being mixed with 10 ml of an acetone/methanol solvent at a ratio of 7:3, followed by 15 ml of hexane. The mixture was then kept below -20°C for 24 hours for optimum extraction. The next day, all the extracted pigments were collected and dried under nitrogen-free oxygen gas to obtain pure crude without the solvent. The dried crude plant pigments were then stored below -20°C prior to analysis.

## Preparation for Antimicrobial Analysis

The antimicrobial potential of the carotenoid, chlorophyll and betalain pigments were analysed using the minimum inhibitory concentration (MIC) and minimum bactericidal concentration (MBC) to detect their ability to inhibit microbial growth. All the antimicrobial analysis procedures were carried out in a Class 2 biological hazard cabinet. The procedure with the cleaning of the surfaces of the cabinet by wiping them with 70% ethanol solution and sterilising them under UV radiation for 20 minutes prior to use. A broth medium was prepared according to the method specified by the manufacturer. Distilled water was used to dissolve the medium, and it was then autoclaved at 121°C for 20 minutes (Mastura & Saiful Azmi, 2023). Two microbes, namely *S. aureus* (ATCC 25923) and *E. coli* (ATCC 35218), were purchased from the American Type Culture Collection (ATCC) to represent gram-positive and gram-negative bacteria, respectively.

The inoculation process involved taking between 5–8 ml of the broth and putting it into sterile test tubes using an aseptic technique and labelling them as inoculum or control samples. At the same time, the microbes that had been stored in cryovials were extracted and added to the inoculum-labelled test tubes. The tubes containing both the inoculum and control samples were then incubated in an incubator shaker set at 37°C for 18–24 hours. The level of turbidity was used as an indication of microbial growth, while a clear control broth would be an indication that no microbial growth had taken place (Mohtar et al., 2009). Next, a standardisation procedure was performed by adjusting the turbidity to obtain similar levels of turbidity, according to the 0.5 McFarland standard (1.5 x 108 CFU/ml). Also, a cell density meter was used to measure the absorbance to ensure that it was between 0.05–0.1 at a wavelength of 600 nm. Then, the dried crude plant pigments were diluted and a certain amount was pipetted into a 96-well microplate that already contained broth, bringing the final volume of the sample and broth mixture to 50 μl. The different amounts of sample and broth mixture produced samples with serial dilutions of 500, 250, 125, 62.5, 31.25, 15.625, and 0 μg/ml, which were ready for exposure to standardised microbes.

## Minimum Inhibitory Concentration (MIC) Analysis

The MIC is a quantitative *in vitro* assay used to evaluate the potential or ability of a pigment to inhibit the activity of selected microbes. Stock solutions of each sample were prepared in 96-well microplates according to the guidelines of the Clinical and Laboratory Standards Institute (CLSI, 2006). Each standardised microbe was added to the sample and broth mixture separately and incubated at 37°C for 24 hours. The turbidity was recorded as an indication of microbial growth, where the lowest concentration that was still clear after a macroscopic evaluation was recorded as the MIC. It can be summarised that the lower the MIC, the higher the ability of the pigment to inhibit microbes.

## Minimum Bactericidal Concentration (MBC) Analysis

The MBC was recorded as the average concentration multiplier. The inhibitory activity status of each sample, whether permanent/cidal or temporary/static, was determined immediately after a macroscopic MIC evaluation. Therefore, 20 µl of the reagent solution, 3-(4,5- dimethylthiazol-2-yl)-2,5-diphenyl tetrazolium bromide, was put into the same microplate that showed the MIC to test the inhibition status of the microbial cells for each sample and incubated at 37°C for 20 minutes. After 20 minutes, observations were made to check for any colour changes. A change from clear yellow to dark blue would indicate that the microbes were still active after the temporary/static inhibition. Meanwhile, if the colour remained clear yellow, it would indicate permanent/cidal inhibition of microbes.

## Results

The antimicrobial potential of the selected pigments was verified through the MBC and MIC. The results were recorded as the mean concentration of triplicates (Table 1), where the mean value of triplicates at a concentration of 5 mg/mL was used as the maximum concentration. Gibbon (2004) recommended the use of MIC concentrations of < 1000 µg/mL for natural compound extracts, including plant pigments (Štumpf et al., 2020; Rogoza et al., 2009). Thus, the results revealed that the highest antimicrobial potential detected from all the tested pigments was for the chlorophyll pigment extracted from the *C. vulgaris* as it was able to inhibit both the *S. aureus* and *E. coli* at 2.5 µg/µL in its bactericidal nature. The pigment with the second highest antimicrobial potential was bixin, extracted from *Bixa orellana* L. seeds, which was able to inhibit both the *S. aureus* at a concentration of 5 µg/µL and the *E. coli* at 2.5 µg/µL in its bactericidal nature. However, 5 µg/µL and more of the *S. androgynus, M. oleifera, O. basilicum,* and *H. polyrhizus* were needed to inhibit the *S. aureus* and *E. coli.*

**Table 1.** Antimicrobial profile of carotenoid, chlorophyll, and betalains extracted from plants

| Bacteria/Candida | Plant Species | MIC (MBC/MCC) (µg/µl) | |
|---|---|---|---|
| | | Permanent/ Cidal | Static/ temporary |
| S aureus (Gram-positive) (ATCC 25923) | Bixa orellana (Annatto) | 5 | 5 |
| | Sauropus androgynus (Cekur manis) | 5 | 5 |
| | Moringa oleifera (Kelor) | 5 | 5 |
| | Ocimum basilicum (Selasih) | 5 | 5 |
| | Chlorella vulgaris (Green algae) | 2.5 | 2.5 |
| | Hylocereus polyrhizus (Dragon fruit) | >5 | >5 |

| | | | |
|---|---|---|---|
| E. coli (Gram-negative) (ATCC 35218) | Bixa orellana (Annatto) | 2.5 | 2.5 |
| | Sauropus androgynus (Cekur manis) | 5 | 5 |
| | Moringa oleifera (Kelor) | 5 | 5 |
| | Ocimum basilicum (Selasih) | >5 | >5 |
| | Chlorella vulgaris (Green algae) | 2.5 | 2.5 |
| | Hylocereus polyrhizus (Dragon fruit) | >5 | >5 |

## Discussion

Alwathnani and Perveen (2017) stated that *C. vulgaris* extracts not only exhibit antibacterial activity but also induce morphological changes in human pathogenic bacteria, thus indicating their potential as antimicrobial agents. The results of a previous study also mentioned that *C. vulgaris* can inhibit *Streptococcus mutans*, another pathogenic bacterium (Jafari et al., 2018). Thus, the available research proves that *C. vulgaris* exhibits inhibitory effects against both the tested bacteria in this study, thereby highlighting its potential as the best natural source of antimicrobial compounds.

Ahmed et al. (2020) focused on investigating the antimicrobial activity of the carotenoid pigment extracted from *B. orellana*. They specifically tested the efficacy of this pigment against two common bacterial strains, *E. coli* and *Bacillus subtilis* (*B. subtilis*). They found that the carotenoid pigment exhibited a high potential to inhibit the growth of both bacterial strains and was effective against them within a certain range, suggesting its potential as an antibacterial agent. This indicates that the pigment could be a promising natural alternative for combating bacterial infections. Abayomi et al. (2014), Alim et al. (2016), and Karmakar et al. (2018) investigated the antimicrobial potential of the carotenoid pigment extracted from *B. orellana* against different bacterial strains. Abayomi et al. found that the pigment showed good antimicrobial activity against *S. aureus* and *B. subtilis*, while Alim et al. and Karmakar et al. reported positive results in inhibiting *E. coli*. These findings collectively suggest that the carotenoid pigment extracted from *B. orellana* possesses significant antimicrobial properties, demonstrating its efficacy against various bacterial strains such as *E. coli, B. subtilis*, and *S. aureus*.

The findings of the antimicrobial potential of *S. androgynus* aligned with previous research on the antimicrobial efficacy of plant extracts. Kuttinath et al. (2019) reported that a concentration of 5 mg/ml of *S. androgynus* extract effectively inhibited the growth of bacterial strains. Notably, they observed similar inhibitory effects against *S. aureus* (5 μg/μl) compared to this finding. However, they reported a lower MIC for *E. coli* (1.25 μg/μl) than this study

(2.5 μg/μl). Kuttinath et al. (2019) employed ethanol, methanol, and aqueous extraction, which, in this context, appeared more effective than the hexane extraction used in this study for maximising antimicrobial inhibition. In addition, the ethyl acetate extract of *S. androgynus* exhibited significant inhibitory activity against both *S. aureus* (1.25 μg/μl) and *E. coli* (2.5 μg/μl) (Kumar and George, 2015).

Several studies have reported the inhibitory effects of *M. oleifera* extract on various microorganisms, including *S. aureus* and *E. coli*. Gull et al. (2016) verified that methanolic *M. oleifera* leaf extract exhibited an MIC of 312.5 μg/mL against *S. aureus* and 625 μg/mL against *E. coli*. Smith (2016) reported that ethanol *M. oleifera* extract showed an MIC and MBC of 0.75 and 3 μg/mL, respectively, against *E. coli*, whereas recent research has shown that *O. basilicum* extract possesses significant antimicrobial activity against both *S. aureus* and *E. coli* (Khalil, 2013). Moghaddam et al. (2011) reported that *O. basilicum* extract displayed an MIC of 18 μg/mL against *S. aureus* and 18.9 μg/mL against *E. coli*. Another study revealed that methanol *O. basilicum* leaves extract showed an MIC of 62.5 μg/mL against *S. aureus* and 125 μg/mL against *E. coli* (Hossain et al., 2010).

Tenore et al. (2012) highlighted that the antimicrobial activity of betalain extracts, particularly from the Cactaceae family like *H. polyrhizus*, is limited. They assessed the antimicrobial efficacy of *H. polyrhizus* extract, both with and without fractionation, against a panel of yeasts, moulds, and bacteria. While the unfractionated extract showed no activity against yeasts and moulds, it exhibited varying degrees of inhibition against bacteria, with an MIC ranging from 125–500 μg/μl. Interestingly, Madadi et al. (2020) demonstrated that polyphenolic fractions from *H. polyrhizus* exhibited significantly enhanced antimicrobial activity, inhibiting all the tested bacteria, yeasts, and moulds at concentrations ranging from 7.8–31.3 μg/μl for bacteria and 125–500 μg/μl for yeasts and moulds. These findings underscore the potential of specific betalain-rich fractions as potent antimicrobial agents. However, despite the growing evidence of the broad-spectrum antimicrobial activity of betalains, research on the specific mechanisms of action, particularly for betalain derived from *H. polyrhizus*, remains scarce. However, previous studies have proven that *S. androgynus, M. oleifera, O. basilicum,* and *H. polyrhizus* also have the potential to be alternative sources of antimicrobial agents against *S. aureus* and *E. coli* infections.

The observed variations in antimicrobial efficacy can be attributed to several factors, including differences in bacterial cell wall structure. Gram-positive bacteria, with their single peptidoglycan layer, are generally more permeable to external substances compared to gram-negative bacteria, which possess an additional lipopolysaccharide layer (Chaturwedi et al., 2023; Ahmed et al., 2020). This structural difference probably explains the increased susceptibility of gram-positive bacteria to phytochemicals, as their cell structure allows for easier penetration (Chaturwedi et al., 2023; Hochma et al., 2021; Yolmeh, 2016). Furthermore, Pérez-López et al. (2008) emphasised the impact of extraction and storage conditions on antimicrobial activity, noting that shorter durations generally yield a lower MIC, thus indicating higher potency. Conversely, factors like solvent volume and agitation speed appear to have a minimal impact on the MIC. Importantly, Quintero Quiroz et al. (2019) attributed the potent antimicrobial activity of pigment extracts to their high polyphenol content, suggesting a key role for these compounds in inhibiting microbial growth.

## Conclusion

In conclusion, plant pigments offer a multitude of benefits that extend far beyond their vibrant colours. Recent research has highlighted their potential as antimicrobial agents that are capable of inhibiting the growth of bacteria and fungi. Interestingly, microbial pigments, such as carotenoids, chlorophyll, and betalains, have shown antimicrobial activity against *S. aureus* and *E. coli*. This discovery has significant implications for various fields, ranging from food science to medicine. Further investigation into the complex relationship between the colour richness of a pigment and its potential antimicrobial role is crucial for advancing our knowledge in this area. A deeper understanding of the specific mechanisms underlying their antimicrobial actions could unlock novel therapeutic strategies for combating infectious diseases.

## References

Abayomi, M., Adebayo, A. S., Bennett, D., Porter, R., Campbell, J. S., & Dawkin, G. (2014). Phytochemical testing and in vitro antibacterial activity of *Bixa orellana* (annatto) seed extract. *British Journal of Pharmaceutical Research, 4*(11), 1387-1399. https://doi.org/10.9734/BJPR/2014/10484

Ahari, H., & Soufiani, S.P. (2021). Smart and active food packaging: Insights in novel food packaging. *Frontiers in Microbiology, 12*, 657233. https://doi.org/10.3389/fmicb.2021.657233

Ahmad, M., & Wajid, M. (2013). Plants as potential source of antimicrobial agents. *Journal of Pharmacy and Alternative Medicine, 2*(3), 18-25.

Ahmed, S., Moni, B.M., Ahmed, S., Gomes, D.J., & Shohael, A.M. (2020). Comparative phytochemical, antioxidant, and antibacterial study of different parts of *Doigota* plants (*Bixa orellana* L.). *Bulletin of the National Research Centre, 44*, 1-10. https://doi.org/10.1186/s42269-020-00483-0

Alappat, B., & Alappat, J. (2020). Anthocyanin pigments: Beyond aesthetics. *Molecules, 25*(23), 5500. https://doi.org/10.3390/molecules25235500

Alim, S., Bairagi, N., Shahriyar, S., Kabir, M. M., & Rahman, M. H. (2016). In vitro antibacterial potential of *Bixa orellana* L. against some pathogenic bacteria and comparative investigation on some standard antibiotics. *Journal of Pharmacognosy and Phytochemistry, 5*(2), 178-181.

Alwathnani, H., & Perveen, K. (2017). Antibacterial activity and morphological changes in human pathogenic bacteria caused by *Chlorella vulgaris* extracts. *Biomed. Res, 28*, 1610-1614.

Benucci, I., Lombardelli, C., Mazzocchi, C., & Esti, M. (2022). Natural colorants from vegetable food waste: Recovery, regulatory aspects, and stability—A review. *Comprehensive Reviews in Food Science and Food Safety, 21*(3), 2715-2737. https://doi.org/10.1111/1541-4337.12838

Bohn, T. (2019). Carotenoids and markers of oxidative stress in human observational studies and intervention trials: Implications for chronic diseases. *Antioxidants, 8*(6), 179. https://doi.org/10.3390/antiox8060179

Breijyeh, Z., Jubeh, B., & Karaman, R. (2020). Resistance of gram-negative bacteria to current antibacterial agents and approaches to resolve it. *Molecules, 25*(6), 1340. https://doi.org/10.3390/molecules25061340

Cai, Y. Z., Sun, M., & Corke, H. (2005). Characterization and application of betalain pigments from plants of the *Amaranthaceae*. *Trends in Food Science & Technology, 16*(9), 370-376. https://doi.org/10.1016/j.tifs.2005.04.011

Câmara, J. S., Albuquerque, B. R., Aguiar, J., Corrêa, R. C., Gonçalves, J. L., Granato, D., Pereira, J. A., Barros, L., & Ferreira, I. C. (2020). Food bioactive compounds and emerging techniques for their extraction: Polyphenols as a case study. *Foods, 10*(1), 37. https://doi.org/10.3390/foods10010037

Chaturwedi, S. B., Mainali, S., & Chaudhary, R. (2023). Antibacterial activity of pigment extracted from pigment producing bacteria. [No publication details available]

Chaudhary, A. (2024). Characterization, antifungal activity and industrial application of pigment produced by *Pseudomonas aeruginosa* (Doctoral dissertation, Kashmir University).

Clifford, T., Howatson, G., West, D. J., & Stevenson, E. J. (2015). The potential benefits of red beetroot supplementation in health and disease. *Nutrients, 7*(4), 2801-2822. https://doi.org/10.3390/nu7042801

Coy-Barrera, E. (2020). Analysis of betalains (betacyanins and betaxanthins). In *Recent Advances in Natural Products Analysis* (pp. 593-619). Elsevier: Netherlands. https://doi.org/10.1016/B978-0-12-818839-0.00037-7

da Silva, J. C., & Lombardi, A. T. (2020). Chlorophylls in microalgae: Occurrence, distribution, and biosynthesis. *Pigments from Microalgae Handbook*, 1-18.

de Andrade Lima, M., Kestekoglou, I., Charalampopoulos, D., & Chatzifragkou, A. (2019). Supercritical fluid extraction of carotenoids from vegetable waste matrices. *Molecules, 24*(3), 466. https://doi.org/10.3390/molecules24030466

De Mejia, E. G., Zhang, Q., Penta, K., Eroglu, A., & Lila, M. A. (2020). The colors of health: Chemistry, bioactivity, and market demand for colorful foods and natural food sources of colorants. *Annual Review of Food Science and Technology, 11*, 145-182. https://doi.org/10.1146/annurev-food-032119-032923

de Oliveira, M. E. P., da Silva, W. M. B., Alves, D. R., de Andrade Neto, J. B., de Morais, S. M., & de Oliveira Pinheiro, S. (2021). Dyes and pigments used in foods: An integrative literature review. *Research, Society and Development, 10*(10), e316101018925-e316101018925. https://doi.org/10.33448/rsd-v10i10.18925

Enaru, B., Drețcanu, G., Pop, T. D., Stănilă, A., & Diaconeasa, Z. (2021). Anthocyanins: Factors affecting their stability and degradation. *Antioxidants, 10*(12), 1967. https://doi.org/10.3390/antiox10121967

Fadiji, T., Rashvand, M., Daramola, M.O., & Iwarere, S.A. (2023). A review on antimicrobial packaging for extending the shelf life of food. *Processes, 11*(2), 590. https://doi.org/10.3390/pr11020590

Favela-González, K. M., Hernández-Almanza, A. Y., & De la Fuente-Salcido, N. M. (2020). The value of bioactive compounds of cruciferous vegetables (*Brassica*) as antimicrobials and antioxidants: A review. *Journal of Food Biochemistry, 44*(10), e13414. https://doi.org/10.1111/jfbc.13414

Fu, X., Wu, Q., Wang, J., Chen, Y., Zhu, G., & Zhu, Z. (2021). Spectral characteristic, storage stability and antioxidant properties of anthocyanin extracts from flowers of butterfly pea (*Clitoria ternatea* L.). *Molecules, 26*(22), 7000. https://doi.org/10.3390/molecules26227000

Gallo, M., Ferrara, L., Calogero, A., Montesano, D., & Naviglio, D. (2020). Relationships between food and diseases: What to know to ensure food safety. *Food Research International, 137*, 109414. https://doi.org/10.1016/j.foodres.2020.109414

Gandía-Herrero, F., Escribano, J., & García-Carmona, F. (2016). Biological activities of plant pigments betalains. *Critical Reviews in Food Science and Nutrition, 56*(6), 937-945. https://doi.org/10.1080/10408398.2013.822376

Ghosh, S., Sarkar, T., Das, A., & Chakraborty, R. (2022). Natural colorants from plant pigments and their encapsulation: An emerging window for the food industry. *LWT, 153*, 112527. https://doi.org/10.1016/j.lwt.2021.112527

Gibbons, S. (2004). Anti-staphylococcal plant natural products. *Natural Product Reports, 21*(2), 263-277. https://doi.org/10.1039/B301323D

Griffiths, J. (1990). Introduction: The evolution of present-day dye technology. *The Chemistry and Application of Dyes*, 1-16.

Guaadaoui, A., Benaicha, S., Elmajdoub, N., Bellaoui, M., & Hamal, A. (2014). What is a bioactive compound? A combined definition for a preliminary consensus. *International Journal of Nutrition and Food Sciences, 3*(3), 174-179. https://doi.org/10.11648/j.ijnfs.20140303.11

Gull, I., Javed, A., Aslam, M. S., Mushtaq, R., & Athar, M. A. (2016). Use of *Moringa oleifera* flower pod extract as natural preservative and development of SCAR marker for its DNA-based identification. *BioMed Research International, 2016*, 7584318. https://doi.org/10.1155/2016/7584318

Hirose, S., Konishi, N., Sato, M., Suzumura, K., Obata, H., Ohtsuka, K., Doi, R., Goto, K., Kai, A., Arai, S., & Hara-Kudo, Y. (2024). Growth and survival of *Escherichia albertii* in food and environmental water at various temperatures. *Journal of Food Protection, 87*(4), 100249. https://doi.org/10.4315/JFP-23-397

Hochma, E., Yarmolinsky, L., Khalfin, B., Nisnevitch, M., Ben-Shabat, S., & Nakonechny, F. (2021). Antimicrobial effect of phytochemicals from edible plants. *Processes, 9*(11), 2089. https://doi.org/10.3390/pr9112089

Hossain, M. A., Kabir, M. J., Salehuddin, S. M., Rahman, S. M., Das, A. K., Singha, S. K., Alam, M. K., & Rahman, A. (2010). Antibacterial properties of essential oils and methanol extracts of sweet basil *Ocimum basilicum* occurring in Bangladesh. *Pharmaceutical Biology, 48*(5), 504-511. https://doi.org/10.3109/13880201003658296

Jafari, S., Mobasher, M. A., Najafipour, S., Ghasemi, Y., Mohkam, M., Ebrahimi, M. A., & Mobasher, N. (2018). Antibacterial potential of *Chlorella vulgaris* and *Dunaliella salina* extracts against *Streptococcus mutans*. *Jundishapur Journal of Natural Pharmaceutical Products, 13*(2).

Jafarzadeh, S., Jafari, S. M., Salehabadi, A., Nafchi, A. M., Kumar, U. S. U., & Khalil, H. A. (2020). Biodegradable green packaging with antimicrobial functions based on the bioactive compounds from tropical plants and their by-products. *Trends in Food Science & Technology, 100*, 262-277. https://doi.org/10.1016/j.tifs.2020.06.004

Jarriyawattanachaikul, W., Chaveerach, P., & Chokesajjawatee, N. (2016). Antimicrobial activity of Thai-herbal plants against food-borne pathogens *E. coli, S. aureus* and *C. jejuni*. *Agriculture and Agricultural Science Procedia, 11*, 20-24. https://doi.org/10.1016/j.aaspro.2016.02.004

Jurić, S., Jurić, M., Król-Kilińska, Ż., Vlahoviček-Kahlina, K., Vinceković, M., Dragović-Uzelac, V., & Donsì, F. (2022). Sources, stability, encapsulation and application of natural pigments in foods. *Food Reviews International, 38*(8), 1735-1790. https://doi.org/10.1080/87559129.2021.1940329

Karmakar, U.K., Sultana, S., Nishi, S., Biswas, N.N., Hossain, L., & Sheikh, S. (2018). Antioxidant, analgesic, antimicrobial, and anthelmintic activity of the dried seeds of *Bixa orellana* (L). *International Journal of Pharmacy, 8*(1), 150-163.

Khalil, A. (2013). Antimicrobial activity of ethanolic extracts of *Ocimum basilicum* leaf from Saudi Arabia. *Biotechnology, 12*(1), 61-64.

Khan, F., Khan, A., & Kazmi, S. U. (2014). Prevalence and susceptibility pattern of multi-drug resistant clinical isolates of *Pseudomonas aeruginosa* in Karachi. *Pakistan Journal of Medical Sciences, 30*(5), 951. https://doi.org/10.12669/pjms.305.5602

Khan, M. I. (2016). Plant betalains: Safety, antioxidant activity, clinical efficacy, and bioavailability. *Comprehensive Reviews in Food Science and Food Safety, 15*(2), 316-330. https://doi.org/10.1111/1541-4337.12177

Khoo, H. E., Azlan, A., Tang, S. T., & Lim, S. M. (2017). Anthocyanidins and anthocyanins: Colored pigments as food, pharmaceutical ingredients, and the potential health benefits. *Food & Nutrition Research, 61*(1), 1361779. https://doi.org/10.1080/16546628.2017.1361779

Kumar, P. R., & George, P. (2015). Antidiabetic effect of *Sauropus androgynus* L. leaves in alloxan-induced diabetic mice. *Journal of Pure and Applied Microbiology, 9*(3), 2565-2570.

Kuttinath, S., Kh, H., & Rammohan, R. (2019). Phytochemical screening, antioxidant, antimicrobial, and antibiofilm activity of *Sauropus androgynus* leaf extracts. *Asian Journal of Pharmaceutical and Clinical Research, 12*(4), 244-250.

Lev-Yadun, S. (2022). The phenomenon of red and yellow autumn leaves: Hypotheses, agreements and disagreements. *Journal of Evolutionary Biology, 35*(10), 1245-1282. https://doi.org/10.1111/jeb.13903

Madadi, E., Mazloum-Ravasan, S., Yu, J.S., Ha, J.W., Hamishehkar, H., & Kim, K.H. (2020). Therapeutic application of betalains: A review. *Plants, 9*(9), 1219. https://doi.org/10.3390/plants9091219

Majid, I., Thakur, M., & Nanda, V. (2018). Innovative and safe packaging technologies for food and beverages: Updated review. *Innovations in Technologies for Fermented Food and Beverage Industries*, 257-287. https://doi.org/10.1201/9780429493280-14

Mancuso, G., Midiri, A., Gerace, E., & Biondo, C. (2021). Bacterial antibiotic resistance: The most critical pathogens. *Pathogens, 10*(10), 1310. https://doi.org/10.3390/pathogens10101310

Manzoor, A., Khan, S., Dar, A.H., Pandey, V.K., Shams, R., Ahmad, S., Jeevarathinam, G., Kumar, M., Singh, P., & Pandiselvam, R. (2023). Recent insights into green antimicrobial packaging towards food safety reinforcement: A review. *Journal of Food Safety, 43*(4), e13046. https://doi.org/10.1111/jfs.13046

Maoka, T. (2020). Carotenoids as natural functional pigments. *Journal of Natural Medicines, 74*(1), 1-16. https://doi.org/10.1007/s11418-020-01423-9

Mastura, M., & Saiful Azmi, J. (2023). Antimicrobial. *A Laboratory Manual on Methods for Bioprospecting of Natural Products. Vol.1*. Forest Research Institute Malaysia (FRIM). Kepong.

Mekapogu, M., Vasamsetti, B. M. K., Kwon, O. K., Ahn, M. S., Lim, S. H., & Jung, J. A. (2020). Anthocyanins in floral colors: biosynthesis and regulation in chrysanthemum flowers. *International Journal of Molecular Sciences, 21*(18), 6537. https://doi.org/10.3390/ijms21186537

Moghaddam, A. M. D., Shayegh, J., Mikaili, P., & Sharaf, J. D. (2011). Antimicrobial activity of essential oil extract of *Ocimum basilicum* L. leaves on a variety of pathogenic bacteria. *Journal of Medicinal Plants Research, 5*(15), 3453-3456

Mohtar, M., Johari, S.A., Li, A.R., Isa, M.M., Mustafa, S., Ali, A.M., & Basri, D.F. (2009). Inhibitory and resistance-modifying potential of plant-based alkaloids against methicillin-resistant *Staphylococcus aureus* (MRSA). *Current Microbiology, 59*, 181-186. https://doi.org/10.1007/s00284-009-9441-9

Molina, A. K., Corrêa, R. C., Prieto, M. A., Pereira, C., & Barros, L. (2023). Bioactive natural pigments' extraction, isolation, and stability in food applications. *Molecules, 28*(3), 1200. https://doi.org/10.3390/molecules28031200

Muthusamy, K., Ramasamy, G., Ravikumar, C., Natesan, S., Muthurajan, R., Uthandi, S., Kalyanasundaram, K., & Tiwari, V. (2024). Exploring bixin from *Bixa orellana* L. seeds: quantification and in silico insights into its anti-cancer potential. *Journal of*

*Biomolecular Structure and Dynamics, 42*(22), 12244-12258. https://doi.org/10.1080/07391102.2024.1987021

Naeem, M., Zhao, W., Ahmad, N., & Zhao, L. (2023). Beyond green and red: unlocking the genetic orchestration of tomato fruit color and pigmentation. *Functional & Integrative Genomics, 23*(3), 243. https://doi.org/10.1007/s10142-023-00944-7

Nassour, R., Ayash, A., & Al-Tameemi, K. (2020). Anthocyanin pigments: Structure and biological importance. *J. Chem. Pharm. Sci, 13*, 45-57.

Ngamwonglumlert, L., Devahastin, S., & Chiewchan, N. (2017). Natural colorants: Pigment stability and extraction yield enhancement via utilization of appropriate pretreatment and extraction methods. *Critical Reviews in Food Science and Nutrition, 57*(15), 3243-3259. https://doi.org/10.1080/10408398.2015.1113847

Nguyen, N.H.K., Tien, H.T.C., Truc, T.T., & Quoc, L.T. (2020, December). Chlorophyll content and antioxidant activity from *folium sauropi* (*Sauropus androgynus* (L.) Merr) with microwave-assisted extraction. *IOP Conference Series: Materials Science and Engineering, 991*(1), 012036. IOP Publishing: England. https://doi.org/10.1088/1757-899X/991/1/012036

Nieto, G., Martínez-Zamora, L., Peñalver, R., Marín-Iniesta, F., Taboada-Rodríguez, A., López-Gómez, A., & Martínez-Hernández, G. B. (2023). Applications of plant bioactive compounds as replacers of synthetic additives in the food industry. *Foods, 13*(1), 47. https://doi.org/10.3390/foods13010047

Okmen, G., Kardas, S., Bayrak, D., Arslan, A., & Cakar, H. (2016). The antibacterial activities of *Crocus sativus* against mastitis pathogens and its antioxidant activities. *World Journal of Pharmacy and Pharmaceutical Sciences, 5*(3), 146-156.

Othman, R. (2009). Biochemistry and genetics of carotenoid composition in potato tubers (Doctoral dissertation, Lincoln University).

Pan, Y., Deng, Z., & Shahidi, F. (2020). Natural bioactive substances for the control of food-borne viruses and contaminants in food. *Food Production, Processing and Nutrition, 2*, 1-19. https://doi.org/10.1186/s43014-020-00004-4

Pattabhiramaiah, M., & Mallikarjunaiah, S. (2023). Foodborne pathogens and food safety regulations. In *Food Microbial and Molecular Biology* (pp. 179-212). Apple Academic Press: Florida, USA.

Pérez-López, A., Orozco-Hayek, M., Rivas-Galindo, V., & De Torres, N.W. (2008). Experimental design to determine the factors affecting the preparation of extracts for antibacterial use. *Natural Product Communications, 3*(3), 1934578X0800300311.

Polturak, G., & Aharoni, A. (2018). "La Vie en Rose": biosynthesis, sources, and applications of betalain pigments. *Molecular Plant, 11*(1), 7-22. https://doi.org/10.1016/j.molp.2017.11.013

Priyadarshi, R., Kumar, B., Deeba, F., Kulshreshtha, A., & Negi, Y.S. (2018). Chitosan films incorporated with apricot (*Prunus armeniaca*) kernel essential oil as active food packaging material. *Food Hydrocolloids, 85*, 158-166. https://doi.org/10.1016/j.foodhyd.2018.07.029

Quintero Quiroz, J., Celis Torres, A., Muñoz Ramirez, L., Silva Garcia, M., Ciro Gomez, G., & Rojas Camargo, J. (2019). Optimization of the microwave-assisted extraction process of bioactive compounds from annatto seeds (*Bixa orellana* L.). *Antioxidants, 8*(2), 37. https://doi.org/10.3390/antiox8020037

Rodriguez-Amaya, D. B. (2016). Natural food pigments and colorants. *Current Opinion in Food Science, 7*, 20-26. https://doi.org/10.1016/j.cofs.2015.10.001

Rodriguez-Concepcion, M., Avalos, J., Bonet, M. L., Boronat, A., Gomez-Gomez, L., Hornero-Mendez, D., ... & Zhu, C. (2018). A global perspective on carotenoids: Metabolism, biotechnology, and benefits for nutrition and health. *Progress in Lipid Research, 70*, 62-93. https://doi.org/10.1016/j.plipres.2018.01.002

Rogoza, L. N., Salakhutdinov, N. F., & Tolstikov, G. A. (2009). Natural and synthetic compounds with an antimycobacterial activity. *Mini-Reviews in Organic Chemistry, 6*(2), 135-151. https://doi.org/10.2174/157019309788184775

Roy, S., & Rhim, J. W. (2021). Anthocyanin food colorant and its application in pH-responsive color change indicator films. *Critical Reviews in Food Science and Nutrition, 61*(14), 2297-2325. https://doi.org/10.1080/10408398.2020.1781767

Sadowska-Bartosz, I., & Bartosz, G. (2021). Biological properties and applications of betalains. *Molecules, 26*(9), 2520. https://doi.org/10.3390/molecules26092520

Sar, T., Kiraz, P., Braho, V., Harirchi, S., & Akbas, M.Y. (2023). Novel perspectives on food-based natural antimicrobials: A review of recent findings published since 2020. *Microorganisms, 11*(9), 2234. https://doi.org/10.3390/microorganisms11092234

Sen, S., Antara, N., & Sen, S. (2021). Factors influencing consumers' decision to take ready-made frozen food. *Current Psychology, 40*, 2634-2643. https://doi.org/10.1007/s12144-019-00430-6

Serwecińska, L. (2020). Antimicrobials and antibiotic-resistant bacteria: A risk to the environment and to public health. *Water, 12*(12), 3313. https://doi.org/10.3390/w12123313

Sharma, M., Usmani, Z., Gupta, V. K., & Bhat, R. (2021a). Valorization of fruits and vegetable wastes and by-products to produce natural pigments. *Critical Reviews in Biotechnology, 41*(4), 535-563. https://doi.org/10.1080/07388551.2020.1837844

Sharma, S., Katoch, V., Kumar, S., & Chatterjee, S. (2021b). Functional relationship of vegetable colors and bioactive compounds: Implications in human health. *The Journal of Nutritional Biochemistry, 92*, 108615. https://doi.org/10.1016/j.jnutbio.2021.108615

Sigwela, V. N. (2020). Extraction, characterization, and application of betalains from cactus pear, beetroot, and amaranth (Doctoral dissertation) University of the Free State.

Smith, B. E. (2016). Anti-bacterial properties of ethanolic *Moringa oleifera* leaf extract and proteomic analysis of its effects on *Escherichia coli* (Doctoral dissertation, Appalachian State University).

Štumpf, S., Hostnik, G., Primožič, M., Leitgeb, M., Salminen, J. P., & Bren, U. (2020). The effect of growth medium strength on minimum inhibitory concentrations of tannins and tannin extracts against *E. coli*. *Molecules, 25*(12), 2947. https://doi.org/10.3390/molecules25122947

Sturm, B., & Hensel, O. (2017). Pigments and nutrients during vegetable drying processes, dried products storage, and their associated color changes. *Handbook of Drying of Vegetables and Vegetable Products*, 257-277. https://doi.org/10.1016/B978-0-12-804242-9.00014-2

Swami, S. B., Ghgare, S. N., Swami, S. S., Shinde, K. J., Kalse, S. B., & Pardeshi, I. L. (2020). Natural pigments from plant sources: A review. *The Pharma Innovation Journal, 9*(10), 566.

Tenore, G.C., Novellino, E., & Basile, A. (2012). Nutraceutical potential and antioxidant benefits of red pitaya (*Hylocereus polyrhizus*) extracts. *Journal of Functional Foods, 4*(1), 129-136. https://doi.org/10.1016/j.jff.2011.10.006

Ullah, F., Ayaz, M., Sadiq, A., Ullah, F., Hussain, I., Shahid, M., Yessimbekov, Z., Adhikari-Devkota, A., & Devkota, H.P. (2020). Potential role of plant extracts and phytochemicals against foodborne pathogens. *Applied Sciences, 10*(13), 4597. https://doi.org/10.3390/app10134597

Venil, C. K., Dufossé, L., & Renuka Devi, P. (2020). Bacterial pigments: Sustainable compounds with market potential for pharma and food industry. *Frontiers in Sustainable Food Systems, 4*, 100. https://doi.org/10.3389/fsufs.2020.00059

Virtanen, O., Constantinidou, E., & Tyystjärvi, E. (2022). Chlorophyll does not reflect green light–how to correct a misconception. *Journal of Biological Education, 56*(5), 552-559. https://doi.org/10.1080/00219266.2022.2053920

Viswanath, B. G., Lakshmi, G. J., Nagamani, K., Reddy, N. V. N., Rao, G. P., & Srinivas, S. S. S. (2017). Emergence of antimicrobial resistance among anaerobic bacteria. *Am J Infect Dis Microbiol, 52*, 87-93. https://doi.org/10.4236/ajidm.2017.52009

Yolmeh, M. (2016). Antimicrobial activity of pigments extracted from *Rhodotorula glutinis* against some bacteria and fungi. *Zahedan Journal of Research in Medical Sciences, 18*(12), e4954.

Zhang, H., Yuan, X., Xiong, T., Wang, H., & Jiang, L. (2020). Bioremediation of co-contaminated soil with heavy metals and pesticides: Influence factors, mechanisms and evaluation methods. *Chemical Engineering Journal, 398*, 125657. https://doi.org/10.1016/j.cej.2020.125657

Zhang, W., Shu, C., Chen, Q., Cao, J., & Jiang, W. (2019). The multi-layer film system improved the release and retention properties of cinnamon essential oil and its application as coating in inhibition to penicillium expansion of apple fruit. *Food Chemistry, 299*, 125109. https://doi.org/10.1016/j.foodchem.2019.125109

Zhao, L., Liu, Y., Zhao, L., & Wang, Y. (2022). Anthocyanin-based pH-sensitive smart packaging films for monitoring food freshness. *Journal of Agriculture and Food Research, 9*, 100340. https://doi.org/10.1016/j.jafr.2022.100340

Zin, M.M., Márki, E., & Bánvölgyi, S. (2020). Conventional extraction of betalain compounds from beetroot peels with aqueous ethanol solvent. *Acta Alimentaria, 49*(2), 163-169. https://doi.org/10.1556/066.2020.00014

# Chapter 8

# Preserving Malay Heritage: The Intersection of Traditional Knowledge and Natural Pigments

**Razanah Ramya[1,*], PhD**
**Nur Atiqah Jamaludin[2], MSc**
**and Rashidi Othman[3], PhD**
[1,2]Institute of the Malay World and Civilization, Universiti Kebangsaan Malaysia, Universiti Kebangsaan Malaysia Bangi, Selangor, Malaysia
[3]Department of Landscape Architecture, Kulliyyah of Architecture and Environmental Design, International Islamic University Malaysia, Kuala Lumpur, Malaysia

## Abstract

Globalization and rapid socioeconomic change have affected indigenous communities in Malaysia. A recent demand for plant-derived bioactive compounds has increased the attention of the researcher to discover phytochemicals content and manage the conservation of natural biodiversity simultaneously. Malay's indigenous knowledge and technology in natural colorants represent a rich heritage deeply rooted in the country's diverse cultural and ecological landscapes. The indigenous communities of Malay have long relied on natural resources for various purposes, including coloring fabrics, food, personal care, and crafts. Indigenous communities possess traditional knowledge passed down through generations regarding the identification, harvesting, and processing of natural colorants. This knowledge often involves an intimate understanding of local flora, including plants, roots, fruits, and even insects, that yield vibrant pigments suitable for dyeing. The use of natural colorants underscores a holistic approach to resource

---

* Corresponding Author's Email: razanah.ramya@ukm.edu.my

In: Natural Pigments: Sources, Properties, and Health Benefits
Editors: Rashidi Othman, Qurratu Aini Mat Ali, Razanah Ramya et al.
ISBN: 979-8-90134-055-4

management and sustainability. Indigenous practices typically emphasize responsible harvesting techniques to ensure the regeneration and preservation of natural dye sources. This sustainable ethos aligns with contemporary environmental conservation efforts and promotes biodiversity conservation. Meanwhile, natural colorants hold profound cultural significance for Malaysia's indigenous communities. The colors are derived from natural sources, which often carry symbolic meanings and are integral to traditional ceremonies, rituals, and artistic expressions. They serve as a tangible link to cultural heritage and identity, fostering a sense of pride and continuity among Indigenous peoples. Therefore, the main goal of this study is to identify and document the Malay indigenous knowledge and technology in natural colorant by focusing on plant resources in the Malay culture, where the ethnoscience behind each usage will be further elaborated. Finally, strategies on the preservation of the Malay indigenous knowledge specifically on natural colorants. In summary, there has been a growing interest in the scientific exploration and commercialization of natural colorants. Researchers and entrepreneurs are investigating the chemical composition, properties, and potential applications of indigenous dye sources for industries such as textiles, cosmetics, and food. This interdisciplinary approach combines traditional knowledge with modern scientific methods to unlock new possibilities for economic development and technological innovation.

**Keywords**: natural colorant, Malay indigenous knowledge, pigment, ethnoscience, technology

## Introduction

For centuries, the indigenous knowledge continues to diminish among North Americans and other Indigenous peoples. However, it is observed that the diminishing rate has dramatically increased in newly industrializing countries in South America and Southeast Asia within the last one or two generations due to the rapid spread of large-scale plantation cultures and trade globalization (Teucher & Sayok, 2018). Indigenous peoples struggle to sustain the traditional method of indigenous knowledge transfer, since it is oral and practical in character. Furthermore, as demonstrated in Figure 1, indigenous knowledge is thought to become extinct sooner or later (Adam et al., 2022). As Indigenous peoples distance themselves from their traditional identities, these challenges make it difficult to effectively transmit indigenous knowledge (Adam et al., 2021). It is critical because if information is not kept, extinction will occur, resulting in the loss of vital knowledge. Indigenous knowledge, such as in agriculture and fishing, directly contributes to a country's biodiversity conservation efforts. However, because of a lack of understanding of the relevance of indigenous systems, this information has been lost or devalued (Halim et al., 2013). Indigenous knowledge is at risk of being lost as a result of colonialism, habitat damage, and modernity. The indigenous knowledge of Orang Asli is naturalistic and ecologically friendly, where its utilization lessens the human influence on the environment while creating a sustainable development (Lambin et al., 2019).

Climate change, global warming, and increased earth's surface temperature have significant global impacts on the nature, which will speed up the process of biodiversity loss. Modern technology poses a serious threat to sustainable development due to their harmful effects on natural resources and the environment (Eyong et al., 2004). Furthermore, global food security is confronting a worldwide crisis in which the global population continues to expand at an exponential pace and is predicted to reach 9 billion by the year 2050; biodiversity is thought to be lost almost as quickly, thus, the human's sources of nourishment would derive from an ever-decreasing number of species (Teucher & Sayok, 2018). In addition to biodiversity loss, such conditions sometimes result in conflict between governments and indigenous communities. Conditions in each nation will impact the lifestyle and quality of Indigenous peoples, as well as the type of ethnobiology study carried out in the individual countries (Hidayati, 2017). Indigenous peoples' rights have been widely accepted since the United Nations Declaration on the Rights of Indigenous Peoples (UN, 2008). Issues on the rights of Indigenous peoples, such as protecting traditional knowledge, human and land rights, and intellectual property, have garnered more attention. Malaysia has yet to develop a specific policy on indigenous knowledge despite the ratified international treaties that recognize the rights of Indigenous peoples to protect their knowledge systems, such as the International Covenant on Economic, Social, and Cultural Rights, the Convention on Biological Diversity, the Rio Declaration on Environment Development, and the United Nations Declaration on the Rights of Indigenous Peoples (Adam et al., 2022). However, Malaysian National Culture Policy was endorsed by the National Culture Congress in 1971. The formation of the National Culture Policy is based on three (3) principles, as well as other factors, as specified by the government (Government Policy, 2019):

1. The national culture must be based on the indigenous culture of this region;
2. Suitable elements from the other cultures may be accepted as part of the national culture;
3. Islam is an important component in the formulation of the national culture.

**Figure 1.** Among the issues encountered on indigenous knowledge.

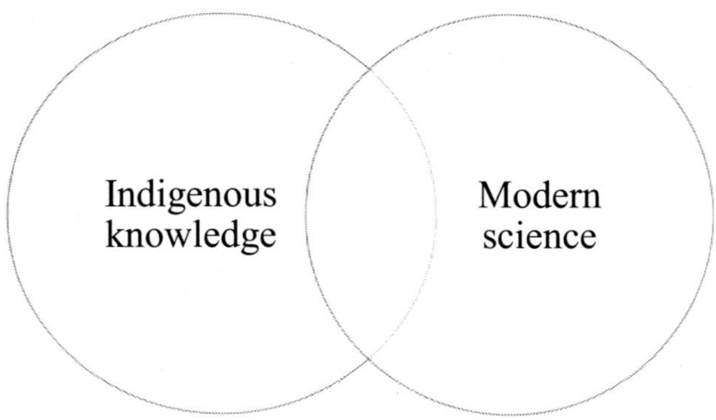


**Figure 2.** Relationship between indigenous knowledge and modern science.

Although contemporary science is widely regarded as having one of the most vital roles today, it cannot address the challenges of the developing world without collaborating with indigenous wisdom. Indigenous knowledge must be encouraged and developed, since it is a very sensitive subject that is linked to the culture and ethnicity of the stakeholders (Haron & Hamiz, 2014a). Because of modernity and progress, the majority of the traditional community no longer consumes indigenous foods. In the past, the community made full use of natural resources to subsist and lived happily and genuinely without excuses (Lambin et al., 2019).

According to Haron & Hamiz (2014b) and Figure 2, the intersection is referred to as the common ground, which can connect contemporary science to indigenous wisdom. The interaction of indigenous knowledge with modern science is one of the acceptable approaches to preserve and protect variety through integration, which benefits both scientists and locals interested in preserving it. The convergence of indigenous knowledge and modern science creates opportunities for collaboration and mutual gain. Supporting and spreading indigenous knowledge is critical because it is inextricably linked to cultural and ethnic identities. Subsequently, the integration of traditional customs with contemporary science can provide a common ground for biodiversity conservation and community sustainability.

## Malay Indigenous Knowledge

Indigenous knowledge is the distinct, traditional, local knowledge that exists within and has evolved around the specific conditions of indigenous women and men in a certain geographical location (Grenier, 1998). It was used interchangeably with traditional and local knowledge to distinguish information acquired by a specific community from knowledge generated by universities, government research institutions, and private enterprise (Warren 1992). The practice of documenting indigenous knowledge across indigenous groups is not new. The cultural applications of plants and animals have been extensively recorded, mostly via ethnobotanical and ethnocultural study (Maden et al., 2008). Ethnobotany was formerly known as aboriginal botany Powers (1874), but Harshberger used the term "ethnobotany" in 1896 to refer to the documentation of plant applications. According to ethnobotanists (Balick & Cox, 2020), "the relationship between plants and people is profound, affecting nearly every aspect of our lives." "Plants have had a significant impact on the human condition from its inception. Indigenous knowledge systems can help in the production of new crop types." In any civilization, the interaction of individuals and their surroundings results in the formation of a system of norms and concepts reflected in vocabulary, and societies often have a far larger vocabulary than scientists might assume (Evans and Levinson, 2009). According to Rajasekaran et al. (1991), indigenous knowledge (IK) varies from Western or scientific knowledge in terms of:

1. Substantive grounds: Differences in the subject matter and characteristics of IK and Western knowledge;
2. Methodological and epistemological grounds: Forms of knowledge employ different methods to investigate reality; and
3. Contextual grounds: Traditional IK is more deeply rooted in its environment.

In 2004, the United Nations Department of Economic and Social Affairs defined indigenous communities, peoples, and nations as, "those which, having a historical continuity with pre-invasion and pre-colonial societies that developed on their territories, consider themselves distinct from other sectors of the societies now prevailing on those territories, or parts of them" (Soh & Omar, 2012). For many centuries, Indigenous peoples have continually observed and confirmed the uses and advantages of plants, whether they were (semi)nomadic hunters and gatherers, farmers, or urbanized. This information has been passed down via generations of oral tradition; nevertheless, bio scientists, botanists, pharmacists, anthropologists, and other researchers have lately begun to thoroughly document Indigenous plant knowledge and usage (Balick & Cox, 2020; Martin, 2012).

The Malay society has a longstanding history spanning thousands of years. In a significant discovery made in 2023 by Assoc. Prof. Dr. Zuliskandar Ramli and his team from the National University of Malaysia, a prehistoric human skeleton was unearthed in Nenggiri, Kelantan, Malaysia. This finding suggests an even older existence than the Silver Man, previously thought to have lived between 10,000 to 11,000 years ago (Bernama, 2023). Notably, in 1991, another significant discovery was made in Perak, Malaysia, at Gua Gunung Runtuh, where a human skeleton dating back to around 11,000 BC was found, indicating the presence of a developed society prior to Western contact. This discovery challenges previous theories suggesting that the earliest settlers were Mesolithic groups arriving from Indochina between 5,000 to 3,000 BC (Soh & Omar, 2012). This recent discovery further solidifies the evidence supporting the Malays' status as the indigenous inhabitants of the Malay Archipelago. Within Malay society, indigenous knowledge encompasses spiritual values (including religion and customs), metaphysical understanding (*ilmu kebatinan*), natural sciences, and the ability to adapt to change and maintain social order. This indigenous knowledge serves to unite the community through a shared normative framework (Soh & Omar, 2012). It is transmitted through various oral traditions, such as folklore, traditional songs and dances, myths, and rituals (Adam et al., 2022). For instance, the poem advocating for following good advice and setting a positive example (*Ikut nasihat dan teladan yang baik*) authored by Saludin & Raihan (2008).

Dibilang nasihat pada yang bermanfaat,
Dibilang pertuah pada yang berfaedah,
Dibilang contoh pada yang senonoh,
Dibilang teladan pada yang sepadan,
Supaya memakai pada yang sesuai,
Supaya makan pada yang tertelan,
Supaya duduk pada yang elok,
Supaya di akhirat beroleh rahmat,
Dicuraikan pertuah dengan amanah,
Dicuraikan pesan dengan amanat,
Supaya hidup beroleh berkat,
Supaya mati beroleh rahmat.

This poem underscores the richness of customary tradition (*Adat Perpatih*), which is brimming with valuable admonitions. Sincere and earnest practice of these traditions promises numerous blessings. It offers abundant advice, exemplars, and practical guidance. Adhering to these traditions enables us to progress while upholding religious laws, etiquette, morality, truth, and consensus. Neglecting traditions leaves one spiritually empty, devoid of genuine love for religion, country, and homeland (Saludin & Raihan, 2008).

The majority of Malays reside in rural areas, engaging in agricultural pursuits and upholding their own customs (*ada*t) and code of conduct (*adab*). Traditionally, they are renowned for their proficiency and expertise in various artisanal crafts, such as woodworking, bamboo and metal crafting, pottery, *batik* printing, and *songket* weaving, among others (Soh & Omar, 2012). The essence of indigenous knowledge in traditional medicine emphasizes a comprehensive approach to wellness, addressing spiritual, emotional, mental, and physical health aspects. This knowledge can also empower traditional communities to adapt to and mitigate future environmental changes through locally relevant adaptation strategies (Adam et al., 2022).

For instance, the Sabah Museum initiated documentation efforts primarily focused on ethnobotanical knowledge among 30 ethnic groups. Another notable indigenous practice extensively documented is the *tagal* or *managal sungai*, a community-led river management system. The fundamental principle of this practice revolves around prohibiting the exploitation of river resources for a designated period, aiming to ensure a sustainable fish population in the river. Each community sets its own rules and regulations within the *tagal* area (Foo, 2011). The success of the *tagal* practice in long-term river conservation led the Sabah Fishery Department to adopt it as a conservation system statewide (Wong, 2009). Sabah's Indigenous peoples possess their own traditional systems, inadvertently contributing to biodiversity conservation (Halim et al., 2013). Traditional medicinal plants hold cultural significance among the Dusun and Bajau communities, prompting them to safeguard their heritage. While many Dusun and Bajau individuals are open to sharing their knowledge, this does not imply consent for non-indigenous exploitation and profit from indigenous resources (Adam et al., 2022).

## Malay Ethnoscience in Natural Colorants

Scholars who have delved into the early history of Malay ethnoscience and ethnotechnology reveal that the Malay World, comprising Malaysia, Indonesia, and the Philippines, boasts a rich heritage in these fields, contributing original insights to the early history of technology (Wan Ramli & Shaharir, 1999). Ethnoscience encompasses the unique set of concepts, propositions, and theories specific to each cultural group worldwide (Meehan, 1980). It is inherently interdisciplinary, fostering collaboration between social sciences and humanities (such as anthropology, sociology, history of science, psychology, and philosophy) with natural sciences, such as biology, ecology, agronomy, climatology, astronomy, and medicine (Ingold, 2000). Ethnoscience encompasses diverse perceptions and categorizations of the world within different cultures. The sustainability aspect of blending traditional and modern inventions profoundly influences the relevance of ethnoscience in preserving cultural beliefs and social cohesion (Kartini Aboo

& Nurshuhada, 2020). Crump (2002) defines science as the systematic pursuit of knowledge about the natural world, developed through speculation, observation, and experimentation, resulting in objective laws and explanations of natural phenomena. Meanwhile, Shaharir (2002), in Introduction to the History and Philosophy of Science, describes science as the systematic, logical, and objective analysis of phenomena using specific methods aimed at establishing reliable knowledge. From the Malay worldview perspective, Malay natural science involves observing phenomena within the Malay environment, leading to the formulation of specific laws through methods or experiments conducted by the Malays (Muhamad Shafiq et al., 2020).

Color holds immense significance among the Malay people and is prominently featured in various artifacts. Natural pigments, sourced from elements found in the Malay community's environment, are commonly utilized to dye materials integrated into daily life, including communication, rituals, ceremonies, and medicine. Dating back to prehistoric times, color has been recognized for its ability to create visual appeal and evoke emotions and ideas (Kumar & Sinha, 2004; Wahdina et al., 2021). It permeates one's surroundings, from clothing to natural and man-made objects, with natural sources such as plants, trees, lichens, insects, and mollusks serving as traditional colorants until the nineteenth century (Samsami et al., 2020).

Ancient Malays placed great importance on the symbolism and qualities of colors. Color plays a significant role in Malay culture, where specific hues are believed to convey security and ward off misfortune. Local traditions have shaped distinctive color palettes among Malays, reflecting deep-rooted meanings and values intertwined with their lifestyle, economy, and environment. Consequently, the Malay community has cultivated a diverse array of traditional handicrafts (Ismail & Othman 2020).

Moreover, color holds pivotal importance across various industries, notably in food, handicrafts, and textile manufacturing. It enhances the aesthetic and market value of products while influencing consumer perception and decision-making based on first impressions (Prayudi Eko et al., 2013).

The term "dye" refers to a coloring substance used to stain various materials. Natural dyes, discovered around 2600 BC, initially involved blending natural pigments with water or oil bases. Fabric dyeing has been integral to human civilization for over 4000 years, with ancient Egyptians notably using various dyes to enhance textiles. The advent of synthetic chemistry led to the discovery of synthetic dyes, offering a broader spectrum of colors and greater affordability. However, concerns arose over their environmental and health impacts due to non-biodegradability and carcinogenic properties, prompting a shift towards eco-consciousness and sustainable alternatives.

Natural dyes have gained traction as ethical and environmentally sustainable options in textile production, offering non-toxic, hypoallergenic, and eco-friendly attributes (Kadolph, 2008). As consumer demand for sustainability rises, the textile industry faces pressure to prioritize natural colorants to mitigate the use of toxic synthetic dyes. Natural dyes find applications beyond textiles, including food substrates, medications, paper, leather, cosmetics, paint, and ink, reflecting Malay local wisdom and resources in the ethnoscience concept of natural dye.

***Natural Colorants in Malay Textile***

The traditional knowledge of utilizing plants for textile dyeing is facing decline due to modernization, including shifts in lifestyles, urbanization, and the widespread use of synthetic colors. Nonetheless, there is a resurgence in textile dyeing with local plants, driven by ecotourism and global interest in natural products (Junsongduang et al., 2017). Color holds significant cultural meaning and application in Malay society, with various hues symbolizing different aspects of identity and status (Lestari & Riyanti, 2017). Natural dyes have evolved into colors or dye sources used by Malay weavers to improve the appeal of basic weaves in textiles, especially *limar* cloth, *cindai*, and *songket*. Natural dyes are not only attractive, but also play a significant role in distinguishing different degrees of hierarchy in Malay social system (Ismail, 2006). The color derived from various sources of plants. Leaf extract from *Artocarpus heterophyllus* (jackfruit) is also used in traditional textiles to give silk threads a light brown color ( Ismail, 2006). Local dyers like to dye their yarn with the red color extract from *Bixa orellana* (annato) seeds (Aziz, 2006). On the other hand, natural dyes, such as green from pandan leaves and yellow from *Curcuma longa* (turmeric), are obtained from nature for textile dyeing (Faridah, 2006). Meanwhile, *gambir* dye on *batik* fabric produces a brown color with different intensity levels depending on the mordant used. Lime mordant and iron treated fabric have stronger color strength than alum treated fabric (Sofyan et al., 2021). Weavers in Sambas and Sintang use 36 dye plant species, including *Psychotria megacoma, Symplocos ophirensis, Litsea angulata, Eusideroxylon zwageri* and *Fibraurea chloroleuca*, which are native to West Kalimantan (Muflihati et al., 2019). For the test of light colorfastness, *Morinda citrifolia* roots are the most successful compared to *Neesia* spp., *Tristaniopsis* spp. and *Psychotria viridiflora* leaves, with a greyscale rating between 4/5 and 5 (Ngo et al., 2013). By adding distinctive value to traditional weaving, the use of natural dyes can boost economic worth, particularly in the international market. In addition, natural fiber particularly wool and silk have better affinity with natural dyes in color absorption than synthetic fiber.

The use of natural dyes adds value to traditional weaving, boosting economic prospects, especially in the international market. Moreover, natural fibers like wool and silk have better affinity with natural dyes, enhancing color absorption compared to synthetic fibers. In 2017, textile-based merchandise contributed significantly to Malaysia's craft industry, with *batik* merchandise leading sales (Malaysian Handicraft Development Corporation, 2017). However, the number of practitioners using natural dyes has dwindled due to the labor-intensive process of extraction, involving chopping, grinding, soaking, and boiling of plant materials (Ngo et al., 2013).

In the Malay dyeing process, suitable parts of raw materials are selected and processed before extraction. Modern technology offers various extraction methods, such as organic solvents and ultrasonic-assisted extraction, which are more efficient than traditional aqueous approaches (Saxena & Raja, 2014). These methods yield greater color extracts in less time, though they may affect dye stability and composition (Hassan & Rahim, 2021). Plant pigments, including chlorophyll, carotenoids, anthocyanins, and betalains, are subject to alteration and degradation based on their chemical structures (Othman et al., 2017). Despite challenges, advancements in extraction processes offer opportunities to streamline production and preserve traditional knowledge of natural dyeing in the Malay culture.

### *Natural Colorants in Malay Gastronomy*

Knowledge of food is acquired through various avenues, including teachings from parents, family members, and community elders, as well as through social and festive gatherings. It encompasses the transmission of skills related to the preparation of traditional foods, deeply rooted in cultural identity, wisdom, practices, and technologies of different tribes (Md. Sharif et al., 2018; Vira et al., 2015). Traditional food serves as a vital cultural element, reflecting local identity, particularly evident in festive foods that hold significant cultural symbolism (Md. Sharif et al., 2018). Malay heritage cuisine draws influence from a diverse range of cultures, including Arab, Indian, Chinese, Siamese, Javanese, and Minangkabau, resulting in dishes characterized by spiciness and the abundant use of aromatic herbs and spices (Raji et al., 2017). Malay food by five characteristics as mentioned in Figure 3 (Ainuddin, 2013).

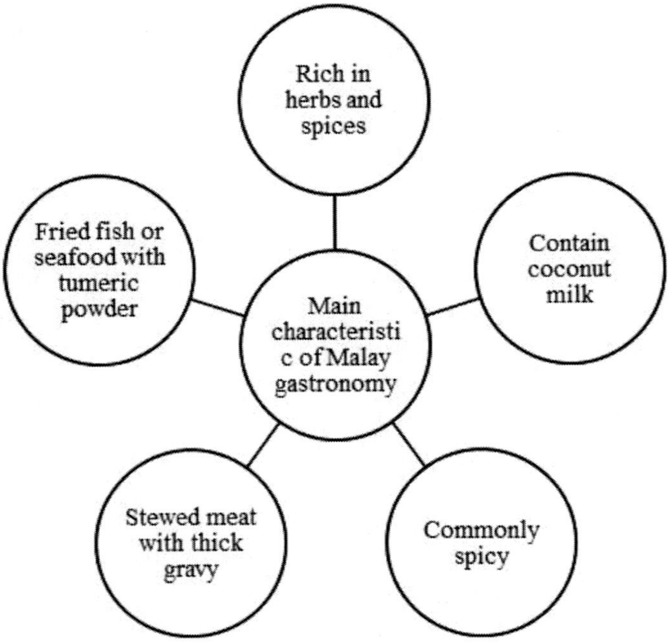


**Figure 3.** Malay gastronomy characteristics.

Various combination of ingredients and color in Malay food may influence the food aesthetic and presentation besides a rich of flavor. The use of kaffir lime leaves, torch ginger, and mint leaves in rice dishes, such as *nasi kerabu* (rice mix with herbs) in Kelantan (Raji et al., 2017). Nutritional studies have indicated that many of these vegetables are rich in carbohydrates, protein, minerals, and vitamins (Ismail, 2000). Some *ulam* is also claimed and traditionally believed to have medical properties. The most common *ulam* varieties consumed by Malay communities include *Centella asiatica (daun pegaga), Oenanthe javanica (daun selom), Cosmos caudatus (ulam raja), Anacardium occidentale (pucuk gajus), Manihot utilissima (pucuk ubi kayu), Piper sarmentosum (daun kaduk), Barringtonia racemose (daun putat)*, and Anacardium occidentale *(pucuk gajus).* Interestingly, in *nasi ulam,* a traditional dish famous in north of Peninsular Malaysia contain almost 40 types of *ulam*. Rebung and umbut are normally cooked in coconut milk, and are called as *masak lemak* (yellow curry) among the Malay communities derived from turmeric may give a natural color to the food. Nor et al. (2012) highlighted the wide use of different rhizomes, namely, *Zingiber officinale* (ginger), *Curcuma longa* (turmeric), *Alpinia officinarum* (galangal), and

*Alpinia officinarum* (lesser galangal), and aromatic leaves such as Persicaria odorata (Vietnamese coriander), *Citrus hystrix* (kaffir lime leaves), *Coriandrum sativum* (coriander leaves), *Pandanus amaryllifolius* (screwpine leaves) and *Murraya koenigii* (curry leaves) in the preparation of Malay dishes.

Vitamin A is an essential nutrient for cell proliferation, the immune system function, and vision (Semba, 1998). Vitamin A deficiency is defined as a public health problem if ≥ 15% of a defined population has a retinol concentration < 0.70 μmol/L (De Pee & Dary, 2002). In Malaysia, Vitamin A nutritional needs are vary; infants (375-400 μg), children (400-500 μg), adolescent (600 μg), adult (600 μg), elderly 600) pregnancy (800 μg) and lactation (850 μg) (Tee et al., 2023). The majority of Malaysian seniors or old folks enjoy eating traditional vegetable or *ulam* as a side dish with rice because of their bitter flavor and utilize them as supplements or alternative medicines to enhance their health (Nor Asiah Awang et al., 2020). About 120 species of herbs and trees are recognized as Malay traditional vegetables (consisting of plants or parts of plants) (Bahri et al., 2019). Traditional vegetables (*ulam*) are rich in polyphenols, antioxidants and fibers which could enhance nutritional status and mood state (You et al., 2020). *Ulam* is fresh green vegetables that are generally consumed in uncooked form or after a short blanch in boiled water (Faujan et al., 2007). The increase in the intake of *ulam* by at least 30 g/day was associated with greater muscle mass, reduced waist circumference, better mood state and cognitive function. The most common *ulam* consumed by Malaysian population were *Centella asiatica* (*pegaga*), *Parkia speciosa* (*petai*), *Oenanthe javanica* (*selom*) and *Cosmos caudatus* (*ulam raja*) (You et al., 2020). In Terengganu, the cinsumption of *ulam* such as *Anacardium occidentale* (gajus), *Syzygium polyanthum* (daun salam*), Leucas zeylanica* (ketumbak)*, Crateva religiosa* (dalur/ kepayang air), *Oroxylum indicum* (beka/ bonglai kayu) and *Musa balbisiana* (pisang abu) as healthy foods in order to obtain the medicinal effects for treating diseases such as diabetes, hypertension, constipation, less food appetite, arthritis and diarrhea (Awang et al., 2016). Nor Asiah Awang et al. (2020) reported *ulam* including are *Cleome gynandra* (maman), *Morinda citrifolia* (mengkudu besar), *Morinda elliptica* (mengkudu kecil), *Clausena excavata* (cemumar) and *Vitex pubescens* (halban) which mostly used for increasing food appetite, anti-aging, postpartum, menstrual tonic, and women contraceptive as well as diabetes and helminthic infection treatments. From 10 *ulam* species (*beko, beluntas, cekur manis, daun selom, kaduk, mengkudu, pegaga, pucuk gajus, tenggek burung and ulam raja*) studied, five main carotenoids identified namely lutein, neoxanthin, violaxanthin, zeaxanthin, and β-carotene (Fatimah et al., 2012). Othman et al. (2017) reported that the highest carotenoid content in 28 *ulam* studied, *cekur manis* is the highest. Lutein and zeaxanthin are xanthophylls known to prevent macular degeneration. The pro-Vitamin A is found in plant products with red, orange, and yellow colors, and converted by an antioxidant into Vitamin A (Chandan et al., 2023).

### *Natural Colorants in Malay Beauty and Personal Care*

For centuries, plant pigments like curcumin, beet anthocyanins, carotenoids from peppers and saffron, and chlorophyll from green leaves have been utilized to add color to both food and cosmetics. In Malaysia, 51 plant species have been identified as sources of natural dyes, predominantly employed for body paints by the Orang Ulu, an informal collective term for various ethnic groups on Borneo Island (Rosli et al., 2015). Thornfeldt (2005) defines cosmetics as encompassing products such as shampoo, facial makeup, skin moisturizer, toothpaste, hair color, deodorant, and any constituent utilized in cosmetic formulations. The designation "natural" in food product labeling is permitted by the U.S. Department of Agriculture if the product contains no artificial flavors, colors, chemical preservatives, or synthetic ingredients (Peng et al., 2014).

Active ingredients sourced from natural origins, including plants, animals, and microorganisms, have long been incorporated into topical creams and lotions (Fowler Jr. et al., 2010). Malaysia, with its rich plant biodiversity, has a tradition spanning decades of utilizing natural ingredients for various purposes, including enhancing beauty. The bioactive compounds found in these natural ingredients offer valuable properties such as antioxidants, photoprotection, anti-aging, and anti-inflammatory effects, making them beneficial for skincare, haircare, and dental care (Mohd-Nasir & Mohd-Setapar, 2018).

*Lawsonia inermis* (henna) is widely used in cosmetology for its dyeing properties, uses as hair and nail color and in the traditional decoration of the soles of the feet and of the palms of the hands. The henna leaves are pounded until smooth, then applied to the nails, and wrapped with cloth about ± 4 hours, before clean it with clean water (Prakash & Majeed, 2015; Styawan et al., 2016). Interestingly, an ointment prepared from the henna leaves can treat wounds and ulcers. *Caesalpinia sappan* (sappanwood) extract can be incorporated to preserve the integrity of formulations and into cosmetic creams, gels, and lotions to provide antibacterial benefits. Further, it lends bright color and aesthetic value to bath soaps, and color cosmetics. According to Styawan et al. (2016), plants such as *Cananga odorata* (*kenanga*) use for body perfume by crushing the leaves until smooth and then applied to the entire body before rinsed with clean water, *Oryza sativa* (rice) applies for facemask by pounding the rice until smooth, mixed with a little water before applied evenly on the face and left about10 minutes, and then rinsed with clean water. Another species namely *Areca catechu* (betel nut) used to whiten the tooth and lipstick. The dried betel nut skin is burned before rubbed on the teeth and cleaned with clean water afterward. Meanwhile for lipstick usage, the betel nut is chopped and then pounded until smooth then mixed with gambier that has been pounded and wrapped with betel leaves that have been smeared with lime before applied to the lips. *Piper betle* (*sirih*) also been use for lipstick by mixing the leaves, lime with betel nut that has been finely pounded with little gambier that has pounded before chewed. *Pandanus amaryllifolius* (*pandan wangi*) use for hair tonic by mixing the leaves with coconut oil. Other than that, *Uncaria gambier* (*gambir*) also can be applied as lipstick, *Cucurma xanthorrhiza* (*temulawak*) to remove black head on the face. *Bixa orellana* (annatto) extract can apply for cosmetic for the coloring of food, pharmaceutical and cosmetic products, to give yellow to red hues (Meena Devi et al., 2013). Based on the study, leaves are the most utilized plant part as cosmetic ingredients, accounting for 57.93%, followed by the fruit (34.92%) while the least utilized organ is rhizome at 7.14% (Styawan et al., 2016).

Furthermore, the market potential for natural colors in cosmetics is expanding globally due to heightened consumer awareness regarding the adverse effects associated with prolonged usage of certain synthetic coloring compounds. Additionally, there is a current trend favoring the incorporation of healthful, natural ingredients in cosmetic formulations.(Prakash & Majeed, 2015). Grand View Research Inc. (2018) forecasted that the global organic personal care market will reach approximately USD 25.11 billion by 2025.

### *Natural Colorants in Malay Craft*

The *batik* industry in Malaysia and Indonesia is dominated by two types of production techniques, that is, hand-drawn *batik* (*batik tjanting*) and *batik* stamp (*batik terap*/blok). During the early 20th century, traditional dyes began to be phased out in favor of chemical dyes imported from Britain and Germany (Syed Shaharuddin et al., 2021). Recent data obtained from Kraftangan Malaysia revealed that the total number of registered *batik* stamp and hand-drawn *batik* entrepreneurs in 2019 was only 651 and is mainly located in Kelantan and Terengganu (Syed Shaharuddin et al., 2021). However, compared to Indonesia, the population of Malaysian *batik* entrepreneurs remains relatively small. Currently, both Indonesia and Malaysia's *batik* industries utilize a combination of natural and synthetic dyes. Traditionally, natural dyes were derived through simple extraction methods from plants like Indigofera tinctoria for indigo dye, *Peltophorum pterocarpum* bark for sogan-yellowish-brown hue, and the root bark of *Morinda citrifolia* for *mengkuda*-dark red hue. Blending these colors, such as indigo with sogan or *mengkuda*, could result in green or purple hues, respectively (Syed Shaharuddin et al., 2021). *M. citrifolia*, has long been employed as a dye for *batik*. The presence of morindin and morindon compounds in the root and bark of *M. citrifolia* makes it a viable source for textile dyes (Walburga Bui & Lapailaka, 2022).

In Pahang, the Semai tribes' way of life is intimately linked with the forest, and they hold their own distinct perspectives on forest resources. They utilize woven bamboo, rattan, or Nipah for roof and wall decorations in their homes. Similarly, the Temuan tribes in Selangor exhibit their unique handicrafts, which reflect the natural environment and emphasize the importance of environmental sustainability (Lambin et al., 2019). This discovery highlights the remarkable skill and knowledge of the Orang Asli tribes in identifying and utilizing raw materials for crafting purposes. They have preserved a tradition of spiritual woodcarving that is of exceptional quality on a global scale. Additionally, the Dayak Iban community primarily employs natural dyes for yarn dyeing in weaving practices. Rattan from *Curculigo latifolia* (known as *lumbah bukit*) is utilized for craft painting, and its products are marketed and utilized in various cultural occasions (Wahdina et al., 2021). The bark from *Xylocarpus* sp (*nyireh*) which inhabit in mangrove area can produce red color or called as tannin for fishing net (Alfa et al., 2018). The skill of weaving Nipah leaves is widespread across the states of Sarawak, Terengganu, and Selangor. Weaving involves the technique of intertwining sheets or strips of leaves by arranging and weaving them together. A variety of plant materials are utilized for weaving, including pandan leaves, *mengkuang* leaves, *sal* leaves, *serdang* leaves, tar leaves, *nipah* sticks, bamboo, *bemban*, rattan, *ribu-ribu*, reeds, *salingkawang*, and others. These weaving techniques are practiced by diverse communities, including the Malay, indigenous, and ethnic groups in Sabah and Sarawak, each showcasing their distinct local identity (Malaysia Handicraft, 2021).

## Strategies in Preserving Malay Indigenous Knowledge on Natural Colorants

There are numerous challenges, limitations, and threats that must be addressed to preserve indigenous knowledge. One of these challenges is the influence of modern technology and education, where new educational systems often overlook the importance of protecting indigenous knowledge (Ngulube, 2002). Additionally, the oral transmission of indigenous knowledge from one generation to the next without proper documentation poses a risk of it fading away (Msuya, 2007). Therefore, it is crucial to undertake proper documentation and conservation measures to preserve the knowledge of natural dye-producing plants (Das & Mondal, 2012). Implementing the Malaysian-Malay-Islamic values system in the teaching and education of science and technology, as well as empowering indigenous scientists and technologists, can help address these challenges (Wan Ramli & Shaharir, 1999). The application of scientific principles alongside indigenous knowledge not only aids in conserving our environment but also contributes to the well-being of nature and society. For instance, indigenous communities in Sabah serve as excellent examples of establishing biodiversity conservation and fulfilling environmental obligations (Halim et al., 2013).

Another challenge pertains to the time-consuming nature of collecting indigenous knowledge, which requires significant labor resources and can be costly to maintain (Lawas & Luning, 1996). Making natural dyes demands time, energy, and patience, as highlighted by Casselman (1993), and achieving consistent colors across dyebaths is often impossible due to the rapid depletion of raw materials and resources (Junsongduang et al., 2017). It is crucial to consistently educate the public on the traditional methods of making *batik* to prevent the loss of familiarity with authentic *batik* fabric, which could diminish its appreciation as an Intangible Cultural Heritage (ICH) (Syed Shaharuddin et al., 2021). Furthermore, the market price of raw materials affects the sustainability of products, as entrepreneurs have limited control over their product costs (Yoshanti & Dowaki, 2017). Traditional techniques are finding it increasingly challenging to attract youths to participate in the local *batik* industry (Syed Shaharuddin et al., 2021). Malaysia is experiencing a decline in craftsmen specializing in the fabrication of *batik* stamps from metals (Lias et al., 2020). Additionally, the deskilling of the younger generation in producing traditional dishes, especially festive food, poses a threat to the continuity of Malay culinary tradition (Md. Sharif et al., 2018).

Furthermore, each aspect of indigenous knowledge requires protection of Intellectual Property (IP). Various intellectual property rights (IPRs) can be applied to achieve this. In Malaysia, the domestic IPRs applicable for safeguarding local *batik* designs include the Copyright Act of 1987 (CA 1987) and the Industrial Designs Act, 1996 (IDA 1996) (Nordin & Bakar, 2012). Recent research has highlighted the beneficial effects of incorporating Fourth Industrial Revolution (IR4.0) technologies such as cloud computing/ manufacturing, cyber-physical systems (CPSs), the Internet of Things (IoT), robotics, big data, additive manufacturing (3D printing), and augmented reality (AR) in fostering the preservation of intangible cultural heritage, industrial processes, and product interaction within a collaborative community (Nagy et al., 2018). Based on Malaysian national Culture Policy (Government Policy, 2019), stated the strategies and implementation of the policy can be achieved through:

1. Restore, preserve, and develop culture towards strengthening national culture through joint research, development, education and culture expansion and connections;
2. Increase and strengthen cultural leadership through training and guidance to interested individuals. Support and mobilize culture as an effective engine of growth;
3. Establish effective communication to instill national awareness and Malaysian nationalism;
4. Fulfil socio-cultural needs; and
5. Improve the standard and quality of arts.

As a result, addressing these research areas and recommendations can significantly contribute to the conservation of Malaysian indigenous knowledge and technology in natural colorants while fostering sustainable development and cultural preservation:

1. *Documentation and preservation*: Thoroughly document indigenous knowledge related to natural colorants, including traditional practices, recipes, and techniques, in collaboration with indigenous communities to ensure accurate representation and preservation of their cultural heritage.
2. *Biological diversity assessment*: Assess the biodiversity of plants and other natural resources used in traditional colorant production,

identifying key species, their ecological significance, and potential threats to their sustainability.

3. *Ecological impact assessment*: Evaluate the ecological impact of traditional colorant extraction methods, determining the sustainability of harvesting practices and their effects on local ecosystems, including potential habitat degradation or loss.
4. *Cultural and socioeconomic impacts*: Investigate the cultural significance of natural colorants within indigenous communities and their role in traditional crafts and ceremonies. Assess the socioeconomic benefits of traditional colorant production, including income generation and community development.
5. *Innovation and technological integration*: Explore opportunities for integrating traditional knowledge with modern technologies to enhance the efficiency and sustainability of natural colorant production. This may involve researching new extraction methods, processing techniques, or product development.
6. *Policy and legal frameworks*: Advocate for the recognition and protection of indigenous intellectual property rights related to natural colorants. Collaborate with policymakers to develop inclusive policies supporting the sustainable use of traditional knowledge and resources.
7. *Capacity building and knowledge exchange*: Facilitate knowledge exchange and capacity-building initiatives between indigenous communities, researchers, and relevant stakeholders. Empower indigenous artisans and practitioners through training programs, workshops, and collaborative research projects.
8. *Market development and fair-trade practices*: Promote the market demand for ethically sourced and sustainably produced natural colorants. Encourage the adoption of fair-trade practices to ensure equitable compensation for indigenous producers and respect cultural values.
9. *Environmental education and awareness*: Raise public awareness about the environmental and cultural significance of natural colorants and the importance of preserving indigenous knowledge. Develop educational materials and outreach programs to engage consumers, businesses, and policymakers.

## Conclusion

The historical evolution of indigenous knowledge among the Malay regarding natural colorants spans various aspects of their lives, encompassing textiles, cuisine, personal care, and handicrafts. This transition from traditional to modern science and technology is evident. While this paper does not delve into the philosophical significance of these practices, it is widely believed that each facet of indigenous knowledge reflects philosophical, social, and cultural values, as well as a commitment to environmental sustainability. Although numerous scientific studies have explored the functions and potential of individual or combined natural phytochemicals, further research is necessary, including clinical trials and formulation studies, to fully harness the benefits of these natural ingredients. Natural colorants consist of secondary metabolite compounds, and a deeper understanding of chemistry, including processes such as fixation, extraction, solution, alum compounds, anthocyanins, and pigments, holds promise for future applications.

Researchers and entrepreneurs are actively exploring the chemical composition, properties, and potential industrial applications of indigenous dye sources in fields like textiles, cosmetics, and food. This interdisciplinary approach harmonizes traditional knowledge with modern scientific methods, opening up new avenues for economic growth and technological advancement. In essence, Malay indigenous knowledge and technology related to natural colorants epitomize a dynamic intersection of tradition, sustainability, culture, and innovation. By acknowledging and preserving these invaluable assets, Malaysia can harness its rich cultural heritage and biodiversity to foster socioeconomic development while safeguarding the environment for future generations.

## References

Adam, A. A., Othman, N., & Ab. Halim, A. (2022). Indigenous knowledge documentation: Perspectives of Dusun and Bajau communities in Kota Belud, Sabah, Malaysia. *International Journal of Academic Research in Business and Social Sciences, 12*(10), 162–175. https://doi.org/10.6007/IJARBSS/v12-i10/14011.

Adam, A. A., Othman, N., & Halim, A. A. (2021). Indigenous knowledge transfer among Dusuns and Bajaus in Kota Belud, Sabah, Malaysia: Approaches and challenges. *International Journal of Academic Research in Business and Social Sciences, 11*(11), 1722–1734. https://doi.org/10.6007/IJARBSS/v11-i11/10461.

Ainuddin, A. H. (2013). *Malaysia cuisine for professional student*. Utusan Publication & Distribution Sdn Bhd.

Alfa, R., M. Sofwan, A., & Muflihati. (2018). Pemanfaatan tumbuhan mangrove oleh masyarakat desa Bakau Besar Laut Kecamatan Sungai Pinyuh Kabupaten Mempawah. *Jurnal Hutan Lestari, 6*(1), 62–70.

Awang, N. A., Mat, N., & Ali, A. M. (2016). Therapeutic values in selected edible plants from Besut, Terengganu, Malaysia. *International Postgraduate Research Conference*, 103.

Awang, Nor Asiah, Mat, N., & Mahmud, K. (2020). Traditional knowledge and botanical description of edible bitter plants from Besut, Terengganu, Malaysia. *Journal of Agrobiotechnology, 11*(1), 32–47.

Aziz, A. (2006). *Rupa & gaya busana Melayu*. Penerbit Universiti Kebangsaan Malaysia.

Baeren, E. J., & Jusilin, H. (2021). The uniqueness of mangrove forests in the state of Sabah as a source of natural color on batik fabric. *Proceedings of the 5th International Conference on Arts Language and Culture (ICALC 2020), 534*, 47–53.

Bahri, A. A., Salleh, W., Lani, M. N., & Abdullah, W. Z. W. (2019). Antimicrobial resistance of *Escherichia coli* isolated from ULAM from supermarkets and wet markets in Kuala Terengganu, Malaysia. *Malaysian Applied Biology, 48*(3), 35–42.

Balick, M. J., & Cox, P. A. (2020). *Plants, people, and culture: The science of ethnobotany*. Garland Science.

Barkeshli, M., Ghani, A., Shan, N. K. S., Meng, K. Y., & Shahari, A. R. (2003). *Textile weaving techniques*. In *Culture Treasures: Textiles of the Malay World*. National Museum.

Bernama. (2023, July). Penemuan rangka manusia wanita pra-neolitik. *Sinar Harian, 1*. https://www.sinarharian.com.my/article/268620/galeri/penemuan-rangka-manusia-wanita-pra-neolitik.

Casselman, K. L. (1993). *Craft of the dyer: Colour from plants and lichens*. Dover Publications Inc.: New York.

Chandan, C., Pardhe, H. A., Nagappan, K., Phani Kumar, G., Sushma, B. V., & Jeyaprakash, M. R. (2023). Formulation and stability enhancement using vitamin A encapsulation in ocular abnormalities: A scientific review. *International Food Research Journal, 30*(3), 564–576. https://doi.org/10.48200/ifrj.2023.30.3.2286.

Crump, P. (2002). *A brief history of science*. Robinson: United Kingdom.

Das, P. K., & Mondal, A. K. (2012). Biodiversity and conservation of some dye yielding plants for justification of its economic status in the local areas of the lateritic zone of West Bengal, India. *Advances in Bioresearch, 3*(1), 43–53. [No DOI available].

De Pee, S., & Dary, O. (2002). Biochemical indicators of vitamin A deficiency: Serum retinol and serum retinol binding protein. *The Journal of Nutrition, 132*(9), 2895S-2901S. https://doi.org/10.1093/jn/132.9.2895S.

Evans, N., & Levinson, S. C. (2009). The myth of language universals: Language diversity and its importance for cognitive science. *Behavioral and Brain Sciences, 32*(5), 429–448. https://doi.org/10.1017/S0140525X0999121l.

Eyong, C. T., Mufuaya, M., & Foy, I. I. (2004). Literature and culture—the sustainability connection from an African perspective. *Regional Sustainable Development Review: Africa, 1*, 1–13.

Faridah, S. (2006). *Natural dye in the art of Malaysian batik.* World Eco Fiber & Textile (Weft). World Eco Textile, Kuching, and Sarawak.

Fatimah, A. M. Z., Norazian, M. H., & Rashidi, O. (2012). Identification of carotenoid composition in selected "ulam" or traditional vegetables in Malaysia. *International Food Research Journal, 19*(2), 527–530.

Faujan, N. H., Abdullah, N., Abdullah Sani, N., & Babji, A. (2007). Antioxidative activities of water extracts of some Malaysian herbs. *ASEAN Food Research, 14*, 61–68.

Foo, J. (2011). Sistem Tagal dalam industri pelancongan: Manfaat kepada komuniti. *GEOGRAFIA Malaysia Journal of Society and Space, 7*(1), 14–25.

Fowler Jr, J. F., Woolery-Lloyd, H., Waldorf, H., & Saini, R. (2010). Innovations in natural ingredients and their use in skin care. *Journal of Drugs in Dermatology: JDD, 9*(6 Suppl), S72–81.

Government Policy. (2019). Malaysia National Culture Policy 2019. http://www.jkkn.gov.my/en/national-culture-policy.

Grand View Research Inc. (2018). Organic personal care market size, share & trends analysis report by application, by region and segment forecasts, 2019–2025. *Reportlinker*, 155. https://www.reportlinker.com/p05479695/Organic-Personal-Care-Market-Size-Share-Trends-Analysis-Report-By-Application-By-Region-And-Segment-Forecasts.html.

Grenier, L. (1998). *Working with indigenous knowledge: A guide for researchers*. IDRC.

Halim, A. A., Othman, N., Ismail, S. R., Jawan, J. A., & Ibrahim, N. N. (2013). Indigenous knowledge and biodiversity conservation in Sabah, Malaysia. *International Journal of Social Science and Humanity, 159–163.*

Haron, H., & Hamiz, M. (2014a). An ontological framework to preserve Malay indigenous health knowledge. *Advanced Science Letters, 20*(1), 226–230. https://doi.org/10.1166/asl.2014.5719.

Haron, H., & Hamiz, M. (2014b). An ontological model for indigenous knowledge of Malay confinement dietary. *Journal of Software, 9*(5), 1302–1312. https://doi.org/10.4304/jsw.9.5.1302-1312.

Harshberger, J. W. (1896). The purpose of ethnobotany. *Botanical Gazette, 21*, 146–154.

Hassan, S. R. B., & Rahim, N. A. B. C. A. (2021). Extraction of natural dye from black tea waste for cotton dyeing application. *Journal of Engineering Science and Technology, 16*(3), 2562–2570. https://jestec.taylors.edu.my/Vol%2016%20Issue%203%20September%202021/16_3_2562_2570.pdf.

Hidayati, S. (2017). The relationship between indigenous languages, traditional knowledge and biodiversity: A case study with the Ba'ie Segan people of Sarawak, Malaysia (Curtin University).

Ingold, T. (2000). *The perception of the environment: Essays on livelihood, dwelling, and skill*. Routledge: United Kingdom.

Ismail, S. (2000). *Sayuran Tradisional Ulam dan Penyedap Rasa*. Penerbit Universiti Kebangsaan Malaysia.

Ismail, S. Z. (2006). *Pakaian Cara Melayu*. Penerbit Universiti Kebangsaan Malaysia.

Junsongduang, A., Sirithip, K., Inta, A., Nachai, R., Onputtha, B., Tanming, W., & Balslev, H. (2017). Diversity and traditional knowledge of textile dyeing plants in Northeastern

Thailand. *Economic Botany, 71*(3), 241–255. https://doi.org/10.1007/s12231-017-9407-4.

Kadolph, S. (2008). Natural dyes: A traditional craft experiencing new attention. *Delta Kappa Gamma Bulletin, 75*(1), 14.

Kartini Aboo, T. K., & Nurshuhada, M. (2020). Midwives and herbal remedies: The sustainable ethnoscience. *Kajian Malaysia, 38*(1), 109–131.

Kumar, J. K., & Sinha, A. K. (2004). Resurgence of natural colourants: A holistic view. *Natural Product Research, 18*(1), 59–84. https://doi.org/10.1080/14786410410001653506.

Lambin, R., Wahab, N. A., Choo, G. S., Mustapha, R., Abdullah, R., & Sharif, R. M. (2019). Transmission and documentation of Orang Asli indigenous knowledge in 'green technology' for sustainable development in Malaysia. *International Journal of Academic Research in Business and Social Sciences, 9*(5), 194–212. https://doi.org/10.6007/IJARBSS/v9-i5/5889.

Lawas, C. M., & Luning, H. A. (1996). Farmers' knowledge and GIS. *International Journal of Geographical Information Systems, 11*(6), 511-522. https://doi.org/10.1080/136588196242116.

Lestari, S., & Riyanti, M. T. (2017). Kajian motif tenun songket Melayu Siak tradisional khas Riau. *Jurnal Dimensi DKV: Seni Rupa dan Desain, 2*(1), 33–48.

Lias, H., Ismail, A. R., & Abd Hamid, H. (2020). Malaysia textile craft industry: Innovation inspired by bamboo for batik block contemporary design. *IOP Conference Series: Earth and Environmental Science, 549*(1), 012087. https://doi.org/10.1088/1755-1315/549/1/012087.

Maden, K., Kongren, R., & Limbu, T. M. (2008). Documentation of indigenous knowledge, skill, and practices of Kirata nationalities with special focus on biological resources. *Research and Development in Sustainable Development, 15*(4), 94–106.

Malaysia Handicraft. (2021). *Annual Report Kraftangan Malaysia 2021.* http://www.kraftangan.gov.my.

Malaysian Handicraft Development Corporation. (2017). *The Malaysian Handicraft Annual Report 2017.*

Martin, D. (2012). Nutrition transition and the public-health crisis: Aboriginal perspectives on food and eating. *Critical Perspectives in Food Studies, 2012*, 208–222.

Md. Sharif, M. S., Abdul Rahman, A. E., Mohd Zahari, M. S., & Mohamad Abdullah, K. (2018). Malay traditional food knowledge transfer. *Asian Journal of Quality of Life, 3*(10), 79–88. https://doi.org/10.21834/ajql.3.10.79.

Meehan, P. (1980). Science, ethnoscience, and agricultural knowledge utilization. In D. M. Warren, D. Brokensha, & O. Werner (Eds.), *Indigenous knowledge systems and development* (pp. 95-112). University Press of America.

Meena Devi, V. N., Ariharan, V. N., & Nagendra Prasad, P. (2013). Annatto: Eco-friendly and potential source for natural dye. *International Research Journal of Pharmacy, 4*(6), 106–108.

Mohd-Nasir, H., & Mohd-Setapar, S. H. (2018). Natural ingredients in cosmetics from Malaysian plants: A review. *Sains Malaysiana, 47*(5), 951–959.

Msuya, J. (2007). Challenges and opportunities in the protection and preservation of indigenous knowledge in Africa. *The International Review of Information Ethics, 7*, 346–353.

Muflihati, Wahdina, Kartikawati, S. M., & Wulandari, R. S. (2019). Natural dye plants for traditional weaving in Sintang and Sambas Regencies, West Kalimantan. *Media Konservasi, 24*(3), 225–236.

Muhamad Shafiq, M. A., Ahmad Bazri, M., & Ros Mahwati, A. Z. (2020). Ilmu etnosains dalam manuskrip alam Melayu (ethnoscience in the Malay manuscripts). *Jurnal Arkeologi Malaysia, 33*(Special Issue), 43–50.

Nagy, J., Oláh, J., Erdei, E., Máté, D., & Popp, J. (2018). The role and impact of Industry 4.0 and the internet of things on the business strategy of the value chain—the case of Hungary. *Sustainability, 10*(3491). https://doi.org/10.3390/su10103491.

Ngo, J. S. K., Ong, W. F., Ahmad, F. B., & Bujang, K. B. (2013). A study of soluble-powdered natural dyes. *Research Journal of Textile and Apparel, 17*(1), 104–112. https://doi.org/10.1108/RJTA-17-01-2013-B010.

Ngulube, P. (2002). Managing and preserving indigenous knowledge in the knowledge management era: Challenges and opportunities for information professionals. *Information Development, 12*(8), 95–102. https://doi.org/10.1177/026666690201200806.

Nor, N. M., Sharif, M. S. M., Zahari, M. S. M., Salleh, H. M., Isha, N., & Muhammad, R. (2012). The transmission modes of Malay traditional food knowledge within generations. *Procedia - Social and Behavioral Sciences, 50*, 79–88. https://doi.org/10.1016/j.sbspro.2012.08.027.

Nordin, R., & Bakar, S. S. A. (2012). Malaysian batik industry: Protecting local batik design by copyright and industrial design laws. *International Journal of Business and Society, 13*(2), 117–132.

Othman, R., Noh, N., Nurrulhidayah, A. F., Anis Hamizah, H., & Jamaludin, M. A. (2017). Determination of natural carotenoid pigments from freshwater green algae as potential halal food colorants. *International Food Research Journal, 24*, 468–471.

Othman, Rashidi, Zaifuddin, F. A. M., & Hassan, N. M. (2017). Characterisation of carotenoid and total retinol equivalent content in ulam and medicinal species as alternative food intervention to combat Vitamin A deficiency. *Journal of Pharmacy and Nutrition Sciences, 7*(3), 81–87. https://doi.org/10.6000/1927-5951.7.3.8.

Peng, W. L., Khanafi, M. A., Setapar, S. H. M., Idham, Z., Yunus, M. A. C., Abbas, M., & Zaini, A. (2014). Development of emulsification containing natural colorant from local plant (Roselle). *Jurnal Teknologi (Sciences and Engineering), 69*(4), 15–17.

Powers, S. (1874). Aboriginal Botany. *California Academy of Sciences, 373–379.*

Prakash, L., & Majeed, M. (2015). Natural actives lend color to cosmetics and more. *CosmeticsDesign-Europe. Com, 1*(1), 16.

Prayudi Eko, S., Yudha, S., & Bakti, J. (2013). Optimalisasi ekstraksi dan uji stabilitas phycocyanin dari mikroalga *Spirulina platensis*. *Jurnal Teknologi Kimia Dan Industri, 2*(2), 61–67.

Rajasekaran, B., Warren, D. M., & Babu, S. C. (1991). Indigenous natural resource management systems for sustainable agricultural development – A global perspective.

*Journal of International Development, 3*(3), 387–401. https://doi.org/10.1002/jid.3380030307.

Raji, M. N. A., Ab Karim, S., Ishak, F. A. C., & Arshad, M. M. (2017). Past and present practices of the Malay food heritage and culture in Malaysia. *Journal of Ethnic Foods, 4*(4), 221–231. https://doi.org/10.1016/j.jfs.2017.08.002.

Rosli, N., Ismail, Z., & Lepun, P. (2015). Plants used as natural dye by the Orang Ulu ethnics in Asao Koyan Belaga Sarawak, Malaysia. *International Journal of Current Research, 7*(08), 19770–19775.

Saludin, M. R., & Raihan, S. H. (2008). Ikut nasihat dan teladan yang baik. *Malaycivilization*. http://malaycivilization.com.my/items/show/101448.

Samsami, S., Mohamadi, M., Sarrafzadeh, M. H., Rene, E. R., & Firoozbahr, M. (2020). Recent advances in the treatment of dye-containing wastewater from textile industries: Overview and perspectives. *Process Safety and Environmental Protection, 143*, 138–163. https://doi.org/10.1016/j.psep.2020.03.018.

Saxena, S., & Raja, A. S. M. (2014). Natural dyes: Sources, chemistry, application and sustainability issues. In *Roadmap to Sustainable Textiles and Clothing* (pp. 37–80). Springer: New York. https://doi.org/10.1007/978-1-4614-6660-2_2.

Semba, R. D. (1998). The role of vitamin A and related retinoids in immune function. *Nutrition Reviews, 56*(1), S38–S48. https://doi.org/10.1111/j.1753-4887.1998.tb01629.x.

Shaharir, B. M. Z. (2002). *Pengenalan Sejarah dan Falsafah Sains*. Penerbit UKM.

Sofyan, S., Failisnur, F., & Silfia, S. (2021). The combination of natural dyes from jengkol POD waste (*Archidendron jiringa*) and liquid waste of gambier (*Uncaria gambir Roxb*) on coloring of cotton fabrics. *4th International Seminar on Chemistry, 2349*, 020061. https://doi.org/10.1063/5.0043117.

Soh, M. B. C., & Omar, S. K. (2012). Small is big: The charms of indigenous knowledge for sustainable livelihood. *Procedia - Social and Behavioral Sciences, 36*, 602–610. https://doi.org/10.1016/j.sbspro.2012.03.068.

Styawan, W., Linda, R., & Mukarlina. (2016). Pemanfaatan Tumbuhan Sebagai Bahan Kosmetik Oleh Suku Melayu Di Kecamatan Sungai Pinyuh Kabupaten Mempawah. *Jurnal Protobiont, 5*(2), 45–52.

Syed Shaharuddin, S. I., Shamsuddin, M. S., Drahman, M. H., Hasan, Z., Mohd Asri, N. A., Nordin, A. A., & Shaffiar, N. M. (2021). A review on the Malaysian and Indonesian batik production, challenges, and innovations in the 21st century. *SAGE Open, 11*(3), 1–19. https://doi.org/10.1177/21582440211031872.

Tee, E. S., Florentino, R. F., Chongviriyaphan, N., Ridwan, H., Appukutty, M., & Mai, T. T. (2023). Review of recommended energy and nutrient intake values in Southeast Asian countries. *Malaysian Journal of Nutrition, 29*(2), 163–241.

Teucher, U., & Sayok, A. K. (2018). Loss of food plants knowledge and identity among indigenous peoples in Malaysia. *Journal of Advanced Research in Social and Behavioural Sciences, 11*(June), 174–188.

Thornfeldt, C. (2005). Cosmeceuticals containing herbs: Fact, fiction, and future. *Dermatologic Surgery, 31*, 873–881. https://doi.org/10.1111/j.1524-4725.2005.31405.x.

United Nation. (2008). United Nations Declaration on the Rights of Indigenous People. [No DOI available] http://www.un.org/esa/socdev/unpfii/documents/DRIPS_en.pdf.

Vira, B., Wildburger, C., & Mansourian, S. (2015). *Forests, trees and landscapes for food security and nutrition: A global assessment report.*

Wahdina, Setiadi, D., Purwanto, Y., & Qayim, I. (2021). Tumbuhan pewarna yang digunakan masyarakat Iban Dayak Dusun Sungai Itik Kalimantan Barat. *Journal of Tropical Ethnobiology, 4*(1), 1–3.

Walburga Bui, A., & Lapailaka, T. (2022). Ekstraksi senyawa turunan antrakuinon dari kulit akar mengkudu (*Morinda citrifolia L.*) asal Pulau Timor aplikasi dye sensitized solar cell (DSSC) berbasis semikonduktor SiO2. *Seminar Nasional Kimia Dan Pendidikan Kimia, 1*(1), 1–7.

Wan Ramli, W. D., & Shaharir, M. Z. (1999). Indigenisation of technology and the challenge of globalization: The case of Malaysia. *Journal of Islamic Science, 15*(1/2), 109–134.

Warren, D. M. (1992). Indigenous knowledge, biodiversity conservation and development: keynote address. Keynote Address for the International Conference on Conservation of Biodiversity in Africa: Local Initiatives and Institutional Roles.

Wong, D. (2009). Wooley and the codification of native customs in Sabah. *Wong, D., 11*(1), 87–105.

Yoshanti, G., & Dowaki, K. (2017). Batik life cycle assessment analysis (LCA) for improving batik small and medium enterprises (SMEs) sustainable production in Surakarta, Indonesia. In M. Matsumoto, K. Masui, S. Fukushige, & S. Kondoh (Eds.), *Sustainability Through Innovation in Product Life Cycle Design* (pp. 997–1008). Springer: New York. https://doi.org/10.1007/978-3-319-55367-1_102.

You, Y. X., Shahar, S., Haron, H., Yahya, H. M., & Che Din, N. (2020). High traditional Asian vegetables (ulam) intake relates to better nutritional status, cognition and mood among aging adults from low-income residential areas. *British Food Journal, 122*(10), 3179–3191. https://doi.org/10.1108/BFJ-06-2020-0495.

You, Yee Xing, Shahar, S., Haron, H., Yahya, H. M., & Che Din, N. (2020). High traditional Asian vegetables (ulam) intake relates to better nutritional status, cognition and mood among aging adults from low-income residential areas. *British Food Journal, 122*(10), 3179–3191. https://doi.org/10.1108/BFJ-06-2020-0495.

Zaki, N. A. M., & Jai, J. (2020). Plant-based pigments: Challenges and future perspective for natural food colourants. *Malaysian Journal of Chemical Engineering and Technology, 3*(1), 44.

# Chapter 9

# Sustainable Extraction Methods for Beta-Carotene

**Ali Haghighi Asl[1,*], PhD**
**Pouya Mottahedin[2], PhD**
**and Maryam Khajenoori[3], PhD**
[1,3]Faculty of Chemical, Petroleum and Gas Engineering, Semnan University, Semnan, Iran
[2]Department of Chemical Engineering, Faculty of Engineering, University of Garmsar, Garmsar, Iran

## Abstract

Beta-carotene is a vital carotenoid that our bodies convert into vitamin A, which is essential for numerous functions. It is plentiful in colorful fruits and vegetables like carrots, sweet potatoes, spinach, and kale. Renowned for its antioxidant qualities, beta-carotene is especially important for preserving eye health, particularly in dim lighting. In addition, it plays a key role in enhancing immune function, safeguarding skin health, promoting heart health, and potentially lowering cancer risks. Recent research also indicates it may help combat cognitive decline by reducing oxidative stress. There are various methods for extracting beta-carotene from plant sources, each with its advantages and drawbacks. Solvent extraction is effective but relies on hazardous, flammable solvents that can leave harmful residues. Cold pressing is a more mechanical process, ideal for oil-rich plants, but often results in lower yields. Steam distillation is non-toxic, though it produces smaller amounts of beta-carotene, and Soxhlet extraction is quite efficient but

[*] Corresponding Author's Email: ahaghighi@semnan.ac.ir

In: Natural Pigments: Sources, Properties, and Health Benefits
Editors: Rashidi Othman, Qurratu Aini Mat Ali, Razanah Ramya et al.
ISBN: 979-8-90134-055-4

requires large amounts of solvents and time. In response to growing environmental concerns, newer "green" extraction methods like supercritical $CO_2$ and subcritical water extraction have emerged as promising alternatives. These methods not only enhance yield but also reduce environmental impact by eliminating the use of harmful solvents. Subcritical water, in particular, is highly effective as it serves as a solvent for non-polar compounds like beta-carotene, although it may require co-solvents to optimize its efficiency. Supercritical $CO_2$ extraction is an eco-friendly and highly efficient technique that selectively extracts non-polar compounds like beta-carotene, operating well under moderate temperatures and high pressure. While these green extraction methods might involve a higher initial cost, they offer significant benefits in terms of safety, sustainability, and the purity of the final product. As more industries focus on health and environmental concerns, these methods are expected to become the preferred option for beta-carotene extraction in the future.

**Keywords**: beta-carotene, green extraction methods, supercritical $CO_2$, subcritical water, health and environmental concerns

## Introduction

### The Imperative for Sustainable Beta-Carotene Extraction

Beta-carotene (β-carotene), a key member of the carotenoid group, is one of the most essential lipophilic pigments in nature, responsible for giving various fruits, vegetables, and microorganisms their bright yellow, orange, and red colors (Vi˜nas-Ospino et al., 2024). The structure and physicochemical properties of β-carotene are summarized in Table 1 (Share Mohammadi et al., 2020). This isoprenoid compound, characterized as a 40-carbon tetraterpenoid, is best known for its vital role as a provitamin A source, which is crucial for human health by supporting vision, immune function, and cellular differentiation (Jiang et al., 2024). In addition to its role as a vitamin A precursor, β-carotene also possesses strong antioxidant and anti-inflammatory properties, thanks to its conjugated double-bond structure (Gazmeh et al., 2025). These qualities have driven extensive research into its potential to lower the risk of chronic diseases, such as various cancers, cardiovascular conditions, eye disorders, and its possible contribution to boosting the immune system against infections, including COVID-19 (Gazmeh et al., 2025). As a result, β-carotene is a highly valued bioactive substance with a wide range of uses in industries like functional foods, nutraceuticals, cosmetics, and animal feed (Kultys & Kurek, 2022). The global carotenoid market is expected to keep expanding significantly, fueled by growing consumer demand for natural ingredients and greater awareness of their health benefits, with a projected market value of USD 2.45 billion by 2034 (Jiang et al., 2024). Figure 1 shows the key health benefits of natural carotenoids and Table 2 provides a detailed description of natural carotenoids (Sereti et al., 2025b).

**Table 1.** Structure and physicochemical properties of β-carotene (Share Mohammadi et al., 2020)

| Formula | Structure | Molecular Weight | CAS number | Melting Temperature |
|---|---|---|---|---|
| C40H56 | 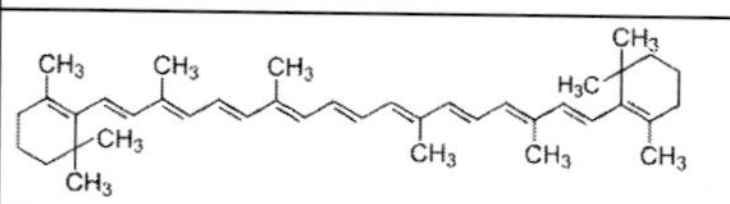 | 536.87 g.mol-1 | 7235-40-7 | 456 K |

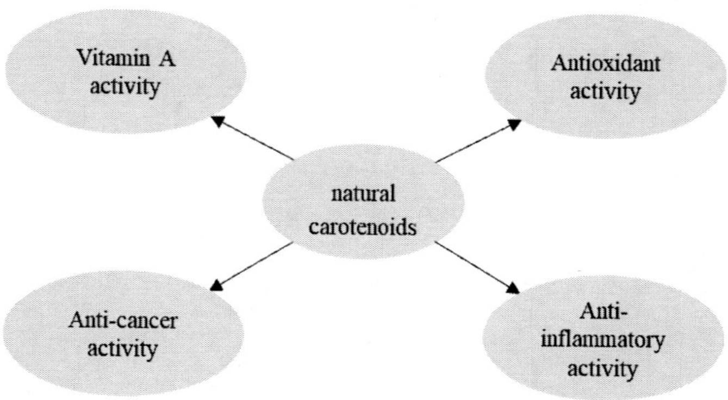


**Figure 1.** Key health benefits of natural carotenoids.

Traditionally, extracting β-carotene from its natural sources has mostly depended on conventional methods using organic solvents like hexane, acetone, tetrahydrofuran, petroleum ether, and chloroform (Morón-Ortiz et al., 2024). While these methods are effective in isolating this hydrophobic compound, they come with notable disadvantages. Many of these solvents are derived from non-renewable petrochemical sources, are highly volatile, and present significant environmental and health risks due to their toxicity and flammability (Morón-Ortiz et al., 2024). The presence of leftover organic solvents in the final extract is a major concern, requiring costly and extensive purification steps to ensure the product is safe and of high quality for use in food and pharmaceuticals (Morón-Ortiz et al., 2024). Moreover, β-carotene is highly prone to degradation when exposed to light, heat, and oxygen during extraction, leading to a loss of its bioactive properties and a decrease in yield (Vi˜nas-Ospino et al., 2024). This instability, combined with the fact that β-carotene is often trapped within complex cellular structures in plant materials or microorganisms, makes it necessary to use advanced extraction methods that can efficiently break open cells without damaging the compound's integrity (Jiang et al., 2024). Table 3 demonstrates the overview of key benefits across various sectors of β-carotene.

The need for sustainable β-carotene extraction arises from these challenges and aligns with the principles of green chemistry and the broader circular economy model (Morón-Ortiz et al., 2024). Green extraction aims to minimize environmental impact by reducing energy usage, waste, and harmful substances, while enhancing efficiency and ensuring the safety and quality of the extracted products (Morón-Ortiz et al., 2024). This shift in approach is driven by increasing regulatory demands, rising consumer preference for natural and eco-friendly products, and the economic opportunity in utilizing agricultural and industrial by-products (Kultys & Kurek, 2022). Instead of treating these by-products as waste, a sustainable approach views them as valuable resources from which high-value compounds like β-carotene can be extracted, helping to improve resource efficiency and reduce ecological footprints (Kultys & Kurek, 2022).

**Table 2.** Main beneficial properties of natural carotenoids

| Activity | Description |
|---|---|
| Vitamin A activity | Carotenoids containing at least one β-ionone ring in their structure (such as β-carotene) possess provitamin A activity. The beneficial effects of Vitamin A involve eyesight effects and embryonic development promotion. |
| Antioxidant activity | Owing to the system of conjugated double bonds found in their polyene chain, carotenoids can act as reactive oxygene species (ROS) scavengers, regulating oxidative stress. |
| Anti-cancer activity | Natural carotenoids are able to regulate the mechanisms controlling cell proliferation. There is a positive effect against various types of cancer, such as against gastric, colon, prostate, breast and liver cancer. |
| Anti-inflammatory activity | The potent antioxidant activity of natural carotenoids has been associated to their anti-inflammatory effect. Carotenoids act against ROS and other inflammatory agents responsible for chronic diseases and inflammation. |

**Table 3.** Applications of β-Carotene: Overview of key benefits across various sectors

| Application Area | Description | Key Benefits | References |
|---|---|---|---|
| Nutrition | B-carotene is widely used as a source of vitamin A in dietary supplements and fortified foods. It is crucial for vision, immune function, and skin health. | Supports immune health, vision, and skin integrity. Prevents vitamin A deficiency. | Buchweitz et al., 2025; Jiang et al., 2024 |
| Cosmetics | B-carotene is used in skincare products due to its antioxidant properties. It helps protect the skin from UV damage and promotes skin rejuvenation. | Protects against oxidative damage, supports skin health, and prevents premature aging. | Sereti et al., 2025a; Morón-Ortiz et al., 2024 |
| Pharmaceuticals | B-carotene is included in pharmaceutical formulations for its antioxidant properties, helping prevent chronic diseases such as heart disease and cancer. | Antioxidant, anti-inflammatory, potential cancer and heart disease prevention. | Gazmeh et al., 2025; Buchweitz et al., 2025 |
| Animal Feed | Used as a supplement in animal feed to improve the health and growth of livestock and poultry, enhancing their immune system and skin color. | Enhances immune function, promotes healthier animals, and improves coloration in poultry. | Kultys & Kurek 2022; Gazmeh et al., 2025 |
| Food Industry | B-carotene is used as a natural food colorant in various processed foods like beverages, snacks, and dairy products. | Provides a natural, safe alternative to synthetic food colorants. Adds nutritional value. | Viñas-Ospino et al., 2023; Jiang et al., 2024 |

In response to this need, there has been a significant increase in research dedicated to developing and refining innovative green extraction methods and bio-based solvents (Morón-Ortiz et al., 2024). These advanced techniques aim to address the limitations of traditional methods by offering gentler extraction conditions, better selectivity, and improved preservation of β-carotene's bioactive properties. Notable green extraction methods include Supercritical Fluid Extraction (SFE), Ultrasound-Assisted Extraction (UAE), Microwave-Assisted Extraction (MAE), Pressurized Liquid Extraction (PLE), Pulsed Electric Field (PEF), Enzyme-Assisted Extraction (EAE), and Accelerated Solvent Extraction (ASE) (Viñas-Ospino et al., 2023). These approaches help to break down cell walls and enhance mass transfer through physical forces or enzymatic processes, often at lower temperatures or shorter times, thus reducing thermal degradation (Staicu et al., 2024).

In parallel with these technologies, considerable advancements have been made in the design and application of novel green and bio-based solvents (Morón-Ortiz et al., 2024). Notable among them is supercritical carbon dioxide ($scCO_2$), a non-toxic, non-flammable solvent that can be easily separated from the extract (Ludwig et al., 2021). Additionally, several vegetable oils, such as sunflower, flaxseed, and corn oil, are food-grade, biodegradable, and capable of acting as protective carriers for lipophilic carotenoids, often eliminating the need for subsequent solvent removal (Vi˜nas-Ospino et al., 2024). Other promising options include ionic liquids (ILs) and deep eutectic solvents (DES), particularly natural deep eutectic solvents (NADES), which are appreciated for their customizable properties, low toxicity, and biodegradability (Vi˜nas-Ospino et al., 2024). Furthermore, bio-derived solvents like 2-methyltetrahydrofuran, cyclopentyl methyl ether, and ethyl lactate are also gaining traction (Morón-Ortiz et al., 2024). The integration of these environmentally conscious solvents with advanced extraction techniques signifies a meaningful transition toward greener and more economical β-carotene extraction methods (Vi˜nas-Ospino et al., 2024).

Although laboratory-scale results show promise, significant challenges remain in scaling up these green extraction methods, addressing their techno-economic feasibility, and ensuring regulatory compliance, all of which require further study (Morón-Ortiz et al., 2024). Future research should aim to refine these techniques for industrial-scale use, conduct detailed life cycle assessments to confirm their environmental benefits compared to traditional methods, and ensure that the extracted β-carotene meets strict food safety and pharmaceutical standards in terms of purity and composition (Morón-Ortiz et al., 2024). The ongoing development of sustainable extraction methods for β-carotene is not just an academic challenge but a critical step in creating a more sustainable food system, fulfilling global nutritional needs, and unlocking the full potential of this valuable bioactive compound. Table 2 demonstrates the applications of β-carotene.

## Sources of Beta-Carotene for Sustainable Extraction

B-carotene, an important member of the carotenoid group, is a highly valued isoprenoid compound known for its bright red-orange color and its notable health benefits (Staicu et al., 2024). In addition to being a natural pigment, it is a vital dietary precursor to vitamin A, which is essential for vision, immune function, and cellular growth, and also exhibits strong antioxidant and anti-inflammatory properties (Sereti et al., 2025a). The rising global demand for functional foods, nutraceuticals, cosmetics, and pharmaceuticals has accelerated the search for more efficient and sustainable sources of β-carotene, moving away from synthetic production and toward natural extracts (Vi˜nas-Ospino et al., 2024). This shift is driven by a consumer preference for natural ingredients and concerns about the safety of chemically synthesized compounds (Jiang et al., 2024; Buchweitz et al., 2025). As a result, finding and optimizing diverse natural sources, especially those that align with circular economy principles and the reuse of waste, has become a key focus in sustainable extraction research.

Historically, β-carotene has been primarily extracted from higher plants, with carrots (Daucus carota) being a prime example due to their high β-carotene content, often reported at 52-60 mg/100g of dry carrot (Kumar, 2024; Gazmeh et al., 2025). Other notable plant sources include pumpkins (Cucurbita moschata), sweet potatoes, spinach, tomatoes, and mangoes, all contributing significantly to the compound's presence in the human diet (Morón-Ortiz et al., 2024). However, the increasing focus on sustainability has prompted researchers to investigate fruit and vegetable by-products as cost-effective and eco-friendly raw materials (Kultys & Kurek, 2022). These by-products, which are often discarded during food processing, hold great potential as untapped sources of bioactive compounds like β-carotene (Kumar, 2024). Examples include carrot pomace, mango pulp and peels, orange peels, and pumpkin peel and pulp (Irianto et al., 2025). Valorizing such agricultural waste not only helps reduce environmental impact but also provides a steady and affordable supply of raw material for extraction (Kultys & Kurek, 2022). However, the β-carotene content in these plant sources can vary considerably depending on factors such as cultivar type, environmental conditions (temperature, sunlight, nutrient availability, soil quality), and post-harvest handling, making consistent large-scale extraction a challenge (Lopes et al., 2024; Matić et al., 2024).

In addition to traditional plant sources, microalgae have emerged as highly promising "cell factories" for β-carotene production due to their capacity for high accumulation and their ability to be cultivated in controlled environments (Sereti et al., 2025a). Dunaliella salina is the most commercially cultivated microalga for β-carotene, with the ability to accumulate over 14% of its dry weight in the compound, and its yields can increase under optimal stress conditions such as high salinity, nitrogen depletion, and intense solar radiation (Ludwig et al., 2021). Other microalgae like Chlorella vulgaris, Phaeodactylum tricornutum, Nannochloropsis salina, and Tetradesmus obliquus also contain valuable carotenoids (Morón-Ortiz et al., 2024). The biotechnological production of β-carotene from microalgae offers a natural and widely accepted alternative to synthetic sources, supporting healthier food products and supplements (Buchweitz et al., 2025). However, β-carotene in microalgae is typically contained within sturdy cellular structures, requiring efficient cell disruption techniques (such as homogenization, ultrasound, or enzymatic treatment) before extraction to optimize yields (Jiang et al., 2024).

In a similar vein, oleaginous yeasts and fungi have emerged as promising biotechnological sources of β-carotene (Sereti et al., 2025a; Thakur, 2018). Species such as Rhodosporidium paludigenum, Rhodosporidium toruloides, Rhodotorula glutinis, and Phaffia rhodozyma are particularly noteworthy for their ability to produce significant amounts of intracellular carotenoids and microbial oils through fermentation (Sereti et al., 2025a). Rhodosporidium paludigenum, for example, has been identified as a versatile microbial factory for synthesizing β-carotene along with other rare carotenoids like torulene and torularhodin, which also have strong antioxidant properties (Sereti et al., 2025a). The β-carotene produced by R. paludigenum can make up a large proportion (70.1% to 93.7%) of its total carotenoid content, depending on the cultivation conditions (Sereti et al., 2025a). The fatty acid composition of microbial oils from these yeasts often mirrors that of conventional vegetable oils such as cocoa butter, making them valuable dual-purpose producers (Buchweitz et al., 2025). Similar to microalgae, the efficient extraction of these intracellular metabolites from yeast biomass requires effective cell disruption methods (Buchweitz et al., 2025).

Choosing a source of β-carotene for sustainable extraction involves a complex decision that balances the compound's natural abundance, its availability within the matrix, and the environmental impact of its cultivation or harvesting (Morón-Ortiz et al., 2024; Kultys & Kurek, 2022). While plant by-products present a strong case for waste valorization and provide readily available raw material, their seasonal availability and compositional inconsistencies can present challenges. In contrast, microbial sources offer controlled production environments and higher yields, but they require advanced upstream (fermentation) and downstream (cell disruption, extraction) processes that need to be optimized for both economic feasibility and environmental sustainability (Morón-Ortiz et al., 2024; Sereti et al., 2025a). Integrating these diverse sources into a biorefinery approach is a key future goal, allowing the co-extraction of multiple valuable compounds (such as microbial oils, polysaccharides, and various carotenoids) from a single biomass. This approach would maximize resource use and improve economic feasibility (Sereti et al., 2025a). Such a comprehensive strategy will be crucial in addressing the growing demand for natural β-carotene in a sustainable way.

## Advanced Green Extraction Technologies for Beta-Carotene

### Supercritical Fluid Extraction (SFE)

Supercritical Fluid Extraction (SFE) plays a pivotal role in the development of sustainable methods for recovering valuable bioactive compounds, such as β-carotene. This technique takes advantage of the unique properties of substances that exist above their critical temperature and pressure, where they show both gas and liquid characteristics (Uwineza & Waśkiewicz, 2020; Sereti et al., 2025a). Carbon dioxide ($CO_2$) is the most commonly used supercritical fluid because of its low critical point (31.1 °C and 7.38 MPa), non-toxic nature, non-flammability, and approval as Generally Recognized As Safe (GRAS) by regulatory agencies (Irianto et al., 2025; Viñas-Ospino et al., 2023). These properties make $CO_2$ an attractive solvent for applications in food and pharmaceuticals, where maintaining product purity and safety is essential. The process involves exposing $CO_2$ to supercritical conditions and passing it through the raw material. The supercritical $CO_2$, with its enhanced diffusivity, low viscosity, and strong solvating power, penetrates the plant matrix like a gas and dissolves the target compounds like a liquid, enabling efficient extraction of non-polar substances, such as triglycerides, fatty acids, and carotenoids (Uwineza & Waśkiewicz, 2020; Irianto et al., 2025; Viñas-

Ospino et al., 2023; Sereti et al., 2025a). Scheme P-T representation of $CO_2$ phase is shown in Figure 2 (Uwineza & Waśkiewicz, 2020).

Using Supercritical Fluid Extraction (SFE) for β-carotene extraction provides significant benefits compared to traditional methods, which often rely on harmful organic solvents like hexane, acetone, or chloroform. These conventional solvents present considerable health and environmental risks due to their toxicity, flammability, and the difficulty of removing solvent residues from the final product (Irianto et al., 2025; Sereti et al., 2025a; Matić et al., 2024). In contrast, SFE eliminates the need for such toxic solvents, resulting in a cleaner extract with little to no solvent residue, which is particularly important for applications in food and cosmetics (Irianto et al., 2025; Viñas-Ospino et al., 2023). Moreover, the gentle temperatures and pressures used in SFE are essential for maintaining the stability of heat-sensitive compounds like β-carotene, reducing the risk of degradation and isomerization that can occur with high-temperature methods (Irianto et al., 2025; Kultys & Kurek, 2022).

Recent advancements in Supercritical Fluid Extraction (SFE) for β-carotene have demonstrated its adaptability across a range of natural sources. It has been effectively used to extract carotenoids from materials like carrot peels, pumpkins, mango peels, and even microbial biomass such as red yeasts and microalgae (de Andrade Lima et al., 2019; Sereti et al., 2025a; Viñas-Ospino et al., 2023). A groundbreaking study highlighted the potential to co-extract β-carotene and microbial oil from red yeast biomass (Rhodosporidium paludigenum) using SFE-$CO_2$, where pretreatment through homogenization optimized cell disruption (Sereti et al., 2025a). Co-solvents play a crucial role in SFE; while $CO_2$ is naturally non-polar, adding a modifier, most commonly ethanol due to its GRAS status and environmentally friendly nature, enhances the system's polarity. This improves the extraction of polar carotenoids and boosts overall yields (Sereti et al., 2025a; Ludwig et al., 2021; Kultys & Kurek, 2022; Irianto et al., 2025). Studies have demonstrated that ethanol is essential for efficiently recovering carotenoids from Rhodosporidium paludigenum and significantly increasing β-carotene yields from Dunaliella salina (Sereti et al., 2025a; Ludwig et al., 2021).

Despite its many advantages and promising results at the laboratory scale, implementing Supercritical Fluid Extraction (SFE) for β-carotene extraction on an industrial scale still faces significant techno-economic challenges.

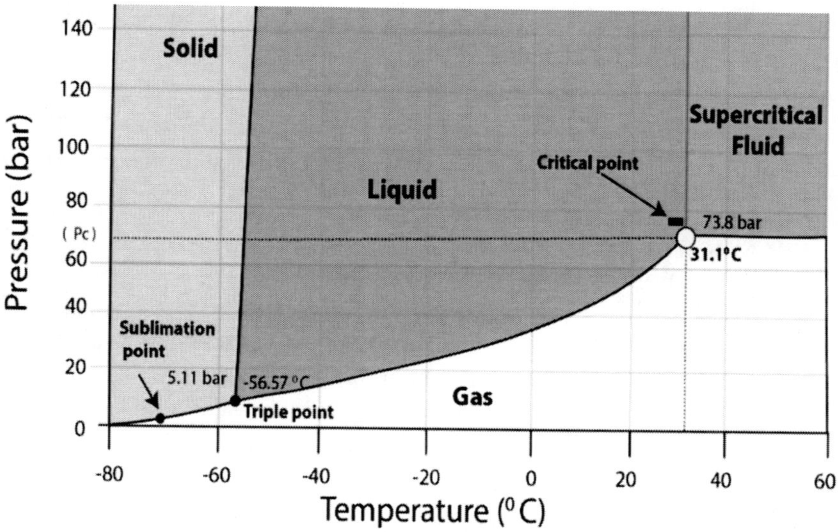


**Figure 2.** Scheme P-T representation of $CO_2$ phase.

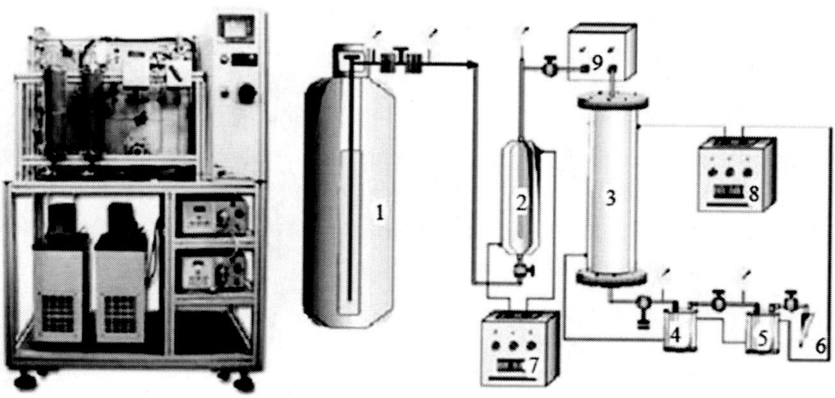


**Figure 3.** Schematic diagram of a supercritical fluid extraction (SFE) system. 1: $CO_2$ cylinder, 2: Surge tank, 3. Extractor, 4, 5: Separators, 6. Collector and gas-measuring device, 7, 8: Thermostatic baths, 9: isocratic pump.

A major point of discussion is the scalability and cost-effectiveness of SFE in comparison to traditional methods. While SFE provides superior extract purity and environmental benefits, it requires substantial capital investment for equipment and high energy consumption, particularly for $CO_2$ compression (Ludwig et al., 2021; Morón-Ortiz et al., 2024; Viñas-Ospino et al., 2023). For example, studies show that SFE-$CO_2$ consumes twice as much energy as n-hexane extraction for β-carotene, although this is often offset by significantly lower solvent costs thanks to efficient $CO_2$ recycling (Ludwig et al., 2021). Additionally, β-carotene's highly lipophilic nature means it is typically extracted alongside other lipophilic compounds. Achieving high selectivity for specific carotenoids from complex matrices remains a challenge and requires additional purification steps, adding both cost and complexity to the process (Morón-Ortiz et al., 2024; Matić et al., 2024; Morón-Ortiz et al., 2024). Figure 3 demonstrates the supercritical fluid extraction (SFE) system used in laboratory settings (Castro-López et al., 2016).

Looking to the future, SFE research for sustainable β-carotene extraction must focus on thorough techno-economic and environmental evaluations, including Life Cycle Assessment (LCA), to fully understand its sustainability advantages over current industrial methods (Morón-Ortiz et al., 2024). Optimizing extraction conditions and creating more cost-effective, scalable equipment are essential for broader industrial adoption, particularly by small and medium-sized enterprises (Irianto et al., 2025). Combining SFE with other green techniques, such as microwave or ultrasound pre-treatment, may further improve efficiency and extraction yields (Irianto et al., 2025). Additionally, exploring the biological properties of SFE-derived β-carotene in cellular assays and incorporating these extracts into new functional food products are critical areas for future research (Sereti et al., 2025a; Matić et al., 2024; Vi˜nas-Ospino et al., 2024). Advancing SFE technology, particularly through the careful selection of GRAS co-solvents and the development of more selective processes, holds great potential to enhance sustainable β-carotene production and meet the growing consumer demand for natural, safe, and functional food ingredients.

## Microwave-Assisted Extraction (MAE)

Microwave-Assisted Extraction (MAE) stands out as a breakthrough in sustainable extraction, offering a highly efficient and eco-friendly technique for obtaining valuable bioactive compounds, particularly carotenoids such as β-carotene (Irianto et al., 2025). As a non-traditional method, MAE uses microwave energy to rapidly heat both the solvent and the plant material, accelerating the extraction of intracellular compounds (Irianto et al., 2025). This approach aligns with green chemistry principles by lowering energy consumption and reducing reliance on conventional organic solvents, which are frequently linked to significant environmental and health risks (Kumar, 2024).

The fundamental concept behind Microwave-Assisted Extraction (MAE) involves the direct interaction of microwave radiation with polar molecules in both the solvent and the plant material (Irianto et al., 2025). This results in localized heating, which initiates a series of effects that enhance extraction efficiency. First, the rapid heating of water within the plant material builds internal pressure, causing cell walls to rupture and releasing intracellular compounds into the solvent (Irianto et al., 2025; Lopes et al., 2024). Second, the increased molecular movement boosts mass transfer, accelerating the diffusion of target compounds (Kumar, 2024). This method typically leads to significantly shorter extraction times and requires less solvent compared to traditional methods such as maceration or Soxhlet extraction (Khajenoori and Haghighi asl, 2014). For instance, research has demonstrated that MAE can extract β-carotene in just a few minutes, whereas conventional methods usually take hours to achieve the same result (Kumar, 2024).

MAE offers several key advantages that strengthen its "green" credentials. Its energy efficiency and reduced solvent use contribute to a lower environmental footprint, making it a more sustainable choice for industrial applications (Irianto et al., 2025; Matić et al., 2024). The ability to precisely control extraction factors such as microwave power, frequency, and time further enhances the method's efficiency and sustainability (Irianto et al., 2025; Viñas-Ospino et al., 2023). Moreover, MAE is highly adaptable, enabling the extraction of a wide variety of natural pigments (Viñas-Ospino et al., 2023). The rapid extraction process and the potential for higher purity extracts increase its appeal, especially for high-value compounds used in the food, cosmetics, and pharmaceutical industries (Irianto et al., 2025; Matić et al., 2024).

Numerous studies have investigated the use of MAE for extracting β-carotene from different natural sources. In the case of carrot waste, MAE has shown promising outcomes, often using edible oils such as flaxseed oil as eco-friendly solvents. One study reported a carotenoid recovery of 77.48% under optimal conditions, which included 165 W microwave power, 9.39 minutes of extraction time, and an oil-to-waste ratio of 8.06:1, with extraction time being the most crucial factor (Morón-Ortiz et al., 2024). To mitigate thermal degradation concerns, intermittent microwave radiation has been explored, leading to the extraction of significant amounts of β-carotene and total carotenoids (289.2 ± 5.4 mg/100 g d.b) from carrot peels in just 5 minutes (Viñas-Ospino et al., 2023). Furthermore, pre-treatments such as blanching with ascorbic acid can enhance MAE efficiency by softening the raw material structure and increasing carotenoid availability (Kultys & Kurek, 2022).

In addition to carrots, MAE has been successfully used to extract β-carotene from other abundant sources. For sea buckthorn pomace, MAE, often paired with edible oils like corn and olive oils, has achieved significant carotenoid yields (26.91–34.35 mg/100 g) (Jiang et al., 2024; Kultys & Kurek, 2022). One innovative method combined ultrasound and microwave (UMAE) with Fatty Acid Ethyl Esters (FAEE) as a green solvent to extract β-carotene from sea buckthorn berries. This UMAE method reportedly increased β-carotene content by 15% compared to ultrasound-assisted extraction (UAE) and 89% compared to MAE alone, all within the same 45-minute extraction at 50°C. The extracted product can be directly used as a food supplement, with the FAEE solvent also supplying valuable omega-3 and omega-6 fatty acids. Adding vitamin E further enhanced the stability of the extract for over 90 days, even with light exposure at room temperature (Staicu et al., 2024). For pumpkin (Cucurbita moschata), MAE with corn oil has been optimized to extract carotenoids, achieving results comparable to other methods. Additionally, research has explored using switchable natural deep eutectic solvents (NADES), such as a caprylic acid:capric acid (3:1) mix, alongside ultrasound-assisted MAE to recover β-carotene from pumpkin by-products (Irianto et al., 2025). MAE has also proven effective for extracting carotenoids from passion fruit peels using olive oil, yielding 1178.54 μg of carotenoids per 100 g of peels under specific conditions (Kultys & Kurek, 2022). Furthermore, recent studies on macroalgae species like Himanthalia elongata demonstrated that optimized MAE using 90% ethanol as a solvent can effectively extract carotenoids, with β-carotene being the primary compound (Lopes et al., 2024). Figure 4 illustrates the schematic of a laboratory-scale microwave-assisted extraction system (Castro-López et al., 2016).

Although MAE provides numerous advantages, it also presents certain challenges. One major issue, particularly with heat-sensitive compounds like carotenoids, is the potential for thermal degradation and cis-trans isomerization (Irianto et al., 2025). This process can negatively affect the bioactivity and quality of the β-carotene extracted. To mitigate this, techniques like intermittent microwave radiation have been developed (Viñas-Ospino et al., 2023; Matić et al., 2024). While MAE significantly reduces solvent usage, some studies still depend on conventional organic solvents, which necessitate extra purification stages and reduce the process's "green" advantages (Matić et al., 2024). Additionally, the economic viability and scalability of MAE for large-scale industrial production remain subjects of ongoing study, requiring comprehensive techno-economic analyses (Irianto et al., 2025; Lopes et al., 2024).

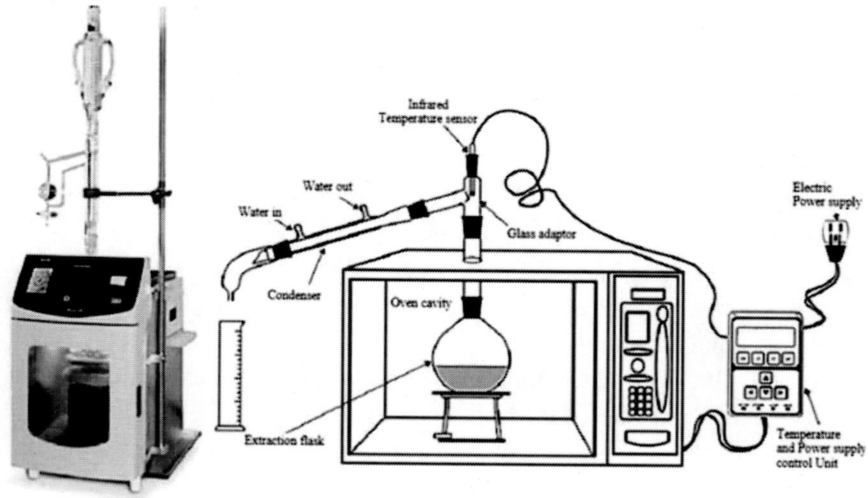


**Figure 4.** Schematic of a laboratory-scale microwave-assisted extraction system.

As global interest in sustainability grows, future research on MAE for β-carotene extraction should concentrate on refining processes that exclusively utilize bio-based and non-toxic solvents, ensuring the extracts are safe for direct application in food products without harmful residues (Viñas-Ospino et al., 2024). Additionally, exploring hybrid extraction techniques, combining MAE with other eco-friendly methods such as ultrasound-assisted extraction (UMAE), holds promise for improving efficiency and increasing yields (Staicu et al., 2024). Conducting detailed life cycle assessments (LCA) is crucial to fully evaluate the environmental impact of MAE compared to conventional industrial methods, providing a comprehensive understanding of their sustainability benefits (Irianto et al., 2025). Continued innovation in equipment design to improve scalability and cost-efficiency will be vital for broader industrial adoption, fostering the sustainable production of high-value natural β-carotene.

## Ultrasound-Assisted Extraction (UAE)

Ultrasound-Assisted Extraction (UAE) has become a leading green technology for extracting bioactive compounds, including β-carotene, from various natural sources, especially fruit and vegetable by-products and microbial biomass (Irianto et al., 2025). This method sets itself apart from traditional techniques by using ultrasonic waves, typically in the range of 20 kHz to 100 kHz, to enhance the extraction process (Alirezapour, et al., 2020). The core mechanism relies on acoustic cavitation, where high-frequency sound waves travel through the medium, causing microscopic bubbles to form, grow, and collapse violently (Irianto et al., 2025; Kultys & Kurek, 2022). This collapse generates brief but intense areas of high temperature (up to 5500 °C) and pressure (up to 50 MPa), producing powerful shear forces (Irianto et al., 2025). The mechanical and thermal effects of these shockwaves efficiently disrupt cell walls, enhance solvent penetration, and improve the transfer of target compounds from the solid matrix into the solvent (Irianto et al., 2025, Khalid et al., 2024). A schematic of the ultrasound-assisted extraction system is presented in Figure 5 (Castro-López et al., 2016).

A key benefit of UAE for β-carotene extraction is its non-thermal nature and high efficiency (Kultys & Kurek, 2022; Viñas-Ospino et al., 2023). Unlike methods that require prolonged high temperatures, UAE can be conducted at lower or ambient temperatures, which is essential for maintaining the sensitive properties and structure of β-carotene, which is easily affected by light, heat, and oxygen (Kultys & Kurek, 2022). This gentle yet effective disruption of cells not only improves extraction yields but also reduces the time and energy needed for the process. For example, traditional stirring techniques might consume around 4.536 × 10^6 J over two hours, whereas UAE can achieve similar results with only 2.999 × 10^6 J in just ten minutes, showcasing its superior energy and time efficiency (Viñas-Ospino et al., 2023). This makes UAE an appealing option for sustainable and eco-friendly processes in the food and pharmaceutical industries (Irianto et al., 2025).

Due to the lipophilic nature of β-carotene, effective extraction requires solvents that can easily dissolve it (Lopes et al., 2024). Traditionally, organic solvents like hexane, acetone, and chloroform have been used, but their toxicity, environmental impact, and difficulty in completely removing them from the final product make them unsuitable for food-grade applications (Viñas-Ospino et al., 2023). This has led to significant research into combining UAE with green and bio-based solvents, which are non-toxic, biodegradable, and often sourced from renewable materials (Viñas-Ospino et al., 2023).

Vegetable oils are widely regarded as excellent green alternatives due to their ability to dissolve lipophilic compounds like carotenoids without causing significant degradation or loss during extraction (Kultys & Kurek, 2022). They also act as an oxygen barrier, further protecting the extracted β-carotene from oxidation (Kultys & Kurek, 2022). Several studies have highlighted the effectiveness of different vegetable oils in UAE for extracting β-carotene from various sources. For example, sunflower oil was successfully used to extract carotenoids from peach palm fruit, achieving a recovery of 163 mg/100 g of dry peel under optimal UAE conditions (Sereti et al., 2025a; Viñas-Ospino et al., 2023; Ordóñez-Santos et al., 2015). Likewise, olive oil has been optimized for UAE to extract carotenoids from passion fruit peels, showing increased yields under optimal conditions of time and temperature (Viñas-Ospino et al., 2023; Vi˜nas-Ospino et al., 2024; Chutia & Mahanta, 2021). Additionally, flaxseed oil and corn oil have been explored for extracting β-carotene from carrot pomace and sea buckthorn pomace, respectively, demonstrating improved yields compared to traditional methods (Kultys & Kurek, 2022). The versatility of vegetable oils also allows for the direct incorporation of the extract into food products, eliminating the need for separate solvent removal, which helps reduce both processing costs and environmental impact (Viñas-Ospino et al., 2023).

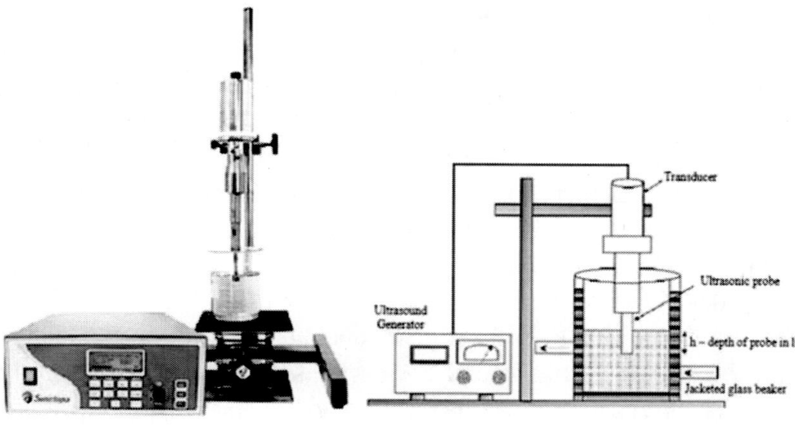


**Figure 5.** Schematic representation of an ultrasound-assisted extraction apparatus.

In addition to vegetable oils, Deep Eutectic Solvents (DES) and Natural Deep Eutectic Solvents (NADES) have gained attention as promising green solvents due to their unique properties such as easy preparation, low cost, biodegradability, and the ability to be customized for specific applications (Vi˜nas-Ospino et al., 2024). Hydrophobic DES, which are made of fatty acids or terpenoids, are especially effective at dissolving naturally occurring compounds with low water solubility, like β-carotene, and they also help protect these compounds during storage (Vi˜nas-Ospino et al., 2024). Research has demonstrated their significant ability to stabilize bioactive compounds, offering opportunities for the development of new functional foods (Vi˜nas-Ospino et al., 2024). For example, UAE combined with specific NADES, such as menthol/camphor, has been explored for extracting carotenoids from orange peels, yielding promising results in both yield and stability (Morón-Ortiz et al., 2024; Viñas-Ospino et al., 2023). Similarly, hydrophobic NADES, like a caprylic acid:capric acid mixture, have proven effective in extracting β-carotene from pumpkin by-products when used with UAE (Kultys & Kurek, 2022; Stupar et al., 2021).

Although UAE shows great promise, its application for β-carotene extraction still faces challenges that require further research. To achieve optimal extraction, it is crucial to carefully optimize parameters such as ultrasound power, intensity, amplitude, frequency, temperature, time, and the solid-to-liquid ratio (Kultys & Kurek, 2022; Viñas-Ospino et al., 2023). If ultrasound intensity is not properly managed, it can lead to the creation of reactive species (such as OH and H radicals) during the cavitation process, which may degrade antioxidant compounds like carotenoids (Kultys & Kurek, 2022; Viñas-Ospino et al., 2023). Thus, it is vital to fine-tune cavitation agents and maintain precise control over these factors to maximize yields and preserve bioactivity (Kultys & Kurek, 2022). Furthermore, while laboratory-scale results are promising, scaling up the UAE process for industrial use remains a significant challenge (Lopes et al., 2024). This scaling-up involves overcoming issues related to the high cost of equipment, energy consumption, and ensuring the consistent quality and purity of the final product (Morón-Ortiz et al., 2024).

Future research on UAE for β-carotene should concentrate on conducting comprehensive techno-economic and life cycle assessments (LCA) to fully understand its sustainability compared to traditional industrial methods (Morón-Ortiz et al., 2024). Further exploration of hybrid extraction techniques that combine UAE with other green technologies, such as microwave-assisted extraction or enzyme-assisted extraction, could lead to synergistic effects, improving both efficiency and yield (Vi˜nas-Ospino et al., 2024; Irianto et al., 2025). Additionally, studies should focus on refining the selectivity of UAE when combined with novel green solvents to target specific carotenoid isomers or mixtures, and evaluating the bioavailability and long-term stability of the extracted compounds for use in functional foods (Morón-Ortiz et al., 2024; Irianto et al., 2025). By addressing these critical areas, UAE can become a key technology for the sustainable and efficient production of natural β-carotene.

## Pressurized Liquid Extraction (PLE)/Accelerated Solvent Extraction (ASE)

Pressurized Liquid Extraction (PLE), also known as Accelerated Solvent Extraction (ASE), is an advanced green extraction technique that utilizes the unique properties of solvents under elevated pressure and temperature to improve the recovery of target compounds from solid or semi-solid materials (Machado et al., 2015; Sarkarat et al., 2023). This method works by increasing the temperature and pressure of the solvent, which significantly changes its physical properties, such as viscosity, surface tension, and diffusivity. This enhancement helps the solvent to penetrate the sample matrix more effectively and dissolve intracellular components more efficiently (Uwineza & Waśkiewicz, 2020; Sarkarat et al., 2023). Traditionally, the extraction of bioactive compounds, including carotenoids, has relied heavily on volatile organic solvents like hexane or acetone. These solvents are not only toxic but also harmful to the environment, consume a lot of energy, and can leave residues in the final product (Yara-Varón et al., 2016; Irianto et al., 2025; Morón-Ortiz et al., 2024). The introduction of PLE/ASE addresses these challenges by aligning with green chemistry principles. It reduces solvent usage, shortens extraction times, and enhances overall efficiency, making it a significant step forward in sustainable extraction practices (Irianto et al., 2025; Sarkarat et al., 2023).

The operational process of PLE/ASE typically involves three stages: static extraction, dynamic extraction, and separation and detection (Machado et al., 2015). During the static phase, a sealed container containing the sample and solvent is heated to a predetermined temperature, which triggers the release of compounds (Machado et al., 2015). In the subsequent dynamic phase, the solvent continuously flows through the extraction vessel, transferring the extracted compounds (Machado et al., 2015). A key feature of PLE/ASE is the precise control of extraction parameters, allowing for the optimization of temperature, pressure, and solvent composition to enhance both yield and purity (Uwineza & Waśkiewicz, 2020). The extraction typically occurs at temperatures ranging from 50 to 200 °C, and pressures between 3.5 and 20 MPa (Machado et al., 2015). These controlled conditions, especially high pressure, help break cell walls and improve solvent penetration, optimizing the transfer of bioactive compounds (Sarkarat et al., 2023). This method offers significant benefits over traditional extraction techniques, including shorter extraction times, more efficient mass transfer, higher yields, reduced solvent use, and the potential for automation (Sarkarat et al., 2023).

Pressurized Liquid Extraction (PLE), also known as Accelerated Solvent Extraction (ASE), provides significant advantages for extracting carotenoids, especially the hydrophobic β-carotene. Carotenoids are often contained within complex cell structures, and their lipophilic nature requires effective solvent penetration and dissolution (Matić et al., 2024). Studies have shown that PLE/ASE can greatly enhance carotenoid extraction efficiency. For example, research by Wang et al. (2023) demonstrated that PLE using dimethyl sulfoxide (DMSO) as a solvent increased carotenoid extraction from dried microalgae (*Chlorella vulgaris*, *Phaeodactylum tricornutum*, and *Spirulina maxima*) by 3 to 5.9 times compared to traditional maceration. Similarly, another study found a 5.6-fold increase in carotenoid content in freeze-dried Spirulina when extracted with DMSO via PLE compared to conventional stirring maceration. This method is particularly advantageous for heat-sensitive compounds like carotenoids, as the short extraction times, despite the use of elevated temperatures, help to minimize degradation, preserving the compound's integrity (Sarkarat et al., 2023).

The choice of solvents plays a crucial role in enhancing the effectiveness of PLE/ASE. While traditional organic solvents like hexane, acetone, and ethyl acetate have been commonly used, the focus in green extraction methods has shifted toward more environmentally friendly alternatives (Saini & Keum, 2018; Morón-Ortiz et al., 2024; Yara-Varón et al., 2016). Ethanol and water mixtures are often preferred for carotenoid extraction in PLE/ASE due to their eco-friendliness and versatility (Matić et al., 2024; Yara-Varón et al., 2016). For instance, Matić et al. (2024) found that a 70% v/v ethanol solution was the most effective solvent for traditional maceration of pumpkin bioactives. They also tested various ethanol concentrations (40% to 80% v/v) in ASE to account for the changes in solvent properties under high pressure and temperature conditions. This method is particularly effective for extracting more polar carotenoids by using solvent mixtures with different polarities. Furthermore, research by Tumbas Šaponjac et al. (2021) found that a 50% v/v ethanol-water mixture was an ideal solvent for carotenoid ASE from carrots, showing significant results for phenolic compound isolation. An optimized ASE method for carrot carotenoids at 80 °C for 5 minutes with a solid-to-liquid ratio of 1:4, using a mixture of 25% acetone and 75% ethanol, yielded 27 mg of total carotenoids per 100 g of raw material. This highlights how even mixtures of traditional solvents can be optimized for better efficiency and a smaller environmental footprint when used in ASE.

Although PLE/ASE offers numerous benefits, it also has its challenges. A major limitation is the high upfront cost for the specialized equipment required

(Sarkarat et al., 2023). This can be a hurdle for small and medium-sized businesses aiming to implement these advanced techniques (Uwineza & Waśkiewicz, 2020). Moreover, while PLE/ASE reduces solvent usage compared to traditional methods, a post-extraction cleaning process may still be necessary, depending on the extracted compounds and the purity required for food-grade applications (Sarkarat et al., 2023). Carotenoids, being sensitive to heat, light, and oxygen, face potential risks of thermal degradation or isomerization when exposed to the high temperatures used in PLE/ASE (Tumbas Šaponjac et al., 2021; Matić et al., 2024). While shorter extraction times can alleviate this issue, high temperatures can still cause the loss of nutritional and functional qualities (Tumbas Šaponjac et al., 2021). Therefore, careful temperature control is essential to optimize yields while preserving the integrity of these heat-sensitive compounds.

Ongoing research in PLE/ASE is focused on optimizing extraction conditions and discovering new applications for extracting β-carotene. The use of predictive modeling tools, like Artificial Neural Networks (ANN), is being explored to fine-tune the process by factoring in variables such as solvent type, temperature, extraction time, and the number of cycles (Matić et al., 2024). This approach aims to produce extracts with high bioactive potential while promoting sustainability (Matić et al., 2024). With its ability to enhance mass transfer, increase extraction yields, and reduce solvent use, PLE/ASE shows promise for large-scale industrial use (Sarkarat et al., 2023). Future research should focus on scaling the process, overcoming techno-economic challenges, and conducting thorough life cycle assessments to confirm the advantages of PLE/ASE over traditional methods (Morón-Ortiz et al., 2024). The revalorization of fruit and vegetable by-products, as seen in studies on pumpkin and carrot wastes, offers a sustainable way to produce high-value carotenoid extracts that can be incorporated into functional food products, addressing consumer demand for natural, safe ingredients while reducing waste (Matić et al., 2024; Viñas-Ospino et al., 2023).

To summarize, PLE/ASE is an efficient and eco-friendly method for extracting β-carotene and other bioactive compounds. It improves extraction efficiency, shortens processing time, and supports the use of sustainable solvents, making it an important technology for advancing sustainable practices in the food and pharmaceutical industries. Ongoing innovation to address economic and scalability challenges will strengthen its position in the future of natural product extraction.

***Pressurized Hot Water Extraction/Subcritical Water Extraction (SWE)***

Pressurized hot water extraction (PHWE), also known as subcritical water extraction (SWE), is a promising and eco-friendly method for extracting valuable bioactive compounds, such as β-carotene, from natural sources. This technique takes advantage of the unique properties of water when heated above its usual boiling point (100°C), while still being kept in a liquid form by applying enough pressure below its critical point (374°C and 22.1 MPa) (Haghighi asl and Khajenoori, 2013). At these specific conditions, water's properties change significantly, transforming from a highly polar solvent at room temperature to one that behaves more like organic solvents such as dimethyl sulfoxide (DMSO) and acetonitrile (Haghighi asl and Khajenoori, 2021). This shift in polarity is due to the weakening of hydrogen bonds and the reduction in water's dielectric constant at higher temperatures, making it more capable of dissolving non-polar compounds (Delkhani, et al., 2024; Badrkhani et al., 2013). A schematic diagram of a subcritical water extraction system is shown in Figure 6.

The core concept behind Superheated Water Extraction (SWE) is the effective penetration of superheated water into the solid structure of the raw material, followed by the dissolution of the target compound (Khajenoori et al., 2015a and 2015b). This process is especially beneficial for extracting β-carotene, a non-polar, fat-soluble compound that is typically insoluble in water at ambient temperatures (Gazmeh et al., 2025). The success of the extraction depends on how well the desired compound dissolves in the solvent, making it essential to accurately determine the solubility of β-carotene in superheated water (Gazmeh et al., 2025).

A key advantage of SWE is its alignment with green chemistry principles, which focus on environmental and human safety (Gazmeh et al., 2024, Mottahedin et al., 2017). Unlike traditional extraction methods that often depend on harmful, toxic, and flammable organic solvents—like hexane, petroleum ether, and tetrahydrofuran—SWE uses water, a non-toxic, inexpensive, widely available, and non-explosive solvent (Khajenoori et al., 2013; Mottahedin et al., 2017, Ghorbani et al., 2024). By eliminating these dangerous solvents, SWE also avoids the expensive and labor-intensive post-extraction purification processes needed to remove leftover contaminants (Gazmeh et al., 2025). As a type of Accelerated Solvent Extraction (ASE) or Pressurized Liquid Extraction (PLE), SWE offers additional benefits such as shorter extraction times, better mass transfer efficiency, higher yields, and less solvent usage, along with the potential for automation (Matić et al., 2024).

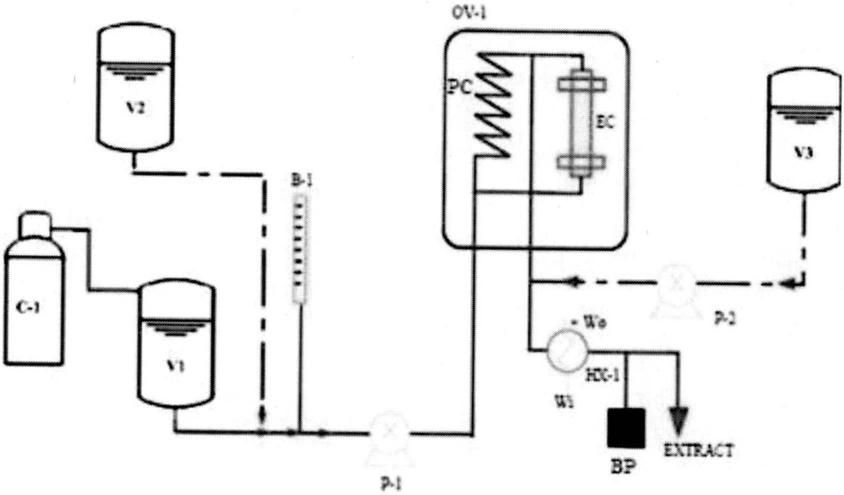


**Figure 6.** Schematic diagram of a subcritical water extraction (SWE) system. B-1: Burette, C-1: Nitrogen capsule, EC: Extraction chamber, PC: Pre-heating coil, OV-1: Oven, P-1: HPLC pump, P-2: Washing pump, V1: Water reservoir, V2: Solvent reservoir, V3: Washing solvent reservoir, HX-1: Heat exchanger, WI: Inlet water, WO: Outlet water, BP: Pressure control regulator.

Research on the use of Superheated Water Extraction (SWE) for extracting β-carotene has shown promising outcomes. Early studies, such as those by Mottahedin et al. (2017), pioneered experimental and modeling work to understand β-carotene solubility in pure and ethanol-modified subcritical water, identifying the optimal conditions for solubility. Building on this, later research, like that by Ebrahimi et al. (2019), further explored β-carotene solubility in pressurized hot water, typically within a temperature range of 70 to 130°C and at consistent pressures of around 2 MPa or 5 MPa. Additional studies by Share Mohammadi et al. (2020) and Gazmeh et al. (2025) have continued to expand our understanding of how β-carotene behaves in superheated water. These experiments have shown a dramatic increase in solubility, with one study reporting an impressive 80-fold increase compared to ambient temperature. The use of co-solvents, especially ethanol at concentrations between 0–10% (w/w or v/v), has been shown to further improve β-carotene solubility in subcritical water. SWE has also been successfully applied to extract β-carotene from a variety of rich sources, such as fruits and vegetables (Share Mohammadi et al., 2020).

Despite its advantages, using Superheated Water Extraction (SWE) for extracting β-carotene comes with some challenges. One major concern is the thermal degradation of β-carotene at high temperatures, which can reduce its solubility and bioactivity once the temperature exceeds an optimal threshold (Mottahedin et al., 2017; Matić et al., 2024). Carotenoids are naturally sensitive to light, heat, and oxygen, which makes extracting them in large quantities while maintaining their integrity a complex process (Matić et al., 2024). While experimental studies help identify the temperatures at which degradation begins, computational methods like molecular dynamics simulations can provide valuable insights into how β-carotene interacts with superheated water, helping to better understand and predict its behavior (Gazmeh et al., 2025). Additionally, like many advanced extraction techniques, the high initial investment in SWE equipment can be a barrier to its widespread industrial use (Matić et al., 2024).

Looking ahead, the focus of SWE development remains on fine-tuning extraction parameters to optimize yield while minimizing degradation. The approach of "elevated temperature extraction for shorter durations," similar to high-temperature short-time (HTST) methods in the food industry, is seen as a promising strategy for scaling up industrial applications (Kumar et al., 2024). Since SWE operates with parameters that are independent of scale and uses safe, water-based solvents, it holds significant potential for wide industrial use in creating β-carotene-enriched food products and nutraceuticals. Future studies will surely continue to refine these processes, balancing economic feasibility with environmental sustainability in the efficient extraction of this important micronutrient.

## Enzyme-Assisted Extraction (EAE)

Enzyme-Assisted Extraction (EAE) is an advanced and eco-friendly technique used for sustainably extracting β-carotene and other carotenoid pigments from various natural sources. Traditionally, extracting these valuable compounds, known for their provitamin A benefits, strong antioxidant properties, and common use as natural colorants in multiple industries, has often depended on harsh organic solvents or energy-consuming mechanical methods (Vi˜nas-Ospino et al., 2024). However, β-carotene and other carotenoids are naturally sensitive to light, heat, and oxygen, making them prone to degradation during intense extraction processes (Kumar, 2024). EAE presents a "green" and gentler solution by using specific enzymes to aid in the release of carotenoids, thus reducing environmental impact while preserving the compounds' structural integrity and bioactivity (Vi˜nas-Ospino et al., 2024). This method supports the increasing demand for natural, safe food ingredients and sustainable industrial practices (Irianto et al., 2025).

The core principle of EAE involves selectively breaking down the complex cellular structures, such as plant or microbial cell walls, that encase the intracellular carotenoids (Jiang et al., 2024). This enzymatic process increases the permeability of the cell structure, significantly enhancing the transfer of carotenoids into the extraction solvent. The primary enzymes used in EAE for carotenoid release are cell-wall-degrading enzymes like pectinases, cellulases, and hemicellulases, as well as lipases, which break down the lipid components of membranes (Jiang et al., 2024). By focusing on these structural components, enzymes reduce the need for harsh physical or chemical treatments, making the process more energy-efficient and environmentally friendly (Jiang et al., 2024). The mild conditions associated with EAE are essential for preserving the heat-sensitive properties and bioactivity of carotenoids, which are vulnerable to degradation under high temperatures and harsh chemicals (Vi˜nas-Ospino et al., 2024).

The effectiveness of EAE has been proven in various carotenoid-rich biomasses, such as microalgae, yeasts, and different plant by-products. For example, in the extraction of β-carotene from the microalga Dunaliella salina, bacterial lipase enzymes have been successfully used. Research has shown that a lipase concentration of 0.5 mg/mL could extract 21.4 μg/mL of β-carotene from the alga after 24 hours under specific salinity and nitrogen conditions. Additionally, when the enzyme concentration was increased to 1 mg/mL, the β-carotene yield (19.5 μg/mL) was similar to that achieved using traditional organic solvent methods with tetrahydrofuran (THF), indicating that lipase is an effective alternative for β-carotene extraction (Sayegh et al., 2025). Enzyme-assisted extraction has also shown promise for microbial yeasts. For instance, β-carotene was successfully extracted from *Rhodotorula glutinis* using this method, highlighting its potential for biorefinery applications in downstream processes (Hladnik et al., 2023).

EAE has also proven effective for carotenoid recovery from plant-derived materials, beyond just microbial sources. Researchers have applied EAE to extract carotenoids from carrots (Kumar et al., 2024), and specifically for β-carotene from carrots using enzyme-assisted green nanoemulsification (Jalali-Jivan et al., 2021). Additionally, research has explored the use of pectinase and cellulase enzymes to extract carotenoid-rich extracts from red capsicum and chili (*Capsicum annuum* L.) (Nath et al., 2016; Sangkharak et al., 2011). These cases highlight the versatility of EAE across various biomass types.

Despite its notable benefits, EAE faces several obstacles that currently hinder its large-scale industrial use and broad adoption. One major issue is the high cost of enzymes and the complexity involved in recovering and recycling them after the extraction, which affects the overall economic feasibility (Morón-Ortiz et al., 2024). While EAE can increase the total yield of carotenoids, it may not always provide the desired selectivity for individual carotenoids when multiple types are present in the raw material, leading to the co-extraction of other lipophilic compounds (Morón-Ortiz et al., 2024). This lack of specific selectivity can make downstream purification steps more difficult, as they are often required to isolate the target compound from other co-extracted cellular components, including the enzymes themselves, thus adding to the complexity and cost of the process (Morón-Ortiz et al., 2024). Additionally, optimizing enzymatic activity demands careful control of several factors, such as enzyme concentration, substrate-to-enzyme ratio, temperature, pH, and incubation time, all of which can vary widely depending on the biomass source and the specific enzyme mixture used (Sayegh et al., 2025). It is also important to note that, compared to other green extraction methods like ultrasound-assisted extraction (UAE) and supercritical fluid extraction (SFE), EAE has historically received less attention in scientific literature, indicating it is a less explored field (Morón-Ortiz et al., 2024; Kumar, 2024).

Future research on EAE for sustainable β-carotene extraction should focus on overcoming these key challenges. Developing more affordable enzyme production methods, including efficient enzyme immobilization techniques, and improving enzyme recovery and recycling processes are essential for enhancing the economic viability of large-scale EAE (Morón-Ortiz et al., 2024). Investigating hybrid extraction methods that combine EAE with other green technologies, like ultrasound-assisted extraction (UAE) or microwave-assisted extraction (MAE), could boost extraction efficiency, reduce processing times, and potentially address cell wall resistance more effectively (Irianto et al., 2025; Hladnik et al., 2023). Additionally, more research is needed to explore the long-term stability and bioavailability of enzymatically extracted β-carotene, especially when directly incorporated into food products, to confirm its effectiveness and market potential (Morón-Ortiz et al., 2024). Tackling the issue of selectivity by developing customizable enzyme systems or enzyme combinations that target specific carotenoids could also make significant progress in this field (Morón-Ortiz et al., 2024). Lastly, conducting comprehensive techno-economic and life cycle assessments is crucial to evaluate the true sustainability and practicality of EAE compared to current industrial standards (Morón-Ortiz et al., 2024). By addressing these challenges, EAE has the potential to become a key player in sustainable carotenoid production, reducing waste, minimizing environmental impact, and creating nutrient-enriched food products that meet the growing consumer demand for natural and safe ingredients.

## Hybrid and Combined Technologies

The growing global demand for natural and safe food ingredients, especially high-value compounds like β-carotene, has driven significant research into more efficient and sustainable extraction techniques (Irianto et al., 2025). While traditional organic solvents are associated with environmental and health risks due to their toxicity and potential residue issues (Irianto et al., 2025), single green extraction methods, despite their advantages, often struggle to achieve optimal yields or address the complex nature of some biomasses. This has led to the development of hybrid and combined technologies, which integrate multiple green extraction methods or pair them with innovative green solvents to improve efficiency, reduce energy use, minimize waste, and maintain the integrity and safety of the extracted compounds (Irianto et al., 2025; Morón-Ortiz et al., 2024; 341, Viñas-Ospino et al., 2023). These combined approaches are a significant step forward in creating truly sustainable biorefineries.

One of the most widely studied hybrid methods involves the combined or sequential use of ultrasound-assisted extraction (UAE) and microwave-assisted extraction (MAE). Both techniques are known for their ability to enhance mass transfer and break down cellular structures, which accelerates the release of intracellular compounds (Viñas-Ospino et al., 2023; Matić et al., 2024). When used together, these methods can outperform their individual applications. For example, in extracting β-carotene from sea buckthorn berries, the combined use of ultrasound and microwave increased β-carotene content by 15% and 89%, respectively, compared to using UAE and MAE alone, all within the same extraction time (Figure 7) (Staicu et al., 2024). This increased efficiency is often attributed to the complementary effects of cell disruption and improved solvent penetration. Importantly, these hybrid methods are typically paired with green solvents such as vegetable oils or fatty acid ethyl esters (FAEE). FAEE, a new green solvent made from renewable resources, is not only environmentally friendly and non-toxic but also safe for consumption, making it suitable for direct use in food supplements (Staicu et al., 2024). The addition of antioxidants like vitamin E can further improve the stability of β-carotene in these extracts (Staicu et al., 2024). Similarly, research on pumpkin and sea buckthorn pomace has shown successful carotenoid extraction using combined UAE and MAE with edible oils like corn and olive oil, demonstrating the dual benefits of enhanced extraction and clean, food-grade products (Kultys & Kurek, 2022; Jiang et al., 2024).

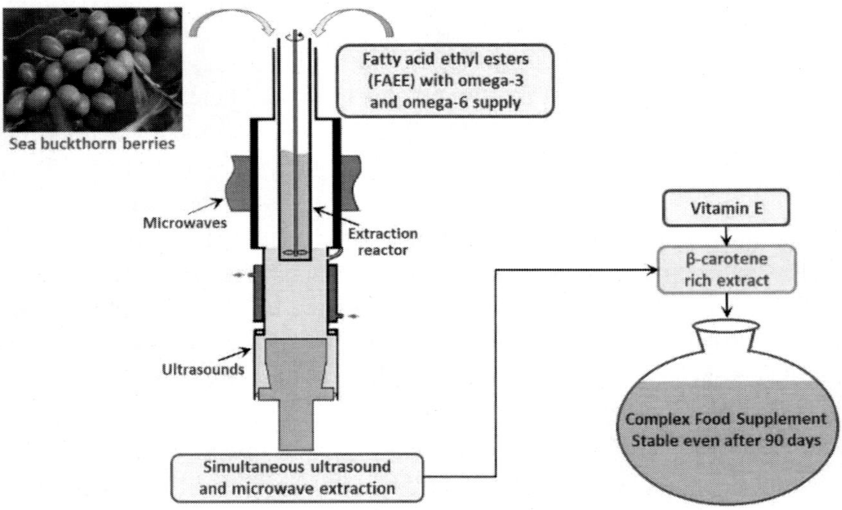


**Figure 7.** The efficient extraction of β-carotene from sea buckthorn berries using a novel solvent, fatty acid ethyl esters, and a combination of ultrasound and microwave.

Supercritical fluid extraction (SFE), especially with $CO_2$, is widely appreciated for its environmental safety and its ability to preserve heat-sensitive compounds due to its mild operating conditions and the ease of solvent removal (Irianto et al., 2025; Viñas-Ospino et al., 2023). Its efficiency is often enhanced by the use of co-solvents that alter $CO_2$'s polarity or increase its solvent capacity. Ethanol is a commonly used co-solvent, playing a critical role in efficiently co-extracting carotenoids and microbial oil from red yeasts like Rhodosporidium paludigenum, achieving high recovery rates (Sereti et al., 2025a). Optimal SFE-$CO_2$ conditions for extracting β-carotene from Dunaliella salina, for example, involved 500 bar, 70 °C, and 10 wt% ethanol, leading to more than 90% pigment recovery (Ludwig et al., 2021). In addition to ethanol, vegetable oils like canola oil have been used as continuous co-solvents in SFE from carrots, improving yields and simultaneously enriching the oil with β-carotene (Irianto et al., 2025; Sun & Temelli, 2006; Kumar, 2024). This integration results in cleaner, food-grade products without the need for extensive downstream purification (Viñas-Ospino et al., 2023). Another green solvent alternative is the use of subcritical water, which, when modified with ethanol, enhances β-carotene solubility with increasing temperature (Share Mohammadi et al., 2023; Gazmeh et al., 2025).

Further innovations in extraction methods include pulsed electric fields (PEF)-assisted extraction and enzyme-assisted extraction (EAE). PEF works by applying electric pulses that create pores in cell membranes, facilitating the release of intracellular compounds and improving the efficiency of subsequent extraction. It has been effectively used to enhance carotenoid extraction from yeasts like Rhodotorula glutinis and to boost the bioaccessibility of carotenoids in carrot purees, particularly when combined with oil (Martínez et al., 2020). EAE utilizes specific enzymes to break down cell walls and matrices, which makes carotenoids more accessible to solvents. This can result in improved extraction efficiency and stability of β-carotene, as demonstrated in enzyme-assisted green nanoemulsification for carrot (Jalali-Jivan et al., 2021), and in lipase enzyme-assisted extraction of β-carotene from Dunaliella salina microalgae (Sayegh et al., 2025).

The strategic use of these technologies in combination with various green solvents is crucial. In addition to vegetable oils and ethanol, deep eutectic solvents (DES) and natural deep eutectic solvents (NADES) offer adjustable properties for targeted extraction. Hydrophobic variants of these solvents can dissolve water-insoluble natural products and help stabilize them during storage (Vi˜nas-Ospino et al., 2024; Morón-Ortiz et al., 2024; Kultys & Kurek, 2022). For instance, a hydrophobic NADES based on fatty acids was effective in extracting β-carotene from pumpkin by-products using ultrasound-assisted extraction (UAE) (Viñas-Ospino et al., 2023). Similarly, ionic liquids (ILs), despite challenges like high viscosity and cost, have shown potential in increasing carotenoid yields through UAE (Morón-Ortiz et al., 2024). Additionally, novel bio-based solvent systems such as 2-methyl tetrahydrofuran (2-MeTHF) and cyclopentyl-methyl-ether (CPME), combined with alcohols and water, are being explored for their effectiveness in extracting hydrophobic compounds like β-carotene and lipids from wet yeast biomass, offering a sustainable alternative to conventional toxic organic solvents (Buchweitz et al., 2025).

**Table 4.** Overview of Sustainable B-carotene extraction methods

| Extraction Method | Solvents | Typical/Optimal Conditions | Advantages | Disadvantages/Challenges | References |
|---|---|---|---|---|---|
| Supercritical Fluid Extraction (SFE) | Supercritical $CO_2$, co-solvents (ethanol, vegetable oils) | $CO_2$ at 31.1°C and 7.38 MPa; process involves supercritical $CO_2$ exposure to raw material to extract non-polar compounds like β-carotene. | Non-toxic, non-flammable, Generally Recognized As Safe (GRAS) certified. Reduces solvent residue, safe for food & pharmaceutical use. Preserves β-carotene. | High energy demand; requires specialized equipment. Solvent recovery and system optimization still challenges at industrial scale. | Irianto et al., 2025; Sereti et al., 2025a; Viñas-Ospino et al., 2023 |
| Pressurized Liquid Extraction (PLE) | Water, ethanol, acetone, and green solvents like ethanol-water mixtures | 50–200°C, 3.5–20 MPa; high-pressure solvent (ethanol, water) to extract bioactive compounds from plant biomass like carrots, sea buckthorn. | Short extraction time, reduced solvent use, high yield, eco-friendly compared to traditional methods, suitable for a wide range of plant materials. | High initial equipment costs; requires precise control of parameters to avoid degradation of heat-sensitive compounds. | Irianto et al., 2025; Morón-Ortiz et al., 2024; Yara-Varón et al., 2016 |
| Subcritical Water Extraction (SWE) | Water, co-solvents (ethanol) | Superheated water at elevated temperatures with co-solvents like ethanol for β-carotene extraction. | Simple and cost-effective; avoids toxic solvents; environmentally friendly. | Thermal degradation of β-carotene at high temperatures; scaling for industrial use and the need for co-solvent optimization. | Share Mohammadi et al., 2020; Gazmeh et al., 2025; mottahedin et al., 2017; Matić et al., 2024 |
| Microwave-Assisted Extraction (MAE) | Vegetable oils, ethanol, water, deep eutectic solvents (DES), fatty acid ethyl esters | Microwave energy used to rapidly heat solvents and raw material, causing efficient extraction of carotenoids in minutes. Optimal at lower temperatures and reduced time. | Faster extraction, lower solvent use, energy efficient, reduces thermal degradation compared to conventional methods. | Potential thermal degradation of sensitive compounds like β-carotene; scalability and high equipment costs remain a challenge. | Irianto et al., 2025; Viñas-Ospino et al., 2024; Morón-Ortiz et al., 2024 |

| Extraction Method | Solvents | Typical/Optimal Conditions | Advantages | Disadvantages/Challenges | References |
|---|---|---|---|---|---|
| Ultrasound-Assisted Extraction (UAE) | Vegetable oils, ethanol, water, natural deep eutectic solvents (NaDES) | 30°C, 500 W, 20 kHz for 80 min; used for carrot pomace, passion fruit peel, etc. | Enhanced mass transfer, energy efficiency, shorter extraction times. Can be combined with green solvents like vegetable oils and NaDES. | Potential thermal degradation if temperature is not controlled; scaling for industrial use can be difficult due to high equipment costs. | Irianto et al., 2025; Morón-Ortiz et al., 2024; Viñas-Ospino et al., 2023 |
| Enzyme-Assisted Extraction (EAE) | Enzymes (pectinases, cellulases, lipases) | Uses specific enzymes to break down cell walls, increasing carotenoid bioavailability. Often used for microalgae or plant by-products. | Green and gentle extraction method, reduces energy and solvent use, preserves bioactivity of carotenoids. | Requires enzyme optimization for different matrices; scaling up for industrial use requires further study. | Viñas-Ospino et al., 2024; Irianto et al., 2025; Jiang et al., 2024 |
| Hybrid and Combined Technologies | Green solvents, Fatty Acid Ethyl Esters (FAEE), vegetable oils, DES, NADES | Combination of ultrasound and microwave-assisted extraction (UAE and MAE) for β-carotene extraction. | Increased efficiency, higher yields, better solvent penetration, non-toxic and biodegradable solvents like FAEE. | Complexity of combining methods; need for optimal process control; higher initial costs for hybrid systems. | Viñas-Ospino et al., 2023; Staicu et al., 2024; Kultys & Kurek, 2022 |

Despite promising results at the laboratory scale, scaling these hybrid technologies for industrial use remains a significant challenge (Irianto et al., 2025; Morón-Ortiz et al., 2024; Morón-Ortiz et al., 2024). Key challenges include high initial investments for specialized equipment, regulatory compliance, and the optimization of parameters for large-scale operations. Future research must focus on refining extraction conditions, enhancing scalability, reducing overall costs, and conducting comprehensive life cycle assessments to fully validate the sustainability of these processes (Irianto et al., 2025; Morón-Ortiz et al., 2024; Morón-Ortiz et al., 2024). Continued innovation in hybrid and combined extraction methods holds great promise for the future of sustainable β-carotene production and its broader applications in functional foods and other industries. Subsections 3.1 to 3.6 are summarized in Table 4.

## Preserving Beta-Carotene Quality and Stability

### Degradation Pathways (Oxidation, Isomerization)

Despite its impressive bioactivity and broad use across the food, cosmetics, and pharmaceutical industries, β-carotene is inherently unstable, presenting significant challenges during its extraction, processing, and storage (Kumar, 2024). This instability is mainly due to its structure, which consists of a long chain of conjugated double bonds, making it highly susceptible to degradation through oxidation and isomerization (Buchweitz et al., 2025).

Oxidation is the primary and most significant process that leads to β-carotene degradation, occurring easily when it's exposed to light, heat, and oxygen (Kumar, 2024). As a potent antioxidant, β-carotene helps neutralize free radicals and deactivate singlet oxygen, which are major contributors to oxidative stress and cellular damage in living organisms (Share Mohammadi et al., 2020; Gazmeh, et al., 2025). However, this antioxidant action also makes β-carotene itself vulnerable to degradation, especially under prolonged or harsh conditions. Incorporating external antioxidants, like vitamin E, has been shown to significantly enhance the stability of β-carotene extracts, preserving their integrity even during prolonged exposure to light and boosting their health benefits (Staicu et al., 2024; Kumar, 2024). Additionally, using green solvents such as vegetable oils during extraction can act as a protective barrier to oxygen, further reducing the oxidative degradation of β-carotene and other fat-soluble bioactive compounds (Kumar, 2024; Kultys & Kurek, 2022).

In addition to oxidation, isomerization, particularly the cis-trans isomerization, plays a significant role in the degradation of β-carotene, impacting both its stability and bioactivity (Adriana Viñas-Ospino et al., 2023; Kumar, 2024). The all-trans isomer is usually the most abundant and stable form found in nature. However, certain processing methods, like microwave-assisted extraction (MAE), can cause it to convert into cis-isomers, such as 9-cis and 13-cis (Viñas-Ospino et al., 2023; Kumar, 2024). Although more research is needed to fully understand the biological significance and functions of these isomers (Kumar, 2024), their presence can affect the quality and nutritional value of the extract. Therefore, it's crucial to manage and control isomerization during processing to preserve the desirable properties of β-carotene extracts.

The impact of temperature on β-carotene degradation is complex, often presenting a trade-off in extraction processes. Higher temperatures can enhance solubility and accelerate mass transfer, improving extraction efficiency (Kultys & Kurek, 2022; Staicu et al., 2024). However, they can also increase the rate of degradation of β-carotene (Staicu et al., 2024). Research has shown that extraction yields improve with temperature up to a certain point (e.g., 85°C in subcritical water or 60°C with ethanol), but temperatures beyond that threshold may cause thermal degradation, which reduces the benefits of improved extraction (Mottahedin et al., 2017; Ludwig et al., 2021). On the other hand, some studies suggest that temperatures as high as 150°C can still be effective for extracting β-carotene into edible oils, as the extraction rate remains high despite some degradation, potentially offering advantages over more costly methods (Kumar, 2024). This highlights the importance of balancing the rate of β-carotene transfer from the matrix into the solvent with its degradation rate (Kumar, 2024). Therefore, selecting and optimizing extraction conditions is key to minimizing degradation and maximizing the recovery of high-quality β-carotene for sustainable applications.

## Strategies for Stability (Antioxidants, Controlled Conditions, Encapsulation)

The previous section, "Degradation Pathways (Oxidation, Isomerization)," discussed the inherent vulnerability of β-carotene due to its structure, making it highly susceptible to oxidative and isomeric degradation when exposed to environmental factors like light, heat, and oxygen. Consequently, developing effective methods to maintain β-carotene's stability and bioactivity is crucial for its successful application in the food, pharmaceutical, and cosmetic industries. To achieve this, strategies typically involve using antioxidants, carefully controlling processing and storage conditions, and applying advanced encapsulation techniques.

One effective approach to reduce β-carotene degradation is by incorporating antioxidant compounds. While β-carotene itself acts as a potent antioxidant, its degradation can accelerate under harsh conditions (Gul et al., 2015). Adding external antioxidants like vitamin E has been shown to greatly improve the stability of β-carotene extracts, preserving their integrity even when exposed to light for extended periods, while enhancing their overall health benefits (Staicu et al., 2024). Additionally, certain natural compounds like L-proline have antioxidant properties that protect carotenoids during extraction, thus maintaining their bioactivity (Viñas-Ospino 2023). Terpenoids and fatty acids, which are often oily, also improve the stability and bioavailability of natural carotenoids by acting as stabilizers (Saini & Keum, 2018).

Managing environmental factors during extraction, processing, and storage is essential for preserving the stability of β-carotene. Temperature is a key factor to consider: while higher temperatures can enhance extraction efficiency by increasing solubility and improving mass transfer, they can also accelerate degradation processes (Kumar et al., 2024). Therefore, optimizing the temperature-time profile is critical to maximizing extraction yields while minimizing degradation. This often involves using high-temperature, short-time (HTST) methods, similar to those used in the food industry, to balance the advantages of better extraction with minimal harm to the β-carotene (Kumar, 2024).

In addition to managing temperature, it is crucial to reduce exposure to light and oxygen to prevent oxidative degradation. Green solvents, such as vegetable oils (e.g., sunflower, corn, and olive oil), serve as effective barriers to oxygen, helping to slow down oxidation during both extraction and storage. These oils are food-grade and non-toxic, making them well-suited for use in food products (Portillo-López et al., 2021; Kultys & Kurek, 2022). Hydrophobic deep eutectic solvents (HDESs) and bio-based solvents like 2-methyltetrahydrofuran (2-MeTHF) and cyclopentyl-methyl-ether (CPME) have also shown effectiveness in dissolving and protecting carotenoids, even at higher temperatures (Stupar et al., 2021; Buchweitz et al., 2025).

Supercritical $CO_2$ extraction is another effective method, using controlled temperature and pressure to achieve a gentle extraction that preserves heat-sensitive compounds like β-carotene while minimizing solvent residues. This is crucial for maintaining the integrity of the product (Irianto et al., 2025). Similarly, subcritical water, which enhances the solubility of β-carotene at higher temperatures by reducing water's polarity, shows promise for both extracting and stabilizing non-polar compounds like β-carotene (Gazmeh, et al., 2025).

Encapsulation stands out as an effective method for protecting β-carotene and improving its stability, water-dispersibility, and bioavailability, especially when integrated into aqueous food matrices. Different encapsulation methods, such as emulsion-based delivery systems and nanoparticles, create protective barriers around the β-carotene molecule, preventing oxidative and isomeric degradation (Chen et al., 2017; Jiang et al., 2024). For instance, encapsulating β-carotene in ferritin nanocages greatly improves its water solubility and thermal stability (Chen et al., 2014).

Encapsulated emulsions, like those made from carrot waste using flaxseed oil, have demonstrated high encapsulation efficiencies, making them ideal for direct incorporation into food products such as flavored milk-based drinks (Tiwari et al., 2021; Tiwari et al., 2022). These systems help ensure that carotenoids stay stable and bioavailable, even in food products that are exposed to light, heat, and oxygen. Future research will be essential in further refining these encapsulation systems to guarantee their long-term effectiveness and maintain high bioaccessibility of the encapsulated β-carotene (Molteni et al., 2022). By combining strategies such as antioxidant integration, precise process control, and advanced encapsulation, we can fully unlock the potential of β-carotene for sustainable, health-boosting applications across various industries.

## Importance of Downstream Processing and Direct Incorporation into Food Products

As previously mentioned, the extraction of β-carotene, has relied on toxic organic solvents such as hexane, acetone, tetrahydrofuran, petroleum ether, diethyl ether, chloroform, and ethyl acetate (Irianto et al., 2025). These solvents pose significant risks to both human health and the environment, as they can leave harmful residues in food products and contribute to pollution during manufacturing and disposal (Irianto et al., 2025; Viñas-Ospino et al.,

2023). As a result, traditional extraction methods require extensive and costly purification processes to remove these toxic residues, which can also lead to the loss or alteration of valuable compounds (Jiang et al., 2024). This challenge highlights the urgent need for innovative extraction methods that either simplify or eliminate complex downstream processing, particularly when the substances are intended for food use (Viñas-Ospino et al., 2023; Vi˜nas-Ospino et al., 2024).

A major shift towards green extraction techniques presents a powerful solution by addressing the challenges of downstream processing and allowing for the direct incorporation of β-carotene into food products. Green solvents, known for their non-toxicity, biodegradability, and renewable origin, help reduce environmental impact while improving safety (Morón-Ortiz et al., 2024; Staicu et al., 2024). Importantly, many of these solvents are non-volatile and are made from Generally Recognized As Safe (GRAS) components, which have been approved by regulatory bodies like the Food and Drug Administration (FDA) and the European Commission for use as food additives (Vi˜nas-Ospino et al., 2024). This means that, unlike traditional solvents, these green alternatives usally do not need to be separated from the extracted β-carotene and can be directly incorporated into bioactive formulations or food products (Viñas-Ospino et al., 2023). This direct incorporation eliminates the need for costly and energy-consuming recovery processes typically involved with harmful solvents, improving both efficiency and cost-effectiveness (Viñas-Ospino et al., 2023; Vi˜nas-Ospino et al., 2024).

Green solvents and extraction techniques have shown great promise for directly incorporating β-carotene extracts into food products. Vegetable oils like sunflower, flaxseed, and corn oils are prime examples. These oils are excellent for extracting lipophilic carotenoids and can be directly added to food products as carriers or natural colorants, boosting both nutritional value and consumer appeal (Vi˜nas-Ospino et al., 2024). For example, β-carotene extracted into sunflower oil can be used as a frying medium for fortified products, helping address nutritional deficiencies. Similarly, flaxseed oil extracts can be turned into emulsions for milk-based drinks, adding valuable omega-3 fatty acids (Kumar, 2024; Viñas-Ospino et al., 2023). Supercritical $CO_2$ extraction (SFE) also produces extracts with little to no solvent residues, making them ready for use as food additives, as demonstrated with mango peel extracts that help protect edible oils from oxidation (Irianto et al., 2025). Deep Eutectic Solvents (DES) and Natural Deep Eutectic Solvents (NADES), particularly the hydrophobic ones, are designed to dissolve water-insoluble natural products like β-carotene, while also protecting them during storage and at high temperatures (Vi˜nas-Ospino et al., 2024). Certain NADES extracts are "ready-to-use" in food products, eliminating the need for complex purification processes and potentially extending shelf life due to their built-in antioxidant properties (Vi˜nas-Ospino et al., 2024). Additionally, fatty acid ethyl esters (FAEE) are another novel green solvent that can be directly incorporated, offering extraction efficiency while adding omega-3 and omega-6 fatty acids to the final product (Staicu et al., 2024).

The ability to directly incorporate β-carotene extracts into food products without the need for extensive downstream processing marks a major step forward in sustainable food production. This approach aligns with the increasing consumer demand for natural, safe, and eco-friendly food ingredients, supporting a "zero waste" industrial model (Viñas-Ospino et al., 2023). However, more research is needed to thoroughly evaluate the purity, compositional selectivity, and sensory effects of these green extracts when incorporated into different food systems (Viñas-Ospino et al., 2023). Additionally, assessing the specific biological properties of these directly incorporated extracts through cellular assays remains a key area for further exploration (Vi˜nas-Ospino et al., 2024). Addressing the techno-economic challenges of scaling these methods to industrial levels will be crucial for their widespread adoption and realizing their full potential in improving global nutrition and health (Morón-Ortiz et al., 2024).

## Challenges and Future Perspectives

### Scalability to Industrial Level: Techno-Economic Considerations and Life Cycle Assessment (LCA)

The growing global demand for natural carotenoids, driven by their health benefits and wide-ranging uses in food, health products, and medicine, highlights the need for more efficient and sustainable extraction methods (Jiang et al., 2024). Although laboratory-scale studies on green extraction techniques often show promising results, scaling these methods to an industrial level remains a significant challenge (Morón-Ortiz et al., 2024). Major technological obstacles, such as the high costs of infrastructure and limitations in production capacity, often hinder this scaling process. Despite more than two decades of research on methods like ultrasound-assisted extraction (UAE) and microwave-assisted extraction (MAE), large-scale commercial applications for β-carotene extraction using these technologies are still largely absent (Kumar, 2024).

From a techno-economic perspective, the economic feasibility of green extraction methods is crucial. While these methods are designed to be cost-effective (Irianto et al., 2025), advanced systems like Accelerated Solvent Extraction (ASE), or Pressurized Liquid Extraction (PLE), often require a significant initial investment (Matić et al., 2024). Additionally, the energy consumption of green extraction methods can vary. Although many green techniques aim to reduce energy usage (Irianto et al., 2025), certain methods like Supercritical Fluid Extraction (SFE) can demand higher energy due to the intense compression needed for carbon dioxide ($CO_2$) (Ludwig et al., 2021). However, these energy costs are often offset by much lower solvent costs, as green solvents are typically recoverable and reusable. For example, SFE requires a much lower solvent-to-biomass ratio than traditional n-hexane extraction, which helps reduce solvent costs overall and could make the extraction process more cost-efficient (Ludwig et al., 2021).

A full assessment of sustainability requires the inclusion of Life Cycle Assessment (LCA), which goes beyond just looking at carbon footprints to include broader environmental impacts throughout the entire process (Morón-Ortiz et al., 2024; Kumar, 2024). Green extraction technologies provide several environmental benefits, such as lower energy consumption, reduced waste production, and the use of non-toxic or biodegradable solvents (Irianto et al., 2025). This helps eliminate hazardous waste and the risk of harmful solvent residues in food products, which are common problems with traditional organic solvents (Irianto et al., 2021). Tools like the Eco-Scale and the Blue Applicability Grade Index (BAGI) are increasingly used to measure the environmental friendliness and feasibility of these extraction methods (Vi˜nas-Ospino et al., 2024, 2024).

Looking to the future, ongoing research is crucial for refining extraction conditions, improving scalability, and ensuring that green technologies meet regulatory standards (Irianto et al., 2025). One ongoing challenge is achieving high purity in the extracts and ensuring a selective carotenoid composition, as these lipophilic compounds often co-extract with other substances (Morón-Ortiz et al., 2024). Innovations like customizable Deep Eutectic Solvents (DES) and Natural Deep Eutectic Solvents (NaDES) may offer a solution to improving selectivity (Morón-Ortiz et al., 2024). The combination of green solvents and advanced extraction methods holds significant potential for the future of carotenoid extraction and for advancing sustainability across various industries (Vi˜nas-Ospino et al., 2024). Future research should also focus on assessing the biological properties of the extracts through cellular assays, ensuring they are effective for their intended uses (Vi˜nas-Ospino et al., 2024).

## Selectivity and Purity of Extracts

The quest for sustainable extraction methods for β-carotene is closely tied to the challenge of achieving high selectivity and purity in the final product. B-carotene, being a highly lipophilic compound, is often found in complex biological mixtures alongside other lipophilic substances like carotenoid isomers, pigments, fatty acids, and polar compounds (Matić et al., 2024; Morón-Ortiz et al., 2024). This complexity often results in the co-extraction of these substances, making it difficult to isolate pure β-carotene (Morón-Ortiz et al., 2024; Matić et al., 2024; Kultys & Kurek, 2022). Traditional organic solvents, like hexane and acetone, have historically been used for extraction but are known for their lack of selectivity, often requiring costly post-extraction purification to remove unwanted co-extracted compounds and toxic solvent residues (Irianto et al., 2025; Morón-Ortiz et al., 2024; Kultys & Kurek, 2022). With the growing global demand for natural and safe food ingredients, there is an increasing need for extraction technologies that naturally provide better selectivity and purity to meet strict regulatory standards and consumer preferences (Irianto et al., 2025; Jiang et al., 2024).

Supercritical Fluid Extraction (SFE), especially with $CO_2$, is considered a “green” technology due to its ability to be finely tuned for selectivity (Morón-Ortiz et al., 2024; Irianto et al., 2025). By adjusting temperature and pressure, the solvent power of supercritical $CO_2$ can be precisely controlled, allowing it to target specific compounds like β-carotene, while reducing the extraction of unwanted impurities (Irianto et al., 2025; Morón-Ortiz et al., 2024). For instance, optimizing the SFE process with 15.5% ethanol as a co-solvent led to efficient recovery of carotenoids from vegetable waste, showing how well it extracts lipophilic compounds (de Andrade Lima et al., 2018; Kultys & Kurek, 2022). One of the main benefits of SFE is the ease with which $CO_2$ is removed, as it volatilizes under normal conditions, leaving minimal solvent residues. This contributes to the purity and safety of the final product, which is essential for food and pharmaceutical applications (Ludwig et al., 2021; Irianto et al., 2025; Viñas-Ospino et al., 2023). However, adding co-solvents, while improving yield, can sometimes reduce the overall selectivity, making it necessary to carefully optimize the process (Ludwig et al., 2021).

Beyond SFE, other innovative green solvents offer unique advantages in enhancing selectivity and purity. Deep Eutectic Solvents (DES) and Natural Deep Eutectic Solvents (NaDES) are gaining attention for their customizable nature, allowing for more precise extraction of specific carotenoids or related compounds (Morón-Ortiz et al., 2024; Kultys & Kurek, 2022). Hydrophobic DES, made from fatty acids or terpenoids, are especially good at dissolving water-insoluble natural products like carotenoids. They can even protect these compounds during storage, which makes downstream processing more straightforward (Morón-Ortiz et al., 2024; Viñas-Ospino et al., 2023). Another group of solvents, vegetable oils, are excellent for extracting lipophilic carotenoids such as β-carotene. The advantage here is that the extract can be directly incorporated into food products, eliminating the need to remove a separate solvent (Viñas-Ospino et al., 2023; Kumar et al., 2024). This method avoids the purification issues associated with solvent residues, although it may not be highly selective among various lipophilic components (Viñas-Ospino et al., 2023). Similarly, new bio-based solvent systems, like 2-methyltetrahydrofuran (2-MeTHF) and cyclopentyl-methyl-ether (CPME), used with ethanol or 1-butanol and water, are being explored for their ability to facilitate a two-step extraction process. This method separates hydrophobic target compounds, including β-carotene, from polar impurities (Buchweitz et al., 2025). The biphasic approach aims to improve purity through phase separation, instead of relying only on the solvent's selectivity.

Preserving the integrity of carotenoids, especially the highly desirable all-trans isomer of β-carotene, is crucial for maintaining extract purity and bioactivity (Viñas-Ospino et al., 2023; Kumar et al., 2024; Morón-Ortiz et al., 2024). B-carotene is highly sensitive to heat, light, and oxygen, which can lead to oxidative degradation and conversion into less active cis isomers, reducing both the purity and nutritional value of the extract (Kultys & Kurek, 2022; Kumar et al., 2024; Staicu et al., 2024). While higher temperatures can improve solubility and mass transfer, it's essential to carefully control the extraction time and temperature to prevent degradation (Kultys & Kurek, 2022; Staicu et al., 2024). For intracellular carotenoids, effective cell disruption before extraction is key to maximizing recovery. This also helps improve the purity of the final product by ensuring better contact between the solvent and the target compound (Sereti et al., 2025a; Mussagy et al., 2019; Sayegh et al., 2025). For example, homogenization is an effective pre-treatment for achieving high yields of carotenoids and microbial oil from yeast biomass, enhancing process efficiency and the concentration of target compounds (Sereti et al., 2025a).

Even with advancements in green extraction technologies, achieving high purity and selective carotenoid composition remains a significant challenge, especially when the raw material contains various co-extracted lipophilic compounds or different carotenoid isomers (Morón-Ortiz et al., 2024; Matić et al., 2024). While laboratory results for green extraction methods show promise, scaling them for industrial use requires further techno-economic and life cycle assessments to ensure their sustainability and cost-effectiveness, particularly when extract purity is a priority (Morón-Ortiz et al., 2024; Irianto et al., 2025). Moving forward, research should focus on developing more advanced separation techniques to improve the selectivity of targeted carotenoids from complex mixtures. Innovations like customizable Deep Eutectic Solvents (DES) and Natural DES (NaDES) could be highly beneficial, enabling more precise targeting of specific β-carotene isomers or even separating colorless carotenoids from the colored ones (Morón-Ortiz et al., 2024). Additionally, evaluating the biological properties of purified extracts through cellular assays is essential to confirm their effectiveness for food, health, and cosmetic applications (Morón-Ortiz et al., 2024). This ongoing research and development will lay the foundation for greener, more efficient, and highly selective β-carotene extraction processes that align with both industrial demands and environmental goals.

## Integration into Circular Economy and Functional Food Development

The growing global demand for natural and safe food ingredients, combined with a deeper understanding of their health benefits, is driving a shift toward sustainable extraction methods for bioactive compounds like β-carotene (Jiang et al., 2024; Morón-Ortiz et al., 2024). This shift supports the principles of the circular economy, focusing on reusing agricultural and food processing by-products to reduce waste and improve resource efficiency (Kumar, 2024). Historically, extracting β-carotene relied on organic solvents that present significant environmental and health risks due to their toxicity, high energy consumption, and the production of hazardous waste (Morón-Ortiz et al., 2024). The move toward green extraction technologies addresses these challenges, offering a safer, more efficient, and cost-effective method of extraction (Vi˜nas-Ospino et al., 2024).

A key element in this shift is the use of green solvents, which include supercritical fluids, vegetable oils, ionic liquids (ILs), deep eutectic solvents

(DES), and terpenoids (Morón-Ortiz et al., 2024). These solvents are favored over traditional organic solvents because they are typically non-toxic, biodegradable, and made from renewable resources (Morón-Ortiz et al., 2024). Their gentle extraction conditions help maintain the structural integrity and bioactivity of carotenoids, which is crucial for industrial applications (Vi˜nas-Ospino et al., 2024). For example, hydrophobic DES, which are made from fatty acids or terpenoids, can dissolve naturally occurring products that don't dissolve well in water and protect them during storage and high-temperature exposure, opening up new opportunities for functional foods (Vi˜nas-Ospino et al., 2024). Similarly, vegetable oils such as sunflower, peanut, and olive oils provide better solubility for lipophilic compounds like carotenoids, with no significant degradation occurring during extraction (Vi˜nas-Ospino et al., 2024). Supercritical $CO_2$ (SFE-$CO_2$) is another widely used green technique, known for its ability to penetrate plant structures like a gas while dissolving compounds like a liquid. This increases efficiency and allows for selective extraction without leaving toxic residues (Irianto et al., 2025).

One of the key benefits of using green solvents for β-carotene extraction is the ability to directly incorporate the extracted compounds into food products, eliminating the need for expensive and time-consuming purification processes to remove solvent residues (Viñas-Ospino et al., 2023). This "ready-to-use" approach simplifies manufacturing and makes food products fortified with natural ingredients more appealing to consumers (Viñas-Ospino et al., 2023). For example, carotenoids extracted with sunflower oil from mandarin peels have been successfully used as natural colorants in bakery products, offering an excellent alternative to synthetic additives like tartrazine (Kumar, 2024). Similarly, β-carotene extracted from carrot by-products using flaxseed oil can be encapsulated in emulsion-based delivery systems and used in flavored milk-based drinks, meeting both nutritional needs and the growing demand for natural functional colorants (Viñas-Ospino et al., 2023). This is particularly important in regions facing Vitamin A Deficiency (VAD), where fortifying staple foods like potato crisps with β-carotene-enriched oil offers a practical solution to improving nutrition (Kumar, 2024). Additionally, carotenoids are known for their health benefits, including their antioxidant properties, immune system support, and ability to protect against various diseases, further highlighting their value in functional food development (Staicu et al., 2024).

While laboratory results for green extraction methods are promising, scaling them up for industrial use remains a significant challenge. This is often due to techno-economic barriers, such as limited production capacity, high infrastructure costs, and the expenses associated with the final product (Morón-Ortiz et al., 2024). More research is essential to optimize extraction conditions, improve scalability, ensure regulatory compliance, and develop cost-effective solutions (Morón-Ortiz et al., 2024; Irianto et al., 2025). Additionally, thorough life cycle assessments are needed to determine if these innovative green extraction methods are truly more sustainable than current industrial practices (Morón-Ortiz et al., 2024). Future research should also focus on hybrid extraction techniques that combine the advantages of different green technologies and refine solvent selectivity to achieve higher purity extracts for specific carotenoids (Morón-Ortiz et al., 2024). Ongoing advancements in green chemistry and engineering are key to unlocking the full potential of β-carotene extraction within a circular economy framework, providing high-value, safe, and sustainable ingredients for the functional food industry (Morón-Ortiz et al., 2024).

## Role of Modeling and Simulation Tools in Process Design

Achieving sustainable β-carotene extraction requires a more advanced approach to process design, moving beyond simple, trial-and-error methods to incorporate advanced modeling and simulation tools (Matić et al., 2024). These computational and statistical techniques are essential for understanding the complex physical and chemical processes involved in extraction, predicting the best conditions, and ultimately improving efficiency and sustainability. By using these tools, researchers can simulate interactions, reduce the need for costly and time-consuming experiments, conserve resources, and speed up the development of processes that are scalable to industrial levels (Buchweitz et al., 2025).

Various modeling techniques play a crucial role in designing an effective extraction process. Kinetic models, for example, are essential for understanding the time-dependent behavior of β-carotene extraction, factoring in both its transfer from the solid material into the solvent and its degradation in both phases. Research on elevated temperature extraction of β-carotene into edible oils has used these models to show that, despite some thermal degradation, higher extraction rates can be achieved by carefully balancing these kinetic factors (Kumar et al., 2024). Thermodynamic models, like the Cubic-Plus-Association Equation of State (CPA EOS) and the modified Apelblat model, are used to predict and correlate the solubility of β-carotene in various solvents, including subcritical water and ethanol-modified subcritical water (Share Mohammadi et al., 2020). Understanding solubility is key as it directly influences extraction efficiency and helps determine the best green solvents to use (Gazmeh, et al., 2025).

In addition to basic modeling, advanced simulation and statistical optimization tools play a crucial role in refining the extraction process. Molecular dynamics (MD) simulations, for example, provide insights into the microscopic interactions between β-carotene and green solvents like superheated water, showing how the solvent properties change at the molecular level to enhance dissolution and diffusion (Gazmeh, et al., 2025). The Conductor-like Screening Model for Real Solvents (COSMO-RS) is also used to predict the solubility of β-carotene, helping to select the best green and bio-based solvent mixtures, which simplifies the solvent screening process (Buchweitz et al., 2025). Additionally, Response Surface Methodology (RSM), often paired with designs like Box-Behnken Design (BBD), is widely applied for experimental optimization, linking variables like temperature, time, and solvent ratio to maximize carotenoid yields (Lopes et al., 2024). Artificial Neural Network (ANN) modeling provides another powerful tool for predicting and optimizing extraction parameters, even with limited data, which is especially valuable for complex green extraction methods like Accelerated Solvent Extraction (ASE) (Matić et al., 2024).

While laboratory results for green extraction methods are promising, scaling these methods for industrial use remains a significant challenge, often due to barriers like infrastructure costs and the expenses associated with the final product (Morón-Ortiz, et al., 2024). Techno-economic assessments and life cycle analyses (LCA) are essential in addressing these challenges. They provide a comprehensive evaluation of a process's sustainability by comparing factors such as energy consumption, solvent costs, and the overall environmental impact to those of traditional methods (Morón-Ortiz, et al., 2024; Ludwig et al., 2021). Moving forward, these tools will be increasingly vital for optimizing hybrid extraction methods, improving solvent selectivity for specific carotenoid isomers, and ensuring that new processes are not only efficient but also truly sustainable over their entire life cycle (Irianto et al., 2025; Morón-Ortiz, et al., 2024).

## Conclusion

Sustainable methods for extracting β-carotene are driving a significant change in how valuable bioactive compounds are sourced. Traditionally, the extraction of β-carotene relied heavily on using toxic, flammable solvents like hexane and acetone, which are harmful to both health and the environment. As concerns grow over these risks, there is an increasing push for safer, eco-friendly alternatives. The demand for natural, safe ingredients in food and cosmetics, along with stricter regulations, has made it crucial to adopt more sustainable methods. The green extraction technologies covered in this chapter, such as supercritical fluid extraction (SFE), ultrasound-assisted extraction (UAE), and microwave-assisted extraction (MAE), offer effective solutions, reducing energy use, waste, and avoiding harmful solvent residues. Among these, supercritical $CO_2$ extraction stands out because it can selectively extract lipophilic compounds like β-carotene with minimal environmental impact. By using $CO_2$, a non-toxic and non-flammable solvent, high-purity β-carotene can be recovered without extensive post-extraction purification. This not only results in a cleaner product but also preserves β-carotene's bioactive properties, which are sensitive to heat and light. Similarly, ultrasound-assisted and microwave-assisted extraction methods speed up mass transfer and improve the efficiency of carotenoid extraction, further boosting the sustainability of the process by reducing time and solvent use. These methods work even better when paired with green solvents like vegetable oils or deep eutectic solvents (DES), which are biodegradable and less toxic.

The shift towards sustainable extraction methods is not only an environmental necessity but also an economic one. Although green extraction technologies may require a higher initial investment, they offer long-term savings by reducing solvent use and cutting energy consumption. However, scalability remains a challenge. While laboratory studies show promising results, scaling these methods for industrial use faces hurdles such as high equipment costs, energy demands, and optimizing extraction conditions. Additionally, achieving high selectivity and purity of β-carotene is still a major hurdle, as carotenoids are often extracted along with other compounds, making purification difficult. Ongoing research needs to focus on refining extraction processes and improving selectivity, potentially through innovations like customizable deep eutectic solvents or co-solvents. Beyond improving extraction methods, it's important to evaluate the broader environmental and economic impacts of these technologies. Life cycle assessments (LCAs) are key to understanding the full environmental footprint of green extraction methods, including energy consumption, water usage, and waste generation. These assessments will help to demonstrate the advantages of green technologies over traditional methods and guide their scalability. The growing focus on circular economy principles, which emphasize reusing waste and by-products, is especially relevant for β-carotene extraction. By utilizing agricultural waste, such as carrot pomace or mango peels, researchers can reduce environmental impacts and provide an affordable, sustainable source of raw materials.

Looking ahead, integrating green extraction methods into a biorefinery model presents an exciting opportunity for more sustainable and resource-efficient β-carotene production. This approach would allow for the extraction of multiple valuable compounds from the same biomass, improving the economic viability of these technologies. For example, extracting carotenoids, essential oils, and other bioactive compounds from plant by-products or microbes could lead to multifunctional food ingredients and nutraceuticals. Additionally, combining green extraction technologies with innovative downstream processing methods, such as enzyme-assisted or hybrid extraction systems, could further increase yield and efficiency. In conclusion, developing sustainable extraction methods for β-carotene is a vital step toward more environmentally responsible and economically viable food production systems. While progress has been made, ongoing research is crucial to overcome challenges related to scalability, selectivity, and cost-effectiveness. By refining these techniques and expanding the use of green solvents and advanced extraction methods, we can unlock the full potential of β-carotene as a natural, bioactive compound that benefits both human health and the environment.

## References

Alirezapour, N., Asl, A. H., & Khajenoori, M. (2020). Ultrasound-assisted extraction of thymol from *Zataria multiflora Boiss.*: Optimization by response surface methodology and comparison with conventional Soxhlet extraction. *Bulgaria Chemical Communications, 52*, 419.

Badrkhani, N., Haghighi Asl, A., & Khajenoori, M. (2013) Subcritical water extraction of Thymol from Ajowan seeds by Taguchi experimental design. *Journal of Applied Chemistry, 7*(25), 51-58.

Buchweitz, V., Dauti, K., Alhadid, A., & Minceva, M. (2025). Evaluation of green and biobased solvent systems for the extraction of β-carotene and lipids from

*Rhodosporidium toruloides*. *ACS Omega, 10*(4), 4132-4142. https://doi.org/10.1021/acsomega.5b00419.

Castro-López, C., Rojas, R., Sánchez-Alejo, E. J., Niño-Medina, G., & Martínez-Ávila, G. C. G. (2016). Phenolic Compound Recovery from Grape Fruit and By- Products: An Overview of Extraction Methods. *InTech*.

Chen, J., Li, F., Li, Z., McClements, D. J., & Xiao, H. (2017). Encapsulation of carotenoids in emulsion-based delivery systems: Enhancement of β-carotene water-dispersibility and chemical stability. *Food Hydrocolloids, 69*, 49-55. https://doi.org/10.1016/j.foodhyd.2017.02.006.

Chen, L., Bai, G., Yang, R., Zang, J., Zhou, T., & Zhao, G. (2014). Encapsulation of β-carotene within ferritin nanocages greatly increases its water-solubility and thermal stability. *Food Chemistry, 149*, 307-312. https://doi.org/10.1016/j.foodchem.2013.09.068.

Chutia, H., & Mahanta, C. L. (2021). Green ultrasound and microwave extraction of carotenoids from passion fruit peel using vegetable oils as a solvent: Optimization, comparison, kinetics, and thermodynamic studies. *Innovative Food Science & Emerging Technologies, 67*, 102547. https://doi.org/10.1016/j.ifset.2020.102547.

de Andrade Lima, M., Charalampopoulos, D., & Chatzifragkou, A. (2018). Optimisation and modelling of supercritical CO2 extraction process of carotenoids from carrot peels. *The Journal of Supercritical Fluids, 133*, 94-102. https://doi.org/10.1016/j.supflu.2017.11.007.

de Andrade Lima, M., Kestekoglou, I., Charalampopoulos, D., & Chatzifragkou, A. (2019). Supercritical fluid extraction of carotenoids from vegetable waste matrices. *Molecules, 24*(3), 466. https://doi.org/10.3390/molecules24030466.

Delkhani, M., Khajenoori, M., & Asl, A. H. (2024). Determination of solubility and thermodynamic parameters of triamterene in pure and ethanol-modified subcritical water at various temperatures. *Arabian Journal for Science and Engineering, 49*(6), 8411-8423. https://doi.org/10.1007/s13369-024-06176-w.

Ebrahimi, N., Asl, A. H., & Mottahedin, P. (2019). Optimization of β-carotene solubility in pressurized hot water using a dynamic method and factorial methodology. *Bulgaria Chemical Communications, 51*, 279-283.

Gazmeh, M., Khajenoori, M., Yousefi-Nasab, S., & Asl, A. H. (2025). Determining the solubility parameter and thermodynamic properties of beta-carotene in superheated water solvent by experimental and MD simulation methods. *Journal of Molecular Liquids, 417*, 126653. https://doi.org/10.1016/j.molliq.2023.126653.

Gazmeh, M., Khajenoori, M., Yousefi-Nasab, S., & Haghighi Asl, A. (2024). Molecular dynamics simulation of extraction of *Curcuma longa L.* extract using subcritical water. *Scientific Reports, 14*(1), 27788. https://doi.org/10.1038/s41598-024-31187-w.

Ghorbani, A. S., Haghihi Asl, A., & Khajenoori, M. (2024). Subcritical water extraction of *Foeniculum vulgare Mill* (Fennel) essential oil. *Progress in Engineering Thermodynamics and Kinetics, 1*(1).

Gul, K., Tak, A., Singh, A. K., Singh, P., Yousuf, B., & Wani, A. A. (2015). Chemistry, encapsulation, and health benefits of β-carotene: A review. *Cogent Food & Agriculture, 1*(1), 1018696. https://doi.org/10.1080/23311932.2015.1018696.

Haghighi Asl, A., & Khajenoori, M. (2013). Subcritical water extraction. In *Mass Transfer-Advances in Sustainable Energy and Environment-Oriented Numerical Modeling* (pp. 459-487). InTech: Rijeka, Croatia. https://doi.org/10.5772/57176.

Haghighi Asl, A., & Khajenoori, M. (2021). *Green extraction in separation technology*. CRC Press: Boca Raton, FL.

Hladnik, L., Vicente, F. A., Košir, A., Grilc, M., & Likozar, B. (2023). Stirred, ultrasound-assisted, and microwave-assisted extraction process of β-carotene from *Rhodotorula glutinis* in biorefinery downstream. *Separation and Purification Technology, 311*, 123293. https://doi.org/10.1016/j.seppur.2023.123293.

Irianto, I., Putra, N. R., Yustisia, Y., Abdullah, S., Syafruddin, S., Paesal, P., Irmadamayanti, A., Herawati, H., Raharjo, B., Agustini, S., Asmaliyah, A., Krismawati, A., & Airlangga, B. (2025). Green technologies in food colorant extraction: A comprehensive review. *South African Journal of Chemical Engineering, 51*(1), 22-34. https://doi.org/10.1016/j.sajce.2024.09.010.

Jalali-Jivan, M., Fathi-Achachlouei, B., Ahmadi-Gavlighi, H., & Jafari, S. M. (2021). Improving the extraction efficiency and stability of β-carotene from carrot by enzyme-assisted green nanoemulsification. *Innovative Food Science & Emerging Technologies, 74*, 102836. https://doi.org/10.1016/j.ifset.2021.102836.

Jiang, Y., Ye, J., Hu, Y., Zhang, J., Li, W., Zhou, X., Yu, M., & Hu, Y. (2024). Extraction and synthesis of typical carotenoids: Lycopene, β-carotene, and astaxanthin. *Molecules, 29*(19), 4549. https://doi.org/10.3390/molecules29194549.

Khajenoori, M., & Haghighi Asl, A. (2014). Review of extraction of plant natural components by microwaves and ultrasounds. *New Food Technologies, 1*(3), 81-91.

Khajenoori, M., Asl, A. H., & Bidgoli, H. N. (2013). Subcritical water extraction of essential oils from *Matricaria chamomilla L. International Journal of Engineering, 26*(5), 489-494.

Khajenoori, M., Asl, A. H., & Eikani, M. H. (2015a). Optimization of subcritical water extraction of *Pimpinella anisum* seeds. *Journal of Essential Oil Bearing Plants, 18*(6), 1310-1320. https://doi.org/10.1080/0972060X.2015.1012996.

Khajenoori, M., Asl, A. H., & Eikani, M. H. (2015b). Subcritical water extraction of essential oils from *Trachyspermum ammi* seeds. *Journal of Essential Oil Bearing Plants, 18*(5), 1165-1173. https://doi.org/10.1080/0972060X.2015.1065443.

Khalid, S., Chaudhary, K., Amin, S., Raana, S., Zahid, M., Naeem, M., Mousavi Khaneghah, A., & Aadil, R. M. (2024). Recent advances in the implementation of ultrasound technology for the extraction of essential oils from terrestrial plant materials: A comprehensive review. *Ultrasonics Sonochemistry, 107*, 106914. https://doi.org/10.1016/j.ultsonch.2024.106914.

Kultys, E., & Kurek, M. A. (2022). Green extraction of carotenoids from fruit and vegetable byproducts: A review. *Molecules, 27*(2), 518. https://doi.org/10.3390/molecules 27020518.

Kumar, R. (2024). Extraction of betalains and β-carotene from beetroot and carrot: A practical approach to food fortification (Doctoral dissertation, University of Reading).

Kumar, R., Oruna-Concha, M. J., Balagiannis, D. P., & Niranjan, K. (2024). Elevated temperature extraction of β-carotene from freeze-dried carrot powder into sunflower oil: Extraction kinetics and thermal stability. *Journal of Food Science, 89*(3), 1642-1657. https://doi.org/10.1111/1750-3841.17856.

Lopes, A., Correia-Sá, L., Vieira, M., Delerue-Matos, C., Soares, C., & Grosso, C. (2024). Sustainable carotenoid extraction from macroalgae: Optimizing microwave-assisted extraction using response surface methodology. *Life, 14*(12), 1573. https://doi.org/10.3390/life14121573.

Ludwig, K., Rihko-Struckmann, L., Brinitzer, G., Unkelbach, G., & Sundmacher, K. (2021). β-Carotene extraction from *Dunaliella salina* by supercritical CO2. *Journal of Applied Phycology, 33*(3), 1435-1445. https://doi.org/10.1007/s10811-021-02491-4.

Machado, A. P. D. F., Pasquel-Reátegui, J. L., Barbero, G. F., & Martínez, J. (2015). Pressurized liquid extraction of bioactive compounds from blackberry (*Rubus fruticosus L.*) residues: A comparison with conventional methods. *Food Research International, 77*(3), 675-683. https://doi.org/10.1016/j.foodres.2015.09.014.

Martínez, J. M., Schottroff, F., Haas, K., Fauster, T., Sajfrtová, M., Álvarez, I., Raso, J., & Jaeger, H. (2020). Evaluation of pulsed electric fields technology for the improvement of subsequent carotenoid extraction from dried *Rhodotorula glutinis* yeast. *Food Chemistry, 323*, 126824. https://doi.org/10.1016/j.foodchem.2020.126824.

Matić, M., Stupar, A., Pezo, L., Ilić, N. Đ., Mišan, A., Teslić, N., Pojić, M., & Mandić, A. (2024). Eco-friendly extraction: A green approach to maximizing bioactive extraction from pumpkin (*Curcubita moschata L.*). *Food Chemistry: X, 22*, 101290. https://doi.org/10.1016/j.fochx.2023.101290.

Molteni, C., La Motta, C., & Valoppi, F. (2022). Improving the bioaccessibility and bioavailability of carotenoids by means of nanostructured delivery systems: A comprehensive review. *Antioxidants, 11*(10), 1931. https://doi.org/10.3390/antiox11101931.

Morón-Ortiz, Á., Mapelli-Brahm, P., & Meléndez-Martínez, A. J. (2024). Sustainable green extraction of carotenoid pigments: Innovative technologies and bio-based solvents. *Antioxidants, 13*(2), 239. https://doi.org/10.3390/antiox13020239.

Mottahedin, P., Asl, A. H., & Lotfollahi, M. N. (2017). Experimental and modeling investigation on the solubility of β-carotene in pure and ethanol-modified subcritical water. *Journal of Molecular Liquids, 237*, 257-265. https://doi.org/10.1016/j.molliq.2017.06.051.

Mottahedin, P., Haghighi Asl, A., & Khajenoori, M. (2017). Extraction of curcumin and essential oil from *Curcuma longa L.* by subcritical water via response surface methodology. *Journal of Food Processing and Preservation, 41*(4), e13095. https://doi.org/10.1111/jfpp.13095.

Mussagy, C. U., Winterburn, J., Santos-Ebinuma, V. C., & Pereira, J. F. B. (2019). Production and extraction of carotenoids produced by microorganisms. *Applied Microbiology and Biotechnology, 103*(3), 1095-1114. https://doi.org/10.1007/s00253-019-09983-4.

Nath, P., Kaur, C., Rudra, S. G., & Varghese, E. (2016). Enzyme-assisted extraction of carotenoid-rich extract from red capsicum (*Capsicum annuum*). *Agricultural Research, 5*(2), 193-204. https://doi.org/10.1007/s40003-016-0229-6.

Ordóñez-Santos, L. E., Pinzón-Zarate, L. X., & González-Salcedo, L. O. (2015). Optimization of ultrasonic-assisted extraction of total carotenoids from peach palm fruit (*Bactris gasipaes*) by-products with sunflower oil using response surface methodology. *Ultrasonics Sonochemistry, 27*, 560-566. https://doi.org/10.1016/j.ultsonch.2015.03.016.

Portillo-López, R., Morales-Contreras, B. E., Lozano-Guzmán, E., Basilio-Heredia, J., Muy-Rangel, M. D., Ochoa-Martínez, L. A., Rosas-Flores, W., & Morales-Castro, J. (2021). Vegetable oils as green solvents for carotenoid extraction from pumpkin (*Cucurbita argyrosperma Huber*) byproducts: Optimization of extraction parameters. *Journal of Food Science, 86*(7), 3122-3136. https://doi.org/10.1111/1750-3841.16197.

Saini, R. K., & Keum, Y. S. (2018). Carotenoid extraction methods: A review of recent developments. *Food Chemistry, 240*, 90-103. https://doi.org/10.1016/j.foodchem.2017.08.033.

Sangkharak, K., Pasitkul, P., Paiyit, N., & Aonchart, P. (2011). Application of lipase treatment on carotenoids extraction from chili (*Capsicum annuum L.*). *African Journal of Microbiology Research, 5*(28), 5057-5061. https://doi.org/10.5897/AJMR11.363

Sarkarat, R., Mohamadnia, S., & Tavakoli, O. (2023). Recent advances in non-conventional techniques for extraction of phycobiliproteins and carotenoids from microalgae. *Brazilian Journal of Chemical Engineering, 40*(2), 321-342. https://doi.org/10.1007/s43153-022-00083-9.

Sayegh, F., Al-naghrani, M. J., Amran, R. H., Jamal, M. T., Nass, N. M., Felemban, W. F., ... & Satheesh, S. (2025). Extraction of beta-carotene from the microalga *Dunaliella salina* using bacterial lipase enzyme and organic solvent under varying stress conditions. *Frontiers in Marine Science, 12*, 1543147. https://doi.org/10.3389/fmars.2025.1543147.

Sereti, F., Alexandri, M., Papapostolou, H., Kachrimanidou, V., Papadaki, A., & Kopsahelis, N. (2025a). Green extraction of carotenoids and oil produced by *Rhodosporidium paludigenum* using supercritical CO2 extraction: Evaluation of cell disruption methods and extraction kinetics. *Food Chemistry, 483*, 144261. https://doi.org/10.1016/j.foodchem.2023.144261.

Sereti, F., Alexandri, M., Papapostolou, H., Papadaki, A., & Kopsahelis, N. (2025b). Recent progress in carotenoid encapsulation: Effects on storage stability, bioaccessibility and bioavailability for advanced innovative food applications. *Food Research International, 115861*. https://doi.org/10.1016/j.foodres.2023.115861.

Share Mohammadi, H. S., Asl, A. H., & Khajenoori, M. (2023). Production of pharmaceutical micro and nano particles by subcritical water-based technologies: A review. *Journal of Drug Delivery Science and Technology, 85*, 104621. https://doi.org/10.1016/j.jddst.2023.104621.

Share Mohammadi, H., Haghighi Asl, A., & Khajenoori, M. (2020). Experimental measurement and correlation of solubility of β-carotene in pure and ethanol-modified subcritical water. *Chinese Journal of Chemical Engineering, 28*(10), 2620-2625. https://doi.org/10.1016/j.cjche.2020.01.015.

Staicu, V., Calinescu, I., Vinatoru, M., Ghimpeteanu, D., Popa, I., & Mason, T. J. (2024). The efficient extraction of β-carotene from sea buckthorn berries using a novel solvent, fatty acid ethyl esters, and a combination of ultrasound and microwave. *Agronomy, 14*(3), 416. https://doi.org/10.3390/agronomy14030416.

Stupar, A., Šeregelj, V., Ribeiro, B. D., Pezo, L., Cvetanović, A., Mišan, A., & Marrucho, I. (2021). Recovery of β-carotene from pumpkin using switchable natural deep eutectic solvents. *Ultrasonics Sonochemistry, 76*, 105638. https://doi.org/10.1016/j.ultsonch.2021.105638.

Sun, M., & Temelli, F. (2006). Supercritical carbon dioxide extraction of carotenoids from carrot using canola oil as a continuous co-solvent. *The Journal of Supercritical Fluids, 37*(3), 397-408. https://doi.org/10.1016/j.supflu.2005.12.010.

Thakur, N. (2018). Heat stability and antioxidant potential of beta-carotene isolated from a fungal isolate. *Bulgarian Journal of Agricultural Science, 24*(5), 891–896. https://doi.org/10.1007/s40003-018-0309-2.

Tiwari, S., Upadhyay, N., & Malhotra, R. (2021). Three-way ANOVA for emulsion of carotenoids extracted in flaxseed oil from carrot bio-waste. *Waste Management, 121*, 67-76. https://doi.org/10.1016/j.wasman.2020.10.013.

Tiwari, S., Upadhyay, N., & Singh, A. K. (2022). Stability assessment of emulsion of carotenoids extracted from carrot bio-waste in flaxseed oil and its application in food model system. *Food Bioscience, 47*, 101631. https://doi.org/10.1016/j.fbio.2022.101631.

Tumbas Šaponjac, V., Kovačević, S., Šeregelj, V., Šovljanski, O., Mandić, A., Ćetković, G., Vulić, J., Podunavac-Kuzmanović, S., & Čanadanović-Brunet, J. (2021). Improvement of carrot accelerated solvent extraction efficacy using experimental design and chemometric techniques. *Processes, 9*(9), 1652. https://doi.org/10.3390/pr9091652.

Uwineza, P. A., & Waśkiewicz, A. (2020). Recent advances in supercritical fluid extraction of natural bioactive compounds from natural plant materials. *Molecules, 25*(17), 3847. https://doi.org/10.3390/molecules25173847.

Vinas-Ospino, A., Jesus, A. R., Paiva, A., Esteve, M. J., Frígola, A., Blesa, J., & López-Malo, D. (2024). Comparison of green solvents for the revalorization of orange by-products: Carotenoid extraction and in vitro antioxidant activity. *Food Chemistry, 442*, 138530. https://doi.org/10.1016/j.foodchem.2023.138530.

Viñas-Ospino, A., López-Malo, D., Esteve, M. J., Frígola, A., & Blesa, J. (2023). Green solvents: Emerging alternatives for carotenoid extraction from fruit and vegetable by-products. *Foods, 12*(4), 863. https://doi.org/10.3390/foods12040863.

Wang, M., Morón-Ortiz, Á., Zhou, J., Benítez-González, A., Mapelli-Brahm, P., Meléndez-Martínez, A. J., & Barba, F. J. (2023). Effects of pressurized liquid extraction with dimethyl sulfoxide on the recovery of carotenoids and other dietary valuable compounds from the microalgae *Spirulina, Chlorella* and *Phaeodactylum tricornutum. Food Chemistry, 405*, 134885. https://doi.org/10.1016/j.foodchem.2023.134885.

Yara-Varon, E., Fabiano-Tixier, A. S., Balcells, M., Canela-Garayoa, R., Bily, A., & Chemat, F. (2016). Is it possible to substitute hexane with green solvents for extraction of carotenoids? A theoretical versus experimental solubility study. *RSC Advances, 6*(33), 27750-27759. https://doi.org/10.1039/C6RA03253B.

## Chapter 10

# The Science and Applications of Natural Red Pigments: From Extraction to Innovation

**Mohd Abhar Haniff Jaafar[1]**
**Eyzzatul Eylia Yunos[2]**
**Wan Syibrah Hanisah Wan Sulaiman[3], PhD**
**and Rashidi Othman[4,*], PhD**

[1,2,3]International Institute for Halal Research and Training,
International Islamic University Malaysia, Kuala Lumpur, Malaysia
[4]Department of Landscape Architecture, International Islamic University Malaysia, Kuala Lumpur, Malaysia

## Abstract

This book chapter explores the significance of natural red pigments, focusing on their sources, extraction methods, applications, challenges, and future prospects. Natural red pigments, derived from various plant, microbial, and animal sources, have been used historically in food, cosmetics, textiles, and pharmaceuticals due to their vibrant hues and potential health benefits. The chapter examines key natural red pigments, such as betalains, anthocyanins, and carotenoids, highlighting their distinct chemical properties and diverse uses across industries. The extraction processes, ranging from traditional methods to advanced technologies like supercritical fluid extraction (SFE) and microwave-assisted extraction (MAE), are also discussed in-depth. These techniques not only improve efficiency but also address challenges related to sustainability, yield, and cost-effectiveness. Moreover, the chapter addresses the growing demand for natural colourants, driven by

* Corresponding Author's Email: rashidi@iium.edu.my

In: Natural Pigments: Sources, Properties, and Health Benefits
Editors: Rashidi Othman, Qurratu Aini Mat Ali, Razanah Ramya et al.
ISBN: 979-8-90134-055-4

consumer preferences for clean-label and eco-friendly products, and the need to replace synthetic dyes that pose health and environmental risks. Despite their advantages, natural red pigments face challenges in terms of stability, cost, and sourcing. Issues like light degradation and temperature sensitivity impact their effectiveness, while the sourcing of raw materials raises ethical and environmental concerns. The chapter provides insight into innovative solutions, such as biotechnological advancements, microbial fermentation, and synthetic biology, which offer promising alternatives for mass production. Looking ahead, the future of natural red pigments is promising, with advancements in biotechnology, sustainable sourcing, and consumer awareness driving innovation. The chapter concludes by exploring the potential for natural pigments in emerging markets, as well as the ongoing research needed to further develop cost-effective, stable, and scalable natural colourants. This work contributes to a deeper understanding of natural red pigments and their growing role in a sustainable, eco-conscious future.

**Keywords**: red pigments, carotenoids, betalains, anthocyanins, natural colourants

## The Vibrant World of Natural Red Pigments

Red, a colour that has captivated human imagination since time immemorial, holds a unique place in culture, art, and science. From the walls of prehistoric caves adorned with crimson ochres to the luxurious fabrics of ancient civilizations dyed with cochineal or madder, red pigments have symbolized power, vitality, and sacredness. Unlike colours that can be easily synthesized today, the earliest red pigments were derived from natural sources (plants, insects, and minerals) reflecting humanity's ingenuity in harnessing nature's palette. This chapter delves into the rich and intricate world of natural red pigments, tracing their historical significance, chemical nature, and contemporary applications, while highlighting the ongoing quest for sustainable and safe alternatives to synthetic dyes.

The significance of red pigments extends beyond aesthetics. In food, cosmetics, textiles, and medicine, red pigments serve as functional molecules that interact with human perception and biology. Their vivid hues are not merely decorative; they often indicate the presence of bioactive compounds, such as antioxidants in anthocyanins or betalains, which confer health benefits (Kalt et al., 2014). This dual role (as visual attractants and biologically active substances) makes red pigments particularly fascinating for researchers, industries, and consumers alike. In modern times, the shift towards natural and sustainable ingredients has revitalized interest in these pigments, as society increasingly demands alternatives to synthetic dyes, which can carry environmental and health concerns (Rahim et al., 2021).

Historically, the journey of red pigments is intertwined with cultural narratives and economic value. Ancient Egyptians prized red ochre (a natural earth pigment primarily composed of hematite ($\alpha$-$Fe_2O_3$), an iron oxide) not only for tomb paintings but also for cosmetics (Beck, 2015). In Mesoamerica, cochineal (a red dye derived from insects) was a symbol of wealth and power, exported extensively during the colonial period (Morris, 2017). Similarly, in Asia, plant-based red pigments like madder root and safflower were central to textile traditions, contributing to the distinctive aesthetics of local cultures (Ueda et al., 2020). These pigments were not only commodities; they were carriers of identity, status, and tradition, embodying the deep human connection to colour and nature.

From a scientific perspective, natural red pigments exhibit remarkable diversity in their chemical composition and biosynthetic pathways. Predominantly, they belong to classes such as anthocyanins, betalains, and carotenoids, each with distinct molecular structures, colour properties, and stability profiles. Anthocyanins, abundant in fruits and flowers, produce vibrant reds that are highly pH-sensitive, shifting hues in response to acidity (Ganguly et al., 2016). Betalains, found in beets and certain cacti, offer a striking crimson or magenta colour, notable for their stability across a range of pH values (Shah et al., 2019). Carotenoids, including lycopene and astaxanthin, contribute deep red tones in tomatoes, watermelon, seaweed, microalgae and shrimp exhibiting both colouring and health-promoting antioxidant properties (Rao & Rao, 2019). Understanding the chemistry of these pigments is critical for their effective extraction, stabilization, and application in diverse industries.

The contemporary relevance of natural red pigments cannot be overstated. With rising consumer awareness of health, sustainability, and environmental impact, industries are increasingly replacing synthetic dyes with naturally derived alternatives. Natural red pigments are now employed in processed foods, beverages, cosmetics, pharmaceuticals, and even in biotechnological applications (Burgos et al., 2017). However, challenges remain, particularly regarding stability, scalability, and cost-effectiveness. Unlike synthetic dyes, natural pigments can degrade under light, heat, and oxygen exposure, limiting their shelf life and industrial utility (Mekonnen et al., 2016). Researchers are actively exploring innovative extraction methods, biotechnological production systems, and formulation strategies to overcome these limitations, making natural red pigments a dynamic field at the intersection of chemistry, biology, and industrial technology.

This chapter aims to provide a comprehensive exploration of natural red pigments, encompassing their historical roots, chemical structures, biological sources, extraction techniques, applications, regulatory considerations, and future prospects. By bridging traditional knowledge with modern scientific insights, it seeks to illuminate not only the beauty and vibrancy of red pigments but also their profound functional and cultural significance. In doing so, it highlights the essential role these pigments play in shaping industries, human health, and the aesthetic experiences that connect us to the natural world.

In essence, understanding natural red pigments is a journey through time, chemistry, culture, and innovation. They are more than mere colours; they are molecules of meaning, linking the biological richness of nature to human creativity and technology. As we progress through this chapter, we will explore the sources, chemistry, applications, and challenges of these pigments, revealing how their enduring allure continues to inspire science, art, and industry in equal measure.

## Chemistry and Biochemistry of Natural Red Pigments

Natural red pigments are a diverse group of compounds that contribute rich, vibrant colours to plants, fruits, flowers, and other biological sources. These pigments are essential for many ecological processes, such as attracting pollinators or deterring herbivores. Chemically, natural red pigments are typically classified into three major groups: anthocyanins, betalains, and carotenoids, each possessing unique chemical structures, properties, and biosynthetic pathways. Understanding their chemistry not only highlights the molecular basis of colour but also provides insight into their applications, stability, and potential for industrial use.

### *Anthocyanins: The Colourful Flavonoids*

Anthocyanins are water-soluble pigments primarily found in the vacuoles of plant cells, responsible for the red, purple, and blue colours of many fruits, flowers, and leaves. They are part of the flavonoid class of compounds, which are derived from the amino acid phenylalanine. The basic structure of anthocyanins consists of an anthocyanidin aglycone (a flavonoid backbone) attached to a sugar molecule or several sugar molecules. The anthocyanidin structure includes a three-ring core with hydroxyl or methoxy groups at various positions, and it is this arrangement that determines the colour. The presence of hydroxyl groups on the aromatic rings tends to enhance the red colour of the pigment, while methoxy groups push the colour toward blue or purple (Ganguly et al., 2016). The colour of anthocyanins is pH-dependent, with acidic environments enhancing red hues and alkaline conditions shifting the colour to purple or blue. This pH sensitivity is a result of the equilibrium between different forms of the pigment, including a flavylium cation, which is responsible for the red colouration, and its colourless, conjugated form at higher pH. There are six distinct anthocyanidins commonly found in plants, which are pelargonidin, cyanidin, peonidin, delphinidin, petunidin, and malvidin, classified according to the number and position of their OH groups. Seven different side groups exist on the flavylium cation. These side groups could be either of the three groups: a hydrogen atom (H) or a methoxy (-OCH3) or a hydroxide (-OH) group.

### *Betalains: A Distinct Class of Pigments*

Betalains, found primarily in the order Caryophyllales (e.g., beets, amaranth, and cacti), are a unique class of red and purple pigments that do not belong to the flavonoid family. Instead, they are derived from tyrosine, an amino acid, and consist of two distinct subclasses: betacyanins (red to violet) and betaxanthins (yellow to orange). Betalains are characterized by their nitrogen-containing structure, with betacyanins possessing a glycerol side chain and betaxanthins having a glucose unit (Shah et al., 2019).

Betalains are particularly interesting because of their stability across a wide range of pH values. Unlike anthocyanins, which can degrade easily at higher pH, betalains maintain their vibrant colour even in alkaline conditions, making them more suitable for certain industrial applications where stability is crucial. Furthermore, betalains are less susceptible to enzymatic degradation, which is a common problem for anthocyanins in food processing (Burgos et al., 2017).

### *Carotenoids: Red and Orange Pigments with Antioxidant Properties*

Carotenoids are a class of pigments that play an essential role in photosynthesis by capturing light energy and protecting plant cells from oxidative damage. They are predominantly found in the chloroplasts of plants, and they provide a wide range of yellow, orange, and red hues. Lycopene, a carotenoid found in tomatoes, watermelon, and red peppers, is the most widely recognized red carotenoid. Similarly, astaxanthin is another red pigment found in marine organisms such as shrimp, salmon, seaweed, and microalgae. Carotenoids are terpenoids composed of isoprene units, and their structure consists of long chains of conjugated double bonds, which are responsible for their characteristic colour (Rao & Rao, 2019). Lycopene's deep red colour arises from the specific arrangement of conjugated double bonds, allowing it to absorb light in the blue and green spectrum while reflecting red light. Astaxanthin, another red carotenoid, also exhibits a similar property and contributes to the red hues in marine organisms such as shrimp and salmon. One of the key features of carotenoids, particularly lycopene and astaxanthin, is their antioxidant activity, which has been linked to various health benefits, including a reduced risk of certain cancers and heart disease. In addition to lycopene and astaxanthin, other carotenoids like beta-carotene, which gives an orange colour but can contribute to red tones at higher concentrations, also play a role in the red and orange colouration of many fruits and vegetables. Carotenoids are also essential for the biosynthesis of vitamin A, a crucial nutrient for human health, further highlighting their biological significance.

### *Biosynthesis Pathways of Natural Red Pigments*

The biosynthesis of natural red pigments involves complex metabolic pathways that begin with the primary metabolites, such as sugars, amino acids, and fatty acids.

Anthocyanins are synthesized via the phenylpropanoid pathway, which starts with phenylalanine and produces a series of intermediate compounds, eventually leading to the formation of anthocyanidin aglycones. These aglycones are then glycosylated to form anthocyanins. The key enzymes involved in this pathway are chalcone synthase (CHS) and flavonoid 3',5'-hydroxylase (F3'H), which control the degree of hydroxylation on the flavonoid backbone, thereby influencing the final colour of the anthocyanin (Ganguly et al., 2016).

Betalains, on the other hand, are synthesized through a distinct pathway from tyrosine, via dopa and dopamine intermediates. The conversion of tyrosine to betacyanin is catalyzed by enzymes like tyrosinase and dopa decarboxylase, which contribute to the formation of the nitrogenous pigments (Shah et al., 2019). This pathway is relatively simpler compared to the anthocyanin biosynthesis and results in highly stable pigments in comparison.

Carotenoids are synthesized through the isoprenoid pathway, beginning with the precursor geranylgeranyl pyrophosphate (GGPP). Enzymes like phytoene synthase and lycopene cyclase then catalyze the formation of carotenoid precursors, including phytoene, zeaxanthin, lycopene, and astaxanthin, with the pathway branching out into the production of lutein and beta-carotene. The specific carotenoid produced depends on the enzymes involved and the metabolic context (Rao & Rao, 2019).

## Stability and Environmental Factors

The stability of natural red pigments is a crucial factor in their application, especially in food, pharmaceuticals, and cosmetics. Anthocyanins are highly sensitive to environmental conditions, especially pH, temperature, and light. They tend to degrade in the presence of oxygen, which can lead to a loss of colour and a reduction in their antioxidant properties. To mitigate this, researchers have been exploring encapsulation techniques and the use of stabilizing agents, such as ascorbic acid and citric acid, to improve their shelf life (Zhao et al., 2020).

Betalains, in contrast, are more stable than anthocyanins in terms of pH, heat, and light, though they are still prone to degradation in the presence of certain enzymes. Their robust stability makes them an attractive option for food and cosmetic formulations, where maintaining colour integrity is essential (Alvarado et al., 2020).

Carotenoids are generally more stable than anthocyanins and betalains, but they can also be affected by light and oxygen. Carotenoids are sensitive to high temperatures, which can lead to the breakdown of their antioxidant properties. However, their stability can be enhanced through microencapsulation and controlled storage conditions (Rao & Rao, 2019).

The chemistry and biochemistry of natural red pigments are integral to their applications in various industries, from food to cosmetics and pharmaceuticals. Understanding the molecular structures, biosynthetic pathways, and stability factors of anthocyanins, betalains, and carotenoids not only provides insight into their colour properties but also opens new doors for innovation. As consumer demand for natural, sustainable alternatives to synthetic dyes increases, the exploration of these pigments' biochemical characteristics will be key to overcoming challenges related to their industrial application and stability.

## Sources of Natural Red Pigments

Natural red pigments are not only prized for their striking colours but also for the ecological roles they play in the organisms that produce them. These pigments are found across a vast array of biological sources, including plants, fungi, bacteria, and even insects. The origins of these pigments are as varied as their applications, which range from food and cosmetics to textiles and pharmaceuticals. This section will explore the primary sources of natural red pigments, including plant-based, microbial, and insect-derived sources, highlighting the geographical and cultural significance of each.

### *Plant-Based Sources*

Plants are by far the most abundant source of natural red pigments, with a wide variety of species contributing to the spectrum of colours available. Among the most notable plant-derived red pigments are anthocyanins, betalains, and carotenoids, each of which is produced by different plant species and plays distinct roles in plant biology.

Anthocyanins are found in a broad range of plants, contributing to the red, purple, and blue colouration of fruits, flowers, and leaves. Some of the most common plant sources of anthocyanins include red cabbage, Butterfly Pea flower, lotus, Canna, cherries, strawberries, blackberries, and grapes (Ganguly et al., 2016). These pigments are produced by plants to attract pollinators, signal ripeness to animals, and protect the plant from UV radiation. The red cabbage is particularly rich in anthocyanins, with its vivid red colouration shifting to purple or blue depending on the pH of the soil (Ganguly et al., 2016).

Betalains, another important class of red pigments, are primarily found in the Caryophyllales order of plants, which includes species like beets, amaranth, and dragon fruit or prickly pear cacti. These pigments are often responsible for the deep red to purple hues of these plants and serve similar ecological functions to anthocyanins, such as attracting animals for seed dispersal (Shah et al., 2019). Beetroot, in particular, is widely known for its betalain content, which contributes to its characteristic deep red colour. Betalains are also known for their health benefits, including antioxidant and anti-inflammatory properties, making them a popular ingredient in functional foods (Rasouli et al., 2019).

Carotenoids, which include lycopene, beta-carotene, and astaxanthin, are another important group of pigments found in plants. These pigments are responsible for the red, orange, and yellow hues of many fruits and vegetables.

Tomatoes, seaweed and microalgae are particularly rich in lycopene and astaxanthin, a carotenoid that gives them their deep red colour (Rao & Rao, 2019). Red peppers and paprika also contain high levels of capsanthin, contributing to their bright red colouration. Carotenoids not only provide colour but also have significant antioxidant properties, which help protect the plant from oxidative stress and contribute to human health when consumed (Rao & Rao, 2019).

### *Microbial Sources*

While plants are the most well-known sources of natural red pigments, microbes, particularly fungi and bacteria, also play an essential role in the production of red-coloured compounds. Monascus species, a group of fungi used in traditional Asian food fermentation, produce monascorubrin, a red pigment that imparts a vibrant red colour to fermented rice products such as red yeast rice. Monascus pigments have been used for centuries in Chinese medicine and food applications, not only for their colour but also for their health benefits, such as cholesterol-lowering properties (Ge et al., 2020).

In addition to fungi, certain bacteria also produce red pigments. *Serratia marcescens*, for example, is known for producing the red pigment prodigiosin, which is produced under certain environmental conditions and has been studied for its antimicrobial properties. This pigment has potential applications in medicine and biotechnology due to its biological activity (Kalt et al., 2014).

Microbial sources of red pigments offer several advantages, including the ability to produce large quantities of pigment in controlled environments. This makes microbial fermentation an attractive alternative to traditional plant-based pigment extraction, especially for industries seeking scalable and sustainable production methods.

### *Insect-Derived Sources*

Insects, although less commonly used today, have historically been an important source of red pigments, particularly for the textile and food industries. Cochineal (*Dactylopius coccus*), a scale insect found primarily in Mexico and South America, is one of the most well-known sources of natural red pigment. The carminic acid derived from cochineal provides a brilliant red dye, carmine, which has been used for centuries in textiles, cosmetics, and food colouring (Morris, 2017). The bright red colour of carmine is due to the carminic acid, which is extracted from the female cochineal insect and processed to produce various forms of dye. Carminic acid has antioxidant properties and is used in food products like beverages, candies, and dairy, as well as in cosmetics and pharmaceuticals (Burgos et al., 2017).

In addition to cochineal, other insects, such as the kermes insect found in Mediterranean regions, have also been used to produce red dyes. Kermes is a genus of gall-like scale insects in the Kermesidae family, which feed on oak sap. The females of these insects produce a red dye, also known as "kermes," which is the source of natural crimson. Although the use of insect-derived pigments has diminished with the rise of synthetic dyes, they remain culturally and historically significant, especially in traditional textile dyeing (Morris, 2017; Rahim et al., 2021).

### *Geographic and Cultural Significance of Red Pigment Sources*

The use of natural red pigments is not only a reflection of the biological diversity of the world but also of the cultural and economic practices of different societies. In the Andes, cochineal was once one of the most valuable commodities traded by the indigenous peoples, and it played a key role in the textile arts of the region. The Maya and Aztec civilizations used carmine for everything from body paint to fabrics, creating vibrant red hues that were emblematic of power and prestige (Morris, 2017).

In Asia, plant-based red pigments like madder (*Rubia tinctorum*) and safflower (*Carthamus tinctorius*) have been integral to traditional dyeing techniques. Madder, in particular, has been used for centuries to create red dyes in textiles, while safflower has been used not only for dyeing but also in traditional Chinese medicine (Ueda et al., 2020). In Malaysia, plants such as *Bixa orellana* (annatto) have also been used for their reddish-orange dye, traditionally utilized in food and textile dyeing. The use of these pigments is a testament to the deep connection between people and nature, as well as the knowledge passed down through generations.

In the modern era, the increasing demand for natural pigments has revitalized the use of these historical sources. As consumers become more conscious of the environmental and health impacts of synthetic dyes, industries in Europe, the Americas, and Asia are turning once again to natural red pigments for applications in food, cosmetics, and textiles (Zhao et al., 2020).

Natural red pigments are derived from a rich variety of sources, ranging from the plant kingdom to microorganisms and even insects. These pigments have played a significant role in human culture, providing not only aesthetic value but also functional and medicinal benefits. As the demand for natural, sustainable alternatives to synthetic dyes continues to grow, the exploration of these sources, along with the development of efficient extraction and production methods, will be essential to meeting the needs of modern industries. Whether from the vibrant fruits of the earth, the fermentation chambers of microbes, or the fascinating world of insects, natural red pigments remain a vital part of our global heritage and a symbol of the ever-evolving relationship between nature and human innovation.

## Extraction and Purification Techniques of Natural Red Pigments

The process of extracting natural red pigments from their biological sources is both an art and a science. Efficient extraction methods are critical for obtaining high yields of pigment while preserving the quality and stability of the colourant. Over the years, various extraction techniques have been developed, ranging from traditional methods such as solvent extraction to more advanced techniques like supercritical fluid extraction and biotechnological approaches. This section will explore the most widely used methods for extracting natural red pigments, including their principles, advantages, and limitations.

### *Solvent Extraction: The Traditional Workhorse*

Solvent extraction has long been the most common method for isolating natural pigments, including red pigments, from plant and microbial sources. The principle behind solvent extraction is to dissolve the pigment in a suitable solvent that can effectively separate the pigment from the plant material or microbial biomass.

Common solvents used in pigment extraction include water, ethanol, methanol, acetone, and hexane. For plant-based red pigments such as anthocyanins and betalains, water or ethanol is typically used, as these solvents are effective in breaking down cell walls and releasing the pigments into solution. For carotenoids, non-polar solvents like hexane or chloroform are preferred due to the lipophilic nature of carotenoid molecules (Jouki et al., 2017).

One of the advantages of solvent extraction is its simplicity and cost-effectiveness. The method is easy to scale up, making it suitable for both small- and large-scale production. However, it also has several drawbacks. First, the solvent extraction process can be time-consuming and often requires large amounts of solvent, which can be both expensive and environmentally harmful. Moreover, some solvents, such as hexane, are toxic, necessitating proper disposal and environmental management.

Optimization of solvent extraction involves adjusting factors such as the solvent type, temperature, extraction time, and pH to achieve the highest yield with minimal degradation of the pigment. The use of microwave-assisted extraction and ultrasound-assisted extraction has gained popularity in recent years to improve the efficiency of solvent extraction, reducing extraction time and enhancing pigment yield (Liu et al., 2019).

### *Supercritical Fluid Extraction: A Green Alternative*

Supercritical fluid extraction (SFE) is an advanced and environmentally friendly method that uses supercritical fluids, typically carbon dioxide ($CO_2$), to extract pigments from plant and microbial sources. In the supercritical state, $CO_2$ exhibits properties of both a gas and a liquid, making it an ideal solvent for extracting non-polar compounds like carotenoids and anthocyanins (Giusti et al., 2017).

The advantages of SFE over traditional solvent extraction are numerous. First, $CO_2$ is non-toxic, environmentally benign, and can be easily removed from the extracted pigment, reducing the need for chemical solvents. Additionally, the use of supercritical $CO_2$ allows for a more selective extraction process, which can help preserve the integrity of the pigment and reduce the co-extraction of unwanted compounds such as lipids or phenolic compounds (Alves et al., 2020).

While supercritical fluid extraction is a promising technique for producing high-quality pigments, it does have its challenges. The high initial setup cost of supercritical $CO_2$ extraction equipment can be a barrier to its widespread adoption, particularly for small-scale producers. Additionally, the process requires precise control of temperature and pressure conditions, which can be difficult to optimize for different types of pigments.

### *Enzyme-Assisted Extraction: Unlocking the Pigments*

Enzyme-assisted extraction (EAE) is a relatively novel technique that uses specific enzymes to break down the plant cell walls and release the pigments into the extraction medium. This method is particularly effective for pigments like anthocyanins and betalains, which are stored in vacuoles and require the disruption of cell walls to be efficiently extracted (Velázquez et al., 2018).

The primary enzymes used in enzyme-assisted extraction are cellulases, pectinases, and hemicellulases, which degrade the cellulose, pectin, and hemicellulose components of the plant cell wall. By combining enzyme-assisted extraction with solvent extraction, higher yields of pigments can be achieved in a shorter period of time, as the enzymes help to break down the cell wall and facilitate the release of pigments into the solution (Choudhury et al., 2018).

One of the advantages of enzyme-assisted extraction is that it is more selective than conventional solvent extraction, reducing the extraction of unwanted compounds such as phenols or polysaccharides. Moreover, enzyme-assisted extraction is considered more environmentally friendly than traditional solvent extraction, as it uses fewer toxic solvents and often operates under milder conditions (Santos et al., 2021).

However, enzyme-assisted extraction is still an emerging technique, and its widespread use is limited by the high cost of enzymes and the need for precise control over extraction parameters. Moreover, enzymes can be deactivated at high temperatures, requiring careful optimization of the extraction process.

### *Ultrasonic-Assisted Extraction: Harnessing Sound Waves*

Ultrasonic-assisted extraction (UAE) is another innovative method that uses ultrasonic waves to disrupt the cellular structure of plant or microbial materials, thereby enhancing the extraction of pigments. The ultrasonic waves create microscopic bubbles within the solvent, which rapidly collapse and create shear forces that break apart plant cell walls and release the pigments into solution (Ashraf et al., 2018).

This technique has been shown to increase the efficiency of pigment extraction by reducing the time and solvent required. UAE is particularly effective for extracting anthocyanins from fruits and flowers and betalains from beets (Ganguly et al., 2016). The advantages of UAE include its rapid extraction time, reduced solvent usage, and the ability to operate at ambient temperatures, preserving the stability of heat-sensitive pigments.

One of the challenges of ultrasonic-assisted extraction is the potential for degradation of sensitive pigments due to the high-intensity sound waves. The efficiency of extraction can also be affected by factors such as the frequency and amplitude of the ultrasound waves, making it necessary to optimize these parameters for each specific pigment source.

### *Biotechnology and Microbial Fermentation: Sustainable Production*

In addition to traditional extraction methods, biotechnological approaches are emerging as a sustainable alternative for producing natural red pigments. Microbial fermentation offers a promising method for producing pigments like monascorubrin (from Monascus fungi) and prodigiosin (from *Serratia marcescens*) in large quantities (Zhao et al., 2020).

Through microbial fermentation, red pigments can be produced in a controlled environment, allowing for high yields without the need for large-scale cultivation of plant or insect sources. This method is particularly attractive for large-scale production, as it eliminates the need for extensive land use and environmental disruption associated with the cultivation of plants or harvesting of insects.

Moreover, microbial fermentation allows for the production of pigments without the need for harmful solvents or chemicals, making it a more environmentally friendly option. However, challenges remain in terms of optimizing fermentation conditions, scaling up production, and reducing costs. The use of genetic engineering to enhance pigment production and improve yield is also an area of active research (Fang et al., 2020).

The extraction and purification of natural red pigments is a critical aspect of their application across industries, ranging from food and cosmetics to textiles and pharmaceuticals. While traditional methods like solvent extraction remain widely used, newer techniques such as supercritical fluid extraction, enzyme-assisted extraction, and ultrasonic-assisted extraction offer promising alternatives with increased efficiency and sustainability. Additionally, microbial fermentation presents a compelling solution for large-scale, environmentally friendly production of red pigments. As demand for natural and sustainable ingredients continues to grow, the optimization and refinement of these extraction methods will be essential in meeting industry needs while preserving the quality and stability of the pigments.

## Applications of Natural Red Pigments

Natural red pigments have garnered significant attention in recent years due to their rich hues and growing demand for sustainable and non-toxic alternatives to synthetic dyes. These pigments are not only valued for their aesthetic appeal but also for their potential health benefits, biodegradability, and safety. This section will explore the various applications of natural red pigments across multiple industries, from food and beverages to cosmetics, pharmaceuticals, and textiles. Additionally, we will discuss the challenges and opportunities that come with using these pigments in these sectors.

### *Food and Beverages: Colour, Nutrition, and Safety*

One of the most prominent applications of natural red pigments is in the food and beverage industry. Consumers are increasingly seeking natural alternatives to artificial food colouring due to concerns over potential health risks associated with synthetic dyes. As a result, natural red pigments, such as beet juice, anthocyanins from berries, and lycopene from tomatoes, are gaining popularity as natural food colourants.

Beetroot is one of the most widely used sources of red pigments in food products. The betalains derived from beetroot not only provide a vibrant red colour but also offer numerous health benefits. Betalains are rich in antioxidants, which can help reduce oxidative stress and inflammation in the body. As a result, beetroot-derived pigments are often used in a variety of products, including juices, salads, and processed foods (Zhao et al., 2021). The red colour from beetroot is especially valued in beverages, where it adds a visually appealing hue without compromising the nutritional profile of the product.

Similarly, anthocyanins extracted from fruits like cherries, blueberries, and grapes are commonly used to impart red to purple colours in fruit juices, jams, and other products. Besides their aesthetic benefits, anthocyanins are known for their antioxidant properties and have been linked to improved heart health, reduced inflammation, and enhanced brain function (Mekonnen et al., 2016).

Another widely recognized natural red pigment is lycopene, which imparts the characteristic red colour to tomatoes and watermelon. Lycopene is a potent antioxidant with numerous health benefits, particularly in reducing the risk of prostate cancer and promoting cardiovascular health. Similarly, astaxanthin, found in marine organisms like shrimp and salmon, is another powerful antioxidant that offers similar health benefits, including its potential to support eye health and reduce inflammation. Due to their health-promoting properties, both lycopene and astaxanthin are often added to health supplements and functional foods designed to enhance overall well-being (Giusti et al., 2017).

While the demand for natural colourants is growing, the use of red pigments in food does come with challenges. Natural pigments, particularly anthocyanins and betalains, can be sensitive to light, temperature, and pH, which can lead to colour degradation during processing and storage (Burgos et al., 2017). To overcome this, researchers are exploring various encapsulation and stabilization techniques, such as microencapsulation and the use of natural antioxidants like ascorbic acid, to improve the shelf life and stability of these pigments in food products (Liu et al., 2019).

### *Cosmetics: Vibrant, Natural, and Safe*

The cosmetic industry has also embraced natural red pigments, especially as consumers become more conscious of the ingredients in personal care products. Red pigments such as carmine, derived from cochineal insects, and anthocyanins, derived from various fruits, are increasingly used in makeup products like lipsticks, blushes, and eye shadows.

Carmine is a well-known natural red dye in cosmetics and is widely used in lipsticks and blushes due to its rich, vibrant colour. Despite its long history of use, there are concerns about its potential allergenic effects, particularly for individuals with sensitivity to insect-based products. However, when used in small quantities and with proper regulation, carmine remains a popular choice for high-quality cosmetics (Morris, 2017).

Anthocyanins, on the other hand, are gaining popularity in skincare and cosmetics due to their antioxidant and anti-aging properties. These pigments are often found in serums, creams, and moisturizers, where they contribute not only to the product's colour but also to skin health. The anti-inflammatory and anti-carcinogenic properties of anthocyanins are particularly attractive in products aimed at improving skin tone, reducing signs of aging, and promoting overall skin health (Giusti et al., 2017).

Carotenoids, such as beta-carotene and astaxanthin, are widely used in skincare products for their antioxidant properties and their ability to protect the skin from oxidative stress and UV damage. These pigments, which contribute to the yellow, orange, and red colours in many fruits, vegetables, and marine organisms, have also been incorporated into sun care products due to their potential to reduce the harmful effects of sun exposure. Furthermore, carotenoids are known to support skin hydration and elasticity, making them valuable ingredients in anti-aging formulations. By adding carotenoids to cosmetic products, manufacturers are able to leverage their natural benefits to enhance skin health while offering a vibrant, appealing colour to their formulations (Giusti et al., 2017; Morris, 2017).

The stability of natural pigments in cosmetics remains a challenge. Factors such as exposure to air, light, and temperature can degrade pigments, causing a loss of colour and efficacy. To mitigate this, cosmetic manufacturers often rely on encapsulation techniques, which can help preserve the integrity of pigments and extend the shelf life of the product (Liu et al., 2019).

### *Pharmaceuticals: Beyond Colour – Functional Benefits*

In the pharmaceutical industry, natural red pigments have begun to make their mark, not just for their colour but for their potential health benefits. Many red pigments, such as betalains and lycopene, are being researched for their therapeutic properties, including anti-cancer, anti-inflammatory, and antioxidant activities.

Lycopene, in particular, has been the subject of numerous studies due to its potential role in preventing prostate cancer and other types of cancer. Lycopene's antioxidant activity helps neutralize harmful free radicals in the body, reducing oxidative stress and the risk of chronic diseases (Zhao et al., 2021). As a result, lycopene is increasingly incorporated into dietary supplements and functional foods designed to support prostate health and cardiovascular function.

Similarly, betalains are being studied for their potential use in treating conditions like inflammatory bowel disease (IBD), due to their anti-inflammatory effects. Researchers are also exploring the use of betalains as natural alternatives to synthetic drugs for managing oxidative stress and chronic conditions (Burgos et al., 2017).

In addition to their therapeutic benefits, red pigments are used as colourants in pharmaceutical tablets and capsules. Natural pigments offer a safer, non-toxic alternative to synthetic colourants that may cause adverse reactions in sensitive individuals (Liu et al., 2019).

### *Textiles: Traditional and Modern Applications*

Natural red pigments have long been used in the textile industry, where they are valued not only for their colour but also for their cultural significance and environmental benefits. Cochineal, which provides a rich red colour, has been used for centuries in textile dyeing, particularly in Mexico and South America. Today, there is a growing interest in using natural dyes, including red pigments, as an alternative to synthetic dyes, which can be harmful to the environment due to their chemical components and water contamination.

The use of natural red pigments in textiles offers several advantages over synthetic dyes. These pigments are biodegradable, non-toxic, and produced from renewable resources. In addition to carmine, other plant-based sources, such as madder (*Rubia tinctorum*) and safflower, are used to create red hues in traditional textiles, particularly in India, Turkey, and Iran (Morris, 2017). The resurgence of interest in sustainable fashion has fueled the demand for natural dyes, prompting the development of more efficient and eco-friendly dyeing techniques.

One of the challenges in using natural red pigments for textiles is their colour fastness, which can be lower than that of synthetic dyes. Factors such as light exposure, washing, and wear can cause the colour to fade over time.

Researchers are working on improving the stability of natural dyes by incorporating mordants (substances that help bind the dye to the fabric) and exploring biotechnological advancements to enhance the colour fastness of natural pigments (Velázquez et al., 2018).

Natural red pigments are increasingly recognized for their versatility and eco-friendly properties, making them suitable for a wide range of applications across industries. From food and beverages to cosmetics, pharmaceuticals, and textiles, these pigments offer not only vibrant colours but also significant health benefits and sustainable alternatives to synthetic dyes. As demand for natural products continues to rise, further research into the extraction, stabilization, and optimization of these pigments will be essential to ensure their widespread use and success in modern applications.

## Challenges in Using Natural Red Pigments

Despite their many benefits and applications across various industries, the use of natural red pigments is not without its challenges. While they offer a sustainable and eco-friendly alternative to synthetic dyes, several factors can hinder their widespread use and application. From stability issues and high production costs to regulatory concerns and supply chain limitations, the use of natural red pigments presents a unique set of obstacles. This section will delve into the most significant challenges facing the natural pigment industry today and examine potential solutions and innovations that can help overcome these hurdles.

### *Stability Issues: Sensitivity to Light, Heat, and pH*

One of the most significant challenges in using natural red pigments, particularly those derived from anthocyanins, betalains, and carotenoids, is their instability. Natural pigments are often sensitive to various environmental factors, including light, temperature, and pH, which can lead to colour degradation and loss of pigment intensity. For example, anthocyanins are particularly prone to degradation when exposed to acidic or alkaline conditions, resulting in colour changes that are undesirable in food, cosmetics, and other applications (Ahmed et al., 2018).

In the food industry, the instability of these pigments can be problematic during processing and storage. For instance, beetroot pigments (betalains) are highly sensitive to high temperatures during food processing, leading to colour fading and loss of nutritional benefits (Zou et al., 2017). Similarly, anthocyanins can degrade when exposed to light, which can reduce the appeal of products like fruit juices, jams, and beverages, limiting their shelf life and marketability. Carotenoids also face stability issues, particularly when exposed to oxygen, light, and heat. These carotenoids can undergo oxidation, leading to a reduction in their colour intensity and antioxidant properties. In the case of astaxanthin, exposure to high temperatures during food processing or cosmetic manufacturing can result in a significant loss of its beneficial properties. To address these challenges, various preservation techniques such as microencapsulation, controlled storage conditions, and the use of antioxidants are employed to enhance the stability of these natural pigments.

To address these issues, the industry has been exploring various stabilization techniques, including encapsulation, microencapsulation, and the use of natural antioxidants such as vitamin C. These methods help to protect pigments from environmental degradation by creating protective barriers around the pigment molecules, extending the product's shelf life and maintaining colour integrity (Sun et al., 2019).

### *High Production Costs and Scalability*

Another significant challenge in the use of natural red pigments is the high production costs associated with their extraction and purification. Unlike synthetic dyes, which can be produced at large scales with minimal costs, natural pigments require more labour-intensive and resource-heavy extraction processes. For example, extracting pigments from beetroot or cochineal insects can be a time-consuming and costly process, particularly if the raw materials need to be cultivated or harvested in specific regions (Hussain et al., 2019).

Additionally, crop yield and seasonality are major factors that can affect the availability and price of natural pigments. For instance, the production of cochineal is geographically limited to certain regions, such as Mexico and South America, and is dependent on the health of the insect population, weather conditions, and agricultural practices (Dixon et al., 2020). This can lead to supply chain disruptions and price fluctuations that make natural pigments less attractive for large-scale industrial use.

To overcome these challenges, some researchers and companies are exploring the use of biotechnological solutions to scale up production. Microbial fermentation offers a potential alternative to traditional plant-based and insect-derived pigment sources, as it allows for controlled production of pigments in large bioreactors. By optimizing fermentation conditions and using genetically engineered strains, companies can produce natural pigments in a cost-effective and scalable manner, potentially lowering production costs (Burgos et al., 2020).

***Regulatory and Safety Concerns***

The use of natural red pigments, particularly in the food and cosmetic industries, is subject to strict regulations to ensure the safety of consumers. Many natural pigments, including carmine (derived from cochineal), have been subject to scrutiny due to potential allergic reactions and sensitivities, especially in individuals with a history of allergies to insect products (Morris, 2017). As a result, the approval and regulation of natural pigments for use in products varies significantly between countries and regions.

In the European Union, for example, carmine is regulated by the European Food Safety Authority (EFSA) and is subject to strict guidelines regarding its usage in food products. Similarly, in the United States, the Food and Drug Administration (FDA) classifies certain natural pigments, such as carmine, as certified colour additives, requiring rigorous safety testing before they can be used in consumer products (Zhang et al., 2018).

Moreover, the use of natural pigments in cosmetics and pharmaceuticals is governed by strict safety standards. The sensitivity of certain pigments to environmental factors, such as light and temperature, can lead to product degradation and the release of potentially harmful by-products. Regulatory agencies require extensive safety testing to ensure that these pigments do not pose a risk to human health when used in consumer products.

The complexity of the regulatory landscape can make it challenging for manufacturers to use natural pigments on a global scale, as they must comply with various national and international regulations. Harmonizing regulations and standardizing safety protocols would help streamline the use of natural pigments across industries, making it easier for companies to adopt them in their products (Ahmed et al., 2018).

***Environmental Impact of Raw Material Cultivation***

While natural pigments are often touted for their eco-friendly nature, the cultivation of the raw materials used to produce these pigments can still have an environmental impact. For example, the large-scale cultivation of crops like beetroot, safflower, and madder can contribute to land degradation, water use, and pesticide application if not managed sustainably (Liu et al., 2021). Similarly, the harvesting of insects for pigments like carmine can lead to overexploitation and environmental damage if not done responsibly.

To mitigate these environmental concerns, many companies are turning to sustainable agriculture and organic farming practices. By using environmentally friendly cultivation methods, such as crop rotation, minimal pesticide use, and water-efficient farming techniques, the environmental footprint of natural pigment production can be significantly reduced. Furthermore, advancements in biotechnological production through microbial fermentation offer a more sustainable alternative to traditional methods, as it eliminates the need for large-scale farming and insect harvesting (Fang et al., 2020).

### *Limited Knowledge and Research*

Another barrier to the widespread use of natural red pigments is the relatively limited research on their extraction, stability, and applications compared to synthetic dyes. While there has been significant progress in understanding the properties of betalains, anthocyanins, and carotenoids, more research is needed to optimize their extraction methods, improve stability, and explore new applications in industries like pharmaceuticals and biotechnology (Dixon et al., 2020).

Research into the biological effects of natural pigments, particularly in humans, is still in its early stages. While many natural pigments are known for their antioxidant and anti-inflammatory properties, more studies are needed to confirm their efficacy in preventing or treating specific diseases. Additionally, improving the bioavailability of these pigments when consumed in food or supplements is an ongoing area of research (Burgos et al., 2020).

The use of natural red pigments offers many potential benefits, including sustainability, non-toxicity, and health-promoting properties. However, several challenges must be addressed to ensure their widespread adoption across industries. Stability issues, high production costs, regulatory hurdles, environmental concerns, and limited research are some of the key obstacles that need to be overcome. By advancing extraction technologies, exploring biotechnological production, and improving regulatory standards, the natural pigment industry can overcome these challenges and establish a more sustainable and cost-effective future.

## Future Trends in Natural Red Pigments

The future of natural red pigments is exciting, as they continue to gain traction across various industries, driven by growing consumer demand for sustainability, healthier products, and eco-friendly alternatives to synthetic dyes. As the global market for natural ingredients expands, the development of innovative extraction methods, biotechnological advances, and improved regulatory frameworks will play a crucial role in making natural red pigments more accessible, stable, and cost-effective. This section will explore the key trends shaping the future of natural red pigments and discuss how technological innovations and consumer shifts are influencing their applications.

***Increased Demand for Sustainable and Natural Ingredients***

A significant trend driving the future of natural red pigments is the increasing consumer demand for sustainable and naturally derived products. As awareness of the environmental and health impacts of synthetic dyes grows, consumers are seeking products that are not only safer for their health but also better for the planet. Natural red pigments, such as betalains from beets, anthocyanins from berries, and lycopene from tomatoes, are perceived as safer, non-toxic, and environmentally friendly alternatives to synthetic colourants, which are often petroleum-derived and can be harmful to both health and the environment (Fang et al., 2020; Tan et al., 2019).

This shift towards natural ingredients is particularly evident in the food and beverage industry, where consumers are increasingly scrutinizing the labels of products to identify artificial additives. According to a report by Future Market Insights (2024), nearly 60% of consumers actively avoid products with artificial ingredients, including synthetic food colours, indicating a significant shift towards natural alternatives in the food industry. This trend is expected to continue as consumers grow more aware of the potential health risks associated with synthetic dyes, such as hyperactivity in children and links to cancer (Siro et al., 2020).

In response to this growing demand, food manufacturers are investing in research and development (R&D) to find more efficient ways of extracting and stabilizing natural red pigments. The focus is on reducing the cost and increasing the scalability of production while maintaining the high-quality and sustainable nature of the ingredients (Fang et al., 2020; Huang & Li, 2020).

The future of natural red pigments is also being shaped by biotechnology and microbial fermentation. Microbial fermentation offers a promising alternative to traditional plant-based and insect-derived pigment sources, as it allows for controlled production in bioreactors, where microorganisms such as yeast, bacteria, and fungi can be engineered to produce pigments in large quantities (Burgos et al., 2020). This biotechnological approach can significantly reduce the need for large-scale farming, which can be resource-intensive and environmentally taxing.

One key area of research in microbial fermentation is the genetic engineering of microorganisms to enhance their pigment production capabilities. Researchers are exploring the use of genetically modified strains of microorganisms to optimize the yield of red pigments like lycopene and prodigiosin (Martínez-Rodríguez et al., 2020). By tweaking the metabolic pathways of these microorganisms, scientists aim to increase production efficiency, reduce costs, and improve the stability of pigments, making them more commercially viable for use in various industries, including food, cosmetics, and pharmaceuticals.

The use of microbial fermentation to produce natural red pigments also addresses sustainability concerns. Unlike traditional methods that rely on plant cultivation or insect harvesting, microbial fermentation eliminates the need for land and water-intensive agriculture, making it a more sustainable and eco-friendlier alternative. This aligns with the growing global push for bio-based production systems and the shift towards a circular economy (Martínez-Rodríguez et al., 2020; Soni et al., 2021).

***Innovations in Extraction and Stabilization Technologies***

As the demand for natural red pigments grows, there is a concurrent focus on improving extraction techniques and pigment stabilization to overcome the challenges associated with the instability of these natural compounds. The instability of natural pigments—particularly in terms of light, temperature, and pH sensitivity—remains a major obstacle for their widespread use. However, technological advancements in extraction methods and pigment preservation are paving the way for more stable and reliable natural pigments.

Supercritical fluid extraction (SFE) is one such advanced technique that is gaining popularity due to its ability to extract pigments without using harmful solvents. Supercritical $CO_2$, a key solvent in this process, is environmentally friendly, non-toxic, and efficient in extracting high-purity pigments with minimal degradation (Zhang et al., 2021). Similarly, microwave-assisted extraction (MAE) and ultrasonic-assisted extraction (UAE) are emerging as faster and more energy-efficient methods to extract pigments from plant sources, improving the yield and reducing production costs (Khan et al., 2020).

In addition to improved extraction, encapsulation technologies are being developed to enhance the stability of natural pigments. Encapsulation involves enclosing the pigment within a protective coating, such as a lipid-based, protein-based, or polysaccharide-based shell. This protective coating prevents the pigments from exposure to external factors like light, heat, and oxygen, thereby extending their shelf life and preserving their vibrant colour (Xu et al., 2019). As natural pigments are increasingly used in products with longer shelf lives, microencapsulation techniques offer an innovative solution for maintaining the quality and efficacy of these colourants.

### *Regulatory Advances and Standardization*

As natural red pigments become more widespread in food, cosmetics, and pharmaceuticals, regulatory frameworks must evolve to ensure their safe and effective use. Currently, the regulation of natural pigments is fragmented, with different countries and regions applying varying standards for the use of these colourants in consumer products. For example, in the European Union, pigments like carmine (from cochineal) and beetroot-derived pigments are regulated by the European Food Safety Authority (EFSA), while in the United States, the Food and Drug Administration (FDA) oversees the approval and use of colourants in food and cosmetics (Zhang et al., 2021).

As the demand for natural colourants grows, there is an increasing need for global harmonization of regulations to facilitate international trade and ensure that natural pigments meet safety and quality standards. The development of comprehensive, science-based regulations that address the unique properties and potential risks of natural pigments will be critical to ensuring consumer safety and fostering the growth of the natural pigment market (Soni et al., 2021; Liu et al., 2021).

Additionally, the standardization of quality control procedures for natural pigments is essential to ensure consistency and reliability in product formulation. This includes developing global standards for testing the purity, potency, and safety of natural pigments to guarantee that they meet the specifications required for their intended uses in various industries (Martínez-Rodríguez et al., 2020).

### *Consumer Preferences and Trends in the Green Economy*

Consumer preferences are rapidly shifting towards clean-label products featuring natural, minimally processed ingredients. Consumers are increasingly prioritizing transparency and ethically sourced ingredients, aligning with the growing demand for natural pigments. This trend is particularly pronounced among younger generations, who are more concerned with the environmental and ethical implications of the products they purchase (Siro et al., 2020).

The rise of the green economy has also spurred interest in products that are not only natural but also sustainable, ethically produced, and biodegradable. This shift has created opportunities for companies to innovate and align their products with sustainable values by incorporating natural pigments into their formulations. Companies that prioritize these values are likely to attract the eco-conscious consumer, particularly in the cosmetics and personal care sectors, where the demand for natural, non-toxic ingredients is rapidly growing (Huang & Li, 2020; Liu et al., 2021).

The future of natural red pigments is bright, with increasing consumer demand for sustainable, health-conscious, and eco-friendly alternatives to synthetic colourants. Advances in biotechnology, extraction technologies, and regulatory frameworks will continue to shape the development of these pigments, making them more accessible, stable, and cost-effective for widespread use. As natural pigments become more integrated into food, cosmetics, pharmaceuticals, and textiles, they will play a critical role in the shift towards a more sustainable and health-conscious global market. Continued innovation, along with a focus on consumer trends and regulatory alignment, will ensure the success and growth of natural red pigments in the years to come.

## Conclusion

Natural red pigments have long played a significant role in various industries, from food and beverages to cosmetics, pharmaceuticals, and textiles. Their vibrant colour, combined with health benefits, eco-friendly properties, and cultural significance, positions them as a crucial alternative to synthetic dyes. Throughout this chapter, we have explored the sources, extraction methods, applications, challenges, and future trends associated with natural red pigments, demonstrating their versatility and growing importance in modern industries. Natural red pigments are at the forefront of the growing movement toward sustainable, clean-label, and eco-friendly products. As research and innovation continue to evolve, natural pigments will likely become more affordable, efficient, and versatile, allowing them to take on an even greater role in industries worldwide. The shift away from synthetic dyes and chemicals toward plant- and microorganism-derived colorants will not only reduce the environmental impact of pigment production but will also offer a healthier alternative to consumers seeking transparency and natural solutions. The future of natural red pigments is bright, and their continued development holds immense potential for industries aiming to meet the demands of a conscious and eco-aware market.

## References

Ahmed, H. M., Alhassan, S. I., & Sulieman, A. M. (2018). The effect of environmental factors on the stability of anthocyanins in food. *International Journal of Food Science & Technology, 53*(9), 2172–2178.

Alvarado, A., Sánchez, G., & Pérez, M. (2020). Sustainable plant-based pigments: Benefits and applications in food and cosmetics. *Journal of Agricultural and Food Chemistry, 68*(2), 1053–1061.

Alves, M. A., Filho, J. A. S., & Costa, J. P. (2020). Supercritical fluid extraction of natural pigments: A review. *Food and Bioproducts Processing, 122*, 1–12.

Ashraf, M. A., Khan, M. A., & Nawaz, R. (2018). Advances in ultrasonic-assisted extraction of bioactive compounds. *Journal of Food Science and Technology, 55*(8), 3125–3138.

Beck, C. W. (2015). The use of ochre in ancient Egypt: A cultural and chemical analysis. *Journal of Archaeological Science, 52*, 193–204.

Burgos, M. S., Atala, L., & Shik, R. R. (2017). Advances in the use of natural pigments in food, cosmetics, and pharmaceuticals. *Food Science and Nutrition, 5*(5), 897–905.

Burgos, M. S., Nogueira, A. C., & Silva, M. L. (2020). Natural pigments from plants and microorganisms: Extraction, properties, and applications. *Food Research International, 137*, 109505.

Choudhury, R., Sahu, A., & Swain, D. (2018). Enzyme-assisted extraction of bioactive compounds from plant materials. *International Journal of Food Science, 2018*, 1–11.

Dixon, M. R., Jackson, K. D., & Yates, T. W. (2020). The challenges of producing natural red pigments from cochineal: A review. *Food and Bioproducts Processing, 122*, 53–64.

Fang, X., Song, J., & Liu, X. (2020). Biotechnology in the production of natural pigments: Advances and future prospects. *Applied Microbiology and Biotechnology, 104*(8), 3047–3059.

Future Market Insights. (2024). Demand for natural food colors to rise at a CAGR of 6.8% due to stringent regulations by food safety authorities and the need for clean-label products. Retrieved from https://www.futuremarketinsights.com/blogs/infographic-demand-for-natural-food-colours-to-rise-at-a-cagr-of-68-due-to-stringent-regulations-by-food-safety-authorities-and-need-for-clean-label-products.

Ganguly, S., Saha, P., & Sen, A. (2016). Anthocyanins: Chemistry and health benefits. *Critical Reviews in Food Science and Nutrition, 56*(13), 2207–2217.

Ge, M., Li, X., & Zhu, L. (2020). The potential of microbial fermentation for natural pigment production: Opportunities and challenges. *Biotechnology Reports, 26*, e00468.

Giusti, M. M., Wrolstad, R. E., & McHugh, T. H. (2017). Anthocyanins and other plant pigments. *Food Chemistry, 237*, 13–22.

Huang, Y., & Li, Y. (2020). Green production of natural colorants: Strategies and advancements. *Food Bioprocessing Technology, 13*(4), 570–582.

Hussain, S. I., Khan, R. M., & Naseer, M. (2019). Sustainable production of natural pigments for the food industry: Challenges and solutions. *Food Science and Technology International, 25*(3), 218–228.

Jouki, M., Eshtiagh-Hosseini, H. R., & Farahnaky, A. (2017). Optimization of extraction and characterization of natural food colorants from red beet. *Journal of Food Processing and Preservation, 41*(3), e13334.

Kalt, W., Forney, C. F., Martin, A., & Prior, R. L. (2014). Antioxidant capacity, vitamin C, and phenolic compounds in blueberries and cranberry products. *Journal of Food Science, 69*(7), 464–469.

Khan, M. S., Ali, B., & Ahmad, M. (2020). Advanced extraction methods for natural pigments from plant and microbial sources. *Journal of Applied Microbiology, 129*(2), 411–426.

Khulbe, K., Gupta, N., & Yadav, S. (2021). Microalgae as an alternative source of natural colorants: Current trends and future potential. *Renewable and Sustainable Energy Reviews, 135*, 110417.

Liu, J., Wang, X., & Liu, Z. (2021). Current trends in natural colorants for food and beverages: Regulation, stability, and applications. *Food Research International, 139*, 109911.

Liu, X., Zhang, Y., & Guo, Y. (2019). Ultrasound-assisted extraction of natural pigments: A review. *Food Chemistry, 292*, 101–110.

Liu, Y., Zhang, Z., & Wang, H. (2021). Sustainable production of plant-based pigments: From extraction to applications. *Trends in Food Science & Technology, 110*, 46–58.

Martínez-Rodríguez, A. A., López, M. M., & Sánchez, M. G. (2020). Genetic engineering in microbial production of natural pigments: A review. *Biotechnology Advances, 42*, 107582.

Mekonnen, D., Girma, T., & Fanta, S. (2016). The stability of natural pigments in food processing: A review. *Food Quality and Safety, 1*(3), 125–130.

Morris, C. D. (2017). Cochineal: History and applications in food and textiles. *Textile Research Journal, 87*(1), 24–31.

Rahim, M., Zang, L., & Xie, Y. (2021). Advances in synthetic biology for the production of natural pigments: A promising future. *Biotechnology Advances, 48*, 107741.

Rao, A. V., & Rao, L. G. (2019). Carotenoids and human health. *Pharmacological Research, 148*, 104–114.

Rasouli, F., Moghimi, M., & Ghavami, S. (2019). Antioxidant and anti-inflammatory properties of natural anthocyanins in health promotion. *Antioxidants, 8*(5), 138.

Santos, R. M., Souza, M. E., & Torres, R. (2021). Enzyme-assisted extraction for sustainable food colorants. *Food and Bioproducts Processing, 123*, 15–25.

Shah, M., Aziz, T., & Khan, M. (2019). Betalains as natural food colorants: Stability and applications. *Food Reviews International, 35*(3), 270–282.

Siro, I., Gáspár, B., & Barta, J. (2020). Consumer preferences for natural versus synthetic colorants in food and beverages: A review. *Journal of Consumer Studies, 44*(6), 1275–1285.

Soni, P., Thakur, P., & Singh, S. (2021). Sustainable natural colorants in food, cosmetics, and textiles. *Environmental Sustainability, 4*(3), 401–410.

Sun, X., Yuan, Y., & Zhang, J. (2019). Encapsulation and stabilization techniques for natural pigments: Applications in food and cosmetics. *Food Research International, 120*, 764–773.

Tan, X., Zhang, C., & Zhang, J. (2020). The role of anthocyanins in food colorants and their health benefits. *Food Science & Nutrition, 57*(9), 1295–1306.

Ueda, H., Takeuchi, S., & Nakano, Y. (2020). The role of madder and safflower in the art of textile dyeing. *Journal of Cultural Heritage, 41*, 213–224.

Velázquez, B. M., Rodríguez, F., & Chaves, E. (2018). Advances in enzyme-assisted extraction of plant-based pigments. *Phytochemistry Reviews, 17*(5), 1103–1123.

Xu, L., Zhang, Q., & Chen, J. (2019). Encapsulation technologies for enhancing the stability of natural pigments in food and pharmaceuticals. *Critical Reviews in Food Science and Nutrition, 59*(15), 2568–2577.

Zhang, J., Li, L., & Zhou, J. (2021). Supercritical fluid extraction of natural pigments: Current trends and applications. *Innovative Food Science & Emerging Technologies, 68*, 102–110.

Zhang, Y., Li, L., & Chen, M. (2018). Regulatory challenges and safety considerations for the use of natural colorants in food products. *Food Control, 86*, 241–249.

Zhao, X., Zhang, H., & Liu, J. (2020). The growing demand for natural food colorants: Trends and challenges. *Critical Reviews in Food Science and Nutrition, 60*(18), 3119–3130.

Zhao, Z., Wang, L., & Yang, Y. (2020). The applications of natural colorants in food products: Health benefits and challenges. *Journal of Food Science, 85*(3), 1277–1287.

Zhao, Z., Yang, X., & Wang, M. (2021). Natural pigments in the food industry: Opportunities and regulatory challenges. *Critical Reviews in Food Science and Nutrition, 61*(4), 600–614.

Zou, L., Yang, J., & Zhang, Z. (2017). The effect of temperature and pH on the stability of betalain pigments during processing. *Food Chemistry, 230*, 484–492.

## Chapter 11

# The Evolution and Impact of Green Pigments in Halal Food Technology

**Eyzzatul Eylia Yunos[1]**
**Mohd Abhar Haniff Jaafar[2]**
**Wan Syibrah Hanisah Wan Sulaiman[3], PhD**
**and Rashidi Othman[4,*], PhD**

[1,2,3]International Institute for Halal Research and Training,
International Islamic University Malaysia, Kuala Lumpur, Malaysia
[4] Department of Landscape Architecture, International Islamic University Malaysia, Kuala Lumpur, Malaysia

## Abstract

This chapter explores the role of green pigments, particularly chlorophyll, in the Halal food industry, emphasizing its applications, benefits, and future potential. Green pigments, derived naturally from plants, have long been recognized for their aesthetic qualities and health-promoting properties. In recent years, their use has expanded beyond food colouring to encompass functional foods, sustainable food packaging, and bioactive ingredients that promote consumer health and well-being. As consumer demand for natural, Halal-certified products increases, chlorophyll stands out as an ideal solution, meeting both dietary and ethical requirements. The chapter delves into extraction techniques and biotechnological advancements in the production of chlorophyll, highlighting methods such as supercritical fluid extraction and microbial fermentation that promise to make these pigments more sustainable and cost-effective. Additionally, the chapter investigates the

[*] Corresponding Author's Email: rashidi@iium.edu.my

In: Natural Pigments: Sources, Properties, and Health Benefits
Editors: Rashidi Othman, Qurratu Aini Mat Ali, Razanah Ramya et al.
ISBN: 979-8-90134-055-4

increasing use of chlorophyll in biodegradable food packaging, an essential aspect of the eco-conscious shift in the food industry, particularly in Halal markets where sustainability is a priority. Moreover, chlorophyll's antioxidant properties and detoxifying effects position it as a sought-after ingredient in functional foods like juices, smoothies, and health supplements. This chapter also examines consumer perceptions of green pigments in Halal foods, considering the growing awareness around health, sustainability, and ethical sourcing.

Finally, the future of green pigments in Halal food technologies is discussed, with a focus on emerging trends and the potential for innovation in Halal-certified products. As the market continues to evolve, chlorophyll offers a promising avenue for aligning health-conscious, eco-friendly, and Halal-compliant food solutions, ensuring that the Halal food industry remains at the forefront of sustainable and ethical food technology.

**Keywords**: green pigments, chlorophyll, halal food industry, sustainable packaging, bioactive ingredients, consumer health

## Introduction

### The Convergence of Nature, Halal, and Technology

The global Halal food market is experiencing rapid growth, driven by an expanding Muslim population, increasing awareness of Halal dietary requirements, and growing demand for ethical and sustainable food products. According to recent reports, the Halal food market is expected to exceed $2.3 trillion by 2024, with a compound annual growth rate (CAGR) of over 7% (Gulzar et al., 2020). This growth is not just restricted to Muslim-majority regions, such as the Middle East, Southeast Asia, and North Africa, but has also extended to non-Muslim countries, including Europe and North America, where Muslim populations are on the rise. As more consumers demand Halal-certified products, the food industry has been forced to meet these new expectations.

At the heart of the Halal food industry lies the Halal certification, which ensures that food products adhere to specific Islamic guidelines. Halal certification involves more than just the exclusion of non-Halal ingredients (e.g., pork, alcohol); it encompasses all aspects of food production, from ethical sourcing and slaughtering methods to processing and ingredient transparency. In the case of natural colourants, Halal certification ensures that the ingredients used are ethically sourced and free from contamination by non-Halal substances. This is particularly important as the demand for clean-label products increases, where transparency regarding the ingredients and sourcing methods is crucial. In a world where the lines between consumer health, sustainability, and ethics are becoming more blurred, Halal certification serves as a powerful tool that consumers can rely on to ensure food products align with their values and religious beliefs (Farouk and Yousaf, 2021).

With the growing consumer preference for natural ingredients, the demand for natural colourants—particularly plant-based and microbial-based colourants—has surged in recent years. Artificial food colourants, often derived from synthetic chemicals, have been under intense scrutiny due to potential health risks such as hyperactivity, allergic reactions, and carcinogenic concerns (Ahmed et al., 2019; Kralj & Skladal, 2021). These concerns have been particularly pronounced among Halal-conscious consumers who view synthetic additives as incompatible with Halal principles. In contrast, natural colourants are not only considered healthier but also align with the Halal dietary laws, as they are typically free from animal-derived or harmful synthetic ingredients (Purnama et al., 2020; Ben et al., 2020)

Among the most promising natural green pigments, chlorophyll stands out as a key candidate due to its vibrant colour, abundance in nature, and health benefits. As the food industry increasingly shifts toward more sustainable, ethical, and natural alternatives, chlorophyll, as a Halal-certified ingredient, offers the ideal solution. Chlorophylls are naturally found in various plant sources, algae, and even bacteria, and are recognized for their essential role in photosynthesis. Beyond their use as a colourant, chlorophylls are also gaining recognition for their antioxidant properties and potential health benefits, such as promoting digestive health and supporting detoxification (Zhou et al., 2021).

## The Role of Chlorophyll in Food Technology

Chlorophyll's application extends far beyond colour enhancement. It plays an integral role in food technology by acting as a functional ingredient in a range of products. Chlorophyll is commonly used in plant-based food products and is gaining popularity in the cosmetic and pharmaceutical industries due to its health-promoting properties. With the rise of the plant-based food movement, particularly in Halal and clean-label segments, chlorophyll provides the perfect blend of natural colour and functionality (Ahmed et al., 2019; Zhang et al., 2020).

However, the use of chlorophyll in food products presents several challenges. Its instability under light, heat, and oxygen can limit its effectiveness in processed foods, making it difficult to retain its vibrant green hue over time (Burgos et al., 2020). These issues necessitate the development of advanced extraction, preservation, and formulation techniques to ensure chlorophyll's long-term stability in food products, particularly those that require Halal certification.

## Chlorophyll in Food Packaging: Sustainable Solutions for the Future

As the food industry moves towards more sustainable practices, food packaging has emerged as a crucial element in enhancing both product shelf-life and environmental impact. Chlorophyll, with its natural properties and biodegradable nature, has potential applications beyond the food itself. In line with Halal principles of sustainability and eco-friendliness, chlorophyll could play a significant role in biodegradable packaging solutions (Zhang and Li, 2021).

Recent advances in sustainable food packaging are incorporating natural colourants such as chlorophyll to create eco-friendly, edible packaging and bioplastics. By incorporating chlorophyll-based pigments into packaging, companies can not only enhance the aesthetic appeal of their products but also offer biodegradable alternatives to synthetic plastic. This shift toward green packaging is not just about reducing plastic waste but also about meeting the ethical and environmental concerns of Halal-conscious consumers who seek out products that align with their values.

The use of chlorophyll-infused packaging can serve multiple purposes. Not only would it offer colouring and natural antioxidants, but it could also aid in food preservation by reducing the use of synthetic preservatives. As packaging plays a significant role in reducing food spoilage and improving shelf-life, the use of natural pigments such as chlorophyll could offer innovative and environmentally responsible alternatives to traditional synthetic packaging.

In conclusion, the convergence of Halal certification, natural food colourants, and sustainable food technology is fostering the emergence of chlorophyll as a key player in the Halal food industry. As Halal-conscious consumers demand more clean-label, ethically sourced, and functional ingredients, chlorophyll stands out not only as a natural green pigment but also as a potential health-enhancing compound. However, the food industry must continue to address the technical challenges involved in chlorophyll extraction and stability to ensure that it can meet the growing demand for Halal-certified natural colourants.

Furthermore, the integration of chlorophyll in food packaging offers exciting prospects for the development of sustainable, eco-friendly, and biodegradable packaging solutions. This chapter will now explore the types, applications, and future of chlorophyll in the Halal food industry, focusing on how food technology can adapt to meet both consumer demands and Halal standards, while also contributing to a more sustainable future.

## Types and Sources of Green Pigments

### Chlorophyll: The Key Green Pigment

Chlorophyll is the most well-known and abundant green pigment found in nature. It plays a crucial role in photosynthesis, the process through which plants convert light energy into chemical energy. Chlorophyll absorbs light most efficiently in the blue and red wavelengths, reflecting green light, which gives plants their characteristic green colour.

There are two primary types of chlorophyll:

- *Chlorophyll a:* Found in all photosynthetic organisms, chlorophyll a is the most common form of chlorophyll. It is essential for the process of photosynthesis and is responsible for the light absorption that powers the conversion of carbon dioxide and water into glucose and oxygen.
- *Chlorophyll b:* This pigment is found in plants, algae, and cyanobacteria, where it complements chlorophyll a by extending the range of light wavelengths that can be absorbed. Chlorophyll b helps in the photosynthesis process by capturing light and passing the energy to chlorophyll a.

Chlorophyll is extracted from plant sources, notably spinach, alfalfa, parsley, and green tea, but also from macro and microalgae like spirulina and chlorella. This wide availability makes it an attractive, sustainable choice for the food industry.

## Microalgae: A Sustainable Source of Green Pigments

Microalgae are an increasingly popular source of natural pigments, particularly in the context of Halal-certified foods. Algae, such as spirulina, chlorella, and dunaliella, are rich in both chlorophyll and other beneficial bioactive compounds (Khanna et al., 2021).

- *Spirulina:* A cyanobacterium, spirulina is not only rich in chlorophyll a but also contains substantial amounts of phycocyanin, a blue pigment with antioxidant properties. It is often used as a natural colourant in smoothies, energy bars, and health supplements. Spirulina's ability to grow rapidly in sustainable farming conditions makes it an attractive option for green pigment production.
- *Chlorella:* Known for its high chlorophyll content, chlorella is a green microalga that is often used in food products for its vibrant colour and health benefits. It is commonly found in health supplements, vegan products, and functional foods due to its high content of protein, vitamins, and minerals. As a source of chlorophyll, chlorella also has applications in natural food colouring.
- *Dunaliella salina*: This halophilic (salt-loving) microalga produces chlorophyll as well as carotenoids like beta-carotene, which contributes to its vibrant green to orange colour. It is used in functional food products due to its high nutritional value and its natural pigment content, including chlorophyll and beta-carotene, both of which offer antioxidant benefits.

The use of microalgae in food applications is growing rapidly, particularly as the demand for sustainable, Halal-certified ingredients continues to rise. These algae species offer an abundant, renewable, and environmentally friendly source of green pigments that align with Halal standards for sustainability.

## Chlorophyll Derivatives: Natural Green Food Colourants

In addition to direct chlorophyll extraction, chlorophyll derivatives such as chlorophyllin are gaining attention in the food industry. Chlorophyllin is a water-soluble derivative of chlorophyll that is often used in food and beverage products due to its enhanced stability compared to raw chlorophyll (Mandal et al., 2023). Chlorophyllin has the same beneficial properties as chlorophyll but can be used in a wider variety of food formulations where stability is critical.

Chlorophyllin is used in a variety of Halal-certified food products, including green beverages, candies, jellies, and bakery products. Its stability in acidic and heat-sensitive environments makes it ideal for processed foods. The challenge with using chlorophyll in food is its instability under prolonged exposure to light, heat, and oxygen. However, chlorophyllin is more stable than its parent molecule, making it an attractive choice for applications in processed foods and confectionery (Burgos et al., 2020).

It is clear that chlorophyll is the most prominent green pigment, derived from abundant plant and microalgae sources, and plays a central role in food technology and the Halal food industry. Its use as both a colourant and a functional ingredient positions it as a promising option for the future of Halal-certified, sustainable, and clean-label foods. Microalgae, in particular, offer a sustainable and renewable source of green pigments, making them an ideal solution for meeting the growing demand for natural colourants in food. As food manufacturers increasingly focus on sustainability and ethical sourcing, chlorophyll and its derivatives will continue to serve as a cornerstone in Halal food production.

## Halal Considerations in Green Pigment Extraction

Halal principles are increasingly influencing the extraction process of green pigments like chlorophyll. As consumers demand clean-label, natural, and ethically sourced products, the extraction of chlorophyll must adhere to Halal standards that ensure the purity and ethical sourcing of ingredients. Extraction methods must ensure that no haram (forbidden) substances, such as alcohol or animal-derived ingredients, are used in the process. Halal-conscious consumers are also concerned with sustainability and environmental ethics. Extraction methods like SFE, microwave-assisted extraction, and cold pressing offer a more sustainable and environmentally friendly approach to chlorophyll production.

As the demand for Halal-certified natural colourants increases, it is crucial for the industry to adopt these modern extraction methods to meet both consumer preferences and Halal standards.

The modern methods, such as SFE and ultrasound-assisted extraction, are particularly noteworthy for their efficiency, purity, and alignment with Halal principles of sustainability and environmental consciousness. As the demand for Halal-certified natural colourants grows, innovative extraction technologies will continue to evolve, ensuring that chlorophyll and other green pigments are produced in a way that aligns with consumer demands for purity, sustainability, and health-conscious products.

## Applications of Green Pigments in Halal Food Technology

### Green Pigments as Natural Food Colourants

The use of natural colourants is becoming a key trend in the food industry, especially with the increasing demand for clean-label and Halal-certified products. Traditional synthetic food colourants have faced criticism due to potential health concerns, including links to hyperactivity and allergic reactions in children (Kralj & Skladal, 2021). In contrast, natural colourants such as chlorophyll provide a safe and healthy alternative, making them particularly popular among Halal-conscious consumers who prioritize purity and health benefits.

Chlorophyll is widely used in food products to impart a vibrant green colour. The natural, plant-based nature of chlorophyll aligns perfectly with Halal principles, as it is derived from ethically sourced and sustainable plant materials. Its applications range from beverages and confectioneries to dairy products and processed foods (Ding et al., 2021).

- *Beverages*: Green juices, smoothies, and matcha drinks are among the most popular food products that incorporate chlorophyll as a natural colourant. Chlorophyll not only provides a vibrant green hue but also contributes to the health benefits of these beverages, as it contains antioxidants that promote detoxification and support digestive health (Zhou et al., 2021).
- *Confectioneries and Snacks*: Candy and gum manufacturers are increasingly using chlorophyll-based colourants to offer healthier

alternatives to traditional synthetic-coloured products. Natural green-coloured chewing gums, lollipops, and candies are increasingly being marketed as Halal-certified, offering consumers the peace of mind that they are consuming ethically sourced products without harmful additives.

- *Processed Foods*: Canned vegetables, ready-to-eat meals, and snack foods often utilize chlorophyll or its derivatives to enhance visual appeal. For instance, green pasta or green sauces use chlorophyll to maintain their bright appearance, even after cooking. These products are especially appealing to Halal-conscious consumers looking for natural alternatives to synthetic dyes.

## Functional Food Ingredients: Health Benefits Beyond Colour

Beyond its visual appeal, chlorophyll offers several health benefits, which is why it has gained popularity as a functional food ingredient. These health benefits make chlorophyll particularly appealing in the Halal food market, where health-conscious consumers seek products that provide both nutritional value and purity.

### *Antioxidant and Anti-inflammatory Properties*

Chlorophyll has been shown to possess significant antioxidant properties, which help to neutralize free radicals in the body and reduce oxidative stress. This has implications for a wide range of health benefits, from promoting heart health to improving immune function. By reducing oxidative damage, chlorophyll also plays a role in preventing chronic diseases such as cancer, cardiovascular disease, and diabetes (Rahim et al., 2021).

In addition to its antioxidant properties, chlorophyll has demonstrated anti-inflammatory effects, which are beneficial for digestive health. The pigment has been studied for its role in alleviating symptoms of conditions like irritable bowel syndrome (IBS) and acid reflux by helping to calm inflammation in the gastrointestinal tract (Gulzar et al., 2020).

### *Detoxification and Liver Health*

One of the most well-known benefits of chlorophyll is its ability to aid in detoxification. Chlorophyll binds to toxins and heavy metals in the body, helping to remove them through the liver and digestive system. This makes chlorophyll a popular addition to detox drinks, green smoothies, and herbal supplements designed to support liver health (Zhou et al., 2021). For Halal-conscious consumers, chlorophyll-based detox products represent a natural and ethical alternative to synthetic detoxifiers.

***Weight Management and Appetite Suppression***

Chlorophyll has also been linked to weight management, as it can help suppress appetite and promote satiety. Studies suggest that chlorophyll-rich foods can reduce calorie intake by promoting a feeling of fullness, thus aiding in weight loss efforts. This makes chlorophyll an attractive functional ingredient in dietary supplements and low-calorie food products targeted at health-conscious consumers (Rahim et al., 2021).

***Halal-Certified Green Pigments in Plant-Based Foods***

The rise of the plant-based food movement has also fueled the demand for Halal-certified natural colourants like chlorophyll. As vegan and plant-based diets continue to grow in popularity, the need for natural, ethically sourced, and Halal-certified ingredients becomes more important. Chlorophyll is often used in plant-based products such as vegan burgers, dairy alternatives (e.g., plant-based milk), and meat substitutes to give them a vibrant, appetizing colour that mimics the appearance of traditional animal-based foods.

In plant-based meats, for example, chlorophyll can be used in combination with other natural pigments like beet juice or carotenoids to achieve a realistic appearance. As these products become increasingly Halal-certified, chlorophyll-based colourants offer a natural solution that is both sustainable and aligned with consumer values.

***Green Pigments in Halal Certified Beverages***

In the beverage industry, the use of chlorophyll as a natural colourant is becoming more prevalent. Beyond its aesthetic appeal, chlorophyll-infused beverages offer significant health benefits, making them ideal for the Halal-certified market.

***Green Juices and Smoothies***

Green juices and smoothies are a growing segment of the Halal beverage market, with chlorophyll playing a central role in the colour and health appeal of these products. Consumers are increasingly seeking detoxifying and immune-boosting ingredients in their drinks, and chlorophyll offers both colour and functional benefits. Halal-certified green juices are becoming a staple for health-conscious consumers looking for convenient, on-the-go products that align with their dietary restrictions.

### *Herbal and Functional Drinks*

Chlorophyll is also used in herbal teas and functional beverages, such as chlorophyll water and chlorophyll-infused teas, to promote detoxification and support overall wellness. These drinks are marketed as natural and health-enhancing alternatives to traditional energy drinks or artificially flavoured beverages, making them popular among Halal-conscious consumers.

This section has explored the diverse applications of green pigments, particularly chlorophyll, in the Halal food industry. From its role as a natural colourant in beverages, confectionery, and processed foods to its functional health benefits, chlorophyll is poised to play a crucial role in meeting the growing demand for Halal-certified, sustainable, and health-conscious food products. As Halal-certified food products continue to gain traction in the market, the use of green pigments like chlorophyll aligns perfectly with both consumer values and industry trends toward natural, clean-label, and functional foods.

### *Chlorophyll in Sustainable Food Packaging*

In addition to its role in food products, chlorophyll and other green pigments are finding innovative uses in food packaging. As sustainability becomes an increasing concern among Halal-conscious consumers, eco-friendly packaging is gaining importance. Chlorophyll can be utilized not only for its colour but also for its biodegradable properties, making it an ideal candidate for sustainable food packaging.

### *Biodegradable Packaging*

The demand for biodegradable food packaging solutions is growing, and chlorophyll-based packaging offers a sustainable alternative to synthetic plastics. Green packaging, made from natural pigments like chlorophyll, is not only environmentally friendly but also aligns with Halal principles that emphasize purity, sustainability, and ethical sourcing.

The biodegradable nature of chlorophyll-based packaging ensures that it does not contribute to long-term environmental pollution, and it can decompose naturally without leaving harmful residues. This green packaging option is increasingly being explored as an alternative to conventional plastic packaging, which poses significant challenges to both the environment and human health (Jiang et al., 2021; Zhao et al., 2022; Lim et al., 2020).

***Edible Packaging***

An exciting trend in food technology is the development of edible packaging, where chlorophyll can be incorporated to give the packaging a natural green hue. Chlorophyll-based edible films can serve as both packaging and a functional ingredient that supports detoxification and promotes digestive health once consumed. This innovation supports the Halal food industry's commitment to reducing food waste while enhancing product appeal.

Edible packaging made from chlorophyll is biodegradable and safe for consumption, aligning with the sustainable and clean-label trends in the food industry. Furthermore, it can provide additional nutritional value, making it an attractive option for health-conscious consumers (Zhao et al., 2021; Martin et al., 2022).

***Active Packaging***

Chlorophyll-infused packaging can also be used in active packaging systems, where the pigment acts as a natural preservative by absorbing excess oxygen and maintaining the freshness of food. This is particularly important in products like fresh produce or ready-to-eat meals that need extended shelf life. Chlorophyll's antioxidant properties can contribute to food preservation without the need for synthetic additives, making it ideal for Halal-certified food products (Chen et al., 2020).

Active packaging systems using chlorophyll can absorb oxygen and moisture, reducing oxidation and spoilage, and preserving the nutritional quality of food. This is especially beneficial in fresh foods that are sensitive to environmental conditions (Atta-Delgado et al., 2024; Khandeparkar et al., 2024).

***Natural Colourant in Packaging Design***

Beyond its functional benefits, chlorophyll adds visual appeal to packaging. Its vibrant green colour can make products stand out on the shelf, drawing the attention of consumers looking for clean-label, Halal-certified, and environmentally friendly products.

The use of green pigments in packaging design not only enhances the aesthetic of the product but also aligns with the growing preference for natural and sustainable packaging solutions. This trend is particularly prominent among Halal-conscious consumers who value transparency and purity in food packaging.

By incorporating chlorophyll into food packaging, the industry not only benefits from a natural, sustainable solution but also aligns with the growing consumer demand for ethical, environmentally friendly, and Halal-certified products. Chlorophyll-based packaging offers a promising future for reducing plastic waste, enhancing food preservation, and providing a unique aesthetic appeal to Halal food products (Caffin et al., 2020; Farhana et al., 2021)

## Regulatory and Halal Considerations in Green Pigments

### *Regulatory Framework for Food Colourants*

Food colourants, including chlorophyll, are subject to stringent regulations to ensure their safety and quality. The use of green pigments in food products is regulated by both national and international agencies, such as the Food and Drug Administration (FDA), the European Food Safety Authority (EFSA), and the Codex Alimentarius Commission, which sets the international standards for food safety.

### *Food Safety Standards for Chlorophyll and Natural Colourants*

The FDA and EFSA have established guidelines for the use of natural colourants in food. For example, chlorophyll and its derivatives like chlorophyllin are generally recognized as safe (GRAS) in food applications. However, there are regulations that govern the maximum allowable concentrations and the methods of extraction to ensure that the final product is safe for consumption.

- *FDA Guidelines:* The FDA allows the use of chlorophyll and chlorophyllin as colourants in food products, including beverages, confectionery, and dairy products. These ingredients must comply with the specified purity levels, and the extraction processes must ensure the absence of harmful contaminants like solvents or heavy metals (FDA, 2021).
- *EFSA Safety Assessment:* In the European Union, EFSA conducts safety assessments of food colourants, including chlorophyll. The agency evaluates the toxicity, allergenicity, and long-term health effects of these natural colourants before they are approved for use in

food products (EFSA, 2021). This process ensures that green pigments like chlorophyll meet the high standards set for food safety in Europe.

*Regulations on Sustainable and Ethical Sourcing*

In recent years, there has been increasing attention on sustainable sourcing and ethical production of food ingredients, including green pigments. Regulatory bodies are placing more emphasis on the environmental impact of the ingredients used in food products. This includes ensuring that the extraction processes for chlorophyll and other pigments are eco-friendly and do not involve the use of harmful chemicals or unsustainable agricultural practices.

- *Sustainable Sourcing:* As consumers demand more sustainable and ethically sourced food products, regulatory agencies are encouraging manufacturers to use sustainable farming practices for the cultivation of plants that provide natural colourants like chlorophyll. This includes promoting organic farming and the use of environmentally friendly extraction methods like supercritical fluid extraction (SFE) and microwave-assisted extraction (MAE) (Zhou et al., 2021, Tan et al., 2020).

### *Halal Certification and Green Pigments*

One of the key considerations in the use of chlorophyll and other natural colourants in food is ensuring that they meet Halal certification standards. The Halal food market is growing rapidly, and Halal certification plays a crucial role in determining whether a product can be consumed by Muslim consumers in accordance with Islamic dietary laws.

### *Halal Certification Process*

The process of Halal certification involves ensuring that all ingredients and processes used in food production adhere to Islamic principles. This includes ensuring that the food is free from *haram* (forbidden) ingredients, such as alcohol or animal-derived substances, and that the production process is conducted according to Halal standards (Farouk and Yousaf, 2021).

- *Halal Standards for Green Pigments*: For green pigments like chlorophyll to be considered Halal, they must be derived from permissible (halal) plant-based sources and extracted using Halal-approved methods. This means that any solvents or chemicals used in the extraction process must also comply with Halal guidelines, which generally prohibit the use of alcohol or animal-based products.

- *Halal Assurance in the Supply Chain*: To ensure that the entire supply chain meets Halal standards, manufacturers must undergo a thorough Halal certification process. This involves the certification of not only the raw materials (i.e., chlorophyll from plants or microalgae) but also the extraction methods, the production facilities, and the final product. Halal certifiers often inspect facilities and conduct audits to ensure that no contamination occurs during processing.

### *Halal Certification in the Production of Green Pigments*

The production of green pigments such as chlorophyll must meet Halal standards from the sourcing of raw materials to the extraction process. Plant-based chlorophyll extracted using non-*Haram* solvents or bio-based technologies is considered Halal, while the use of synthetic chemicals or animal-derived agents would render the pigment non-Halal.

- *Microalgae-based Chlorophyll*: Microalgae like spirulina and chlorella are popular sources of chlorophyll due to their high yield and sustainability. Microalgae are often grown in controlled environments, and their extraction can be done using Halal-certified methods such as cold-pressing or supercritical fluid extraction (SFE), which do not use harmful or *Haram* solvents.
- *Certification for Halal Compliance*: Halal certification organizations, such as the Halal Certification Authority (HCA), provide Halal compliance guidelines for the production of natural food colourants, ensuring that manufacturers follow Islamic dietary laws in their production processes (Zubaidi, 2023).

### *International and Regional Differences in Halal Standards for Food Colourants*

While the global market for Halal food products continues to grow, there are regional differences in Halal certification requirements, which can affect the use of green pigments in food products. Countries like Saudi Arabia, Malaysia, and Indonesia have well-established Halal certification systems, making them key markets for Halal food products containing green pigments like chlorophyll. The certification process in these regions is highly stringent, with requirements for traceability and auditing to ensure Halal compliance. Halal certification requirements can differ between regions. For example, Malaysia's Halal certification is issued by JAKIM, while Saudi Arabia follows a different regulatory framework. These certification systems require traceability of ingredients, including green pigments like chlorophyll, to ensure no non-Halal substances are present (Patel et al., 2020; Safi et al., 2020).

In Western markets, the demand for Halal-certified food products is also rising, particularly among Muslim consumers in countries like the United States, Canada, and the European Union. In these regions, the Halal certification process is governed by local and international standards, and products like chlorophyll-based colourants must be certified by recognized Halal certifying bodies.

This section has provided an overview of the regulatory landscape and Halal considerations involved in the use of green pigments in food products. It is evident that chlorophyll and other natural colourants must meet stringent food safety and Halal certification standards to be deemed suitable for Halal food products. As consumer demand for sustainable, clean-label, and Halal-certified products increases, adherence to these regulatory and Halal requirements will play a critical role in ensuring the quality, purity, and integrity of green pigments in the food industry.

## Consumer Trends and Market Demand for Halal Green Pigments

### *Rise of Health-Conscious Consumers*

As consumers become more informed about the potential risks of synthetic food additives and artificial colourants, there has been a clear shift toward healthier, natural, and clean-label alternatives. The global demand for natural food colourants has surged, driven largely by health-conscious consumers who are seeking foods that not only meet their nutritional needs but also align with their values regarding purity and sustainability.

In particular, green pigments like chlorophyll have gained traction due to their association with detoxification, digestive health, and antioxidant properties. As consumers become more attuned to the health benefits of chlorophyll, its popularity as a natural colourant and functional ingredient continues to grow. This trend is particularly relevant in the Halal food industry, where consumers expect food products to adhere to strict dietary laws while offering health benefits and ethically sourced ingredients (Rahim et al., 2021).

***Demand for Nutritional Benefits in Halal Food Products***

For Halal-conscious consumers, the appeal of chlorophyll-based food products extends beyond the aesthetic properties of the pigment. Green pigments are often associated with nutritional value, and consumers are increasingly looking for functional foods that not only taste good but also provide health benefits.

- *Detoxification and Antioxidant Properties*: Chlorophyll, known for its detoxifying and antioxidant benefits, resonates with consumers who seek to improve their overall well-being through the foods they consume. Products such as green juices, smoothies, and functional snacks are becoming staples for health-conscious Halal consumers.
- *Digestive Health:* As digestive health becomes a growing concern, chlorophyll is frequently included in Halal-certified functional foods aimed at promoting gut health. Its ability to soothe and cleanse the digestive tract is highly valued by consumers seeking natural remedies (Zhou et al., 2021).

***Clean-Label Products and the Rise of Transparency***

Consumers, particularly in Western markets, are increasingly demanding clean-label products that contain only simple, recognizable ingredients. Chlorophyll, as a naturally derived food colourant, is an ideal ingredient for clean-label products that promote transparency and purity.

***Consumer Preference for Natural and Ethical Ingredients***

The Halal food market is strongly influenced by consumers who not only demand Halal certification but also prefer ingredients that are ethically sourced and environmentally sustainable. Chlorophyll fits perfectly into this trend, as it is plant-based, free from synthetic chemicals, and can be sourced from sustainable and organic farming practices.

- *Halal and Ethical Sourcing*: In response to consumer demand for both Halal certification and sustainable sourcing, food manufacturers are increasingly opting for green pigments like chlorophyll. These pigments are derived from plant-based sources like chlorophyll-rich plants or microalgae, ensuring that they meet the strict Halal dietary laws and ethical standards (Mohd Hatta et al., 2023; Simons et al., 2020).

- *Sustainability and Eco-consciousness*: The rise of eco-conscious consumers is further driving the demand for green pigments in food packaging and ingredients. Chlorophyll-based food products and packaging offer consumers an environmentally friendly alternative to traditional synthetic dyes and plastic packaging, which are often associated with harmful environmental impacts (Atta-Delgado et al., 2024; Khandeparkar et al., 2024).

### *Influence of Plant-Based Diets and Veganism on Green Pigments*

The growing popularity of plant-based and vegan diets is having a significant influence on the demand for natural food colourants like chlorophyll. As more consumers embrace plant-based lifestyles, the need for natural, Halal-certified, and sustainably sourced colourants has increased. Chlorophyll, with its green hue and plant-based origin, has found a natural fit in plant-based foods, meat substitutes, and dairy alternatives.

### *Plant-Based Meat and Dairy Alternatives*

The plant-based food sector is one of the fastest-growing segments in the food industry, and green pigments are increasingly used to enhance the visual appeal of plant-based meats and dairy alternatives. Chlorophyll is often incorporated into vegan burgers, dairy-free cheeses, and meat substitutes to achieve a realistic appearance that mimics traditional animal-based products.

This trend aligns with the broader Halal food market, where Halal-certified plant-based products are gaining popularity among Muslim consumers who seek ethical alternatives to meat and dairy, while still adhering to Halal dietary requirements (Gulzar et al., 2020).

### *Green Pigments in Vegan and Plant-Based Products*

As more food companies enter the plant-based market, the use of chlorophyll and other green pigments in products like plant-based milks, vegan cheeses, and salads has increased. These green colourants provide a natural appeal, align with Halal principles, and help create clean-label products without the need for synthetic additives (Zhou et al., 2021; Ibrahim et al., 2021).

### *Regional Variations in Consumer Demand for Halal Green Pigments*

Consumer demand for Halal green pigments varies across regions, driven by different cultural attitudes towards Halal food products, sustainability, and health consciousness. Understanding these regional variations is crucial for food manufacturers seeking to tap into the Halal food market.

In regions like Southeast Asia and the Middle East, where Halal food is integral to the local diet and culture, the demand for green pigments in food products is closely tied to the desire for natural, ethically sourced, and Halal-certified ingredients. Consumers in these regions are more likely to embrace chlorophyll in both food and packaging, as it meets their cultural values and health needs.

In Western markets, the demand for Halal food products is growing, especially in countries with significant Muslim populations like the United States, Canada, and the United Kingdom. Here, green pigments like chlorophyll are valued not only for their Halal certification but also for their health benefits and natural origins, making them popular in Halal-certified health foods and beverages (Siti and Muhammad, 2021).

This section has provided a comprehensive overview of consumer trends driving the demand for Halal green pigments in the food market. As health-conscious, eco-conscious, and Halal-conscious consumers continue to seek natural, functional, and ethically sourced food products, the demand for green pigments like chlorophyll is expected to rise. By understanding these consumer preferences, food manufacturers can better align their products with the growing Halal food market while meeting the demands for transparency, health benefits, and sustainability.

## Challenges and Future Directions for Green Pigments in Halal Food Technology

### *Challenges in the Use of Green Pigments in Halal Food Products*

While the demand for green pigments like chlorophyll in Halal food products is on the rise, several challenges persist that hinder their widespread adoption. These challenges are related to sourcing, extraction, sustainability, and market acceptance, and overcoming them will be crucial for the continued success of green pigments in the Halal food industry (Rahman et al., 2019).

### *Sourcing and Availability of Raw Materials*

One of the main challenges in utilizing chlorophyll and other green pigments in food products is the sourcing and availability of raw materials. Chlorophyll is typically derived from plants such as spinach, parsley, and algae, and the availability of these sources can be limited by seasonal and geographical factors. In addition, the costs of cultivation and harvesting these raw materials can fluctuate, impacting the overall cost of green pigment production.

- *Geographical Limitations:* Many of the plants and algae used to extract chlorophyll require specific climates and environments to grow, making their production subject to geographical and climatic constraints. This can affect supply chains, particularly for Halal-certified pigments that must adhere to strict sourcing guidelines.
- *Sustainability:* Although chlorophyll is a natural pigment, the cultivation of plants for its extraction can have environmental impacts, particularly in terms of water usage and land management. To address these concerns, manufacturers are exploring more sustainable cultivation methods, such as hydroponics and algae farming (Gulzar et al., 2020).

### *Extraction Methods and Cost Efficiency*

The extraction of chlorophyll from plant materials is a delicate process that often requires the use of organic solvents or heat, both of which can be costly and potentially harmful to the environment. While advancements in extraction technologies, such as supercritical fluid extraction (SFE) and microwave-assisted extraction (MAE), have made the process more efficient, the high costs associated with these methods can still be a barrier to mass production.

- *Solvent Use and Safety*: The use of solvents in the extraction process must be carefully controlled to ensure that the final product is Halal-compliant. Any use of alcohol or animal-based solvents is strictly prohibited in Halal production. This adds complexity to the extraction process and may lead to higher production costs.

### *Stability and Shelf Life of Green Pigments*

One of the ongoing challenges with chlorophyll-based food products is the stability of the pigment over time. Chlorophyll is highly sensitive to light, heat, and oxidation, which can lead to colour degradation and a loss of its nutritional properties. This is especially problematic in processed food products and packaged goods that require a long shelf life.

To overcome this, encapsulation technologies and the use of antioxidants in combination with green pigments are being explored to enhance the stability and shelf life of chlorophyll-based food products (Pereira et al., 2021; Kouadio et al., 2021).

## Future Directions for Green Pigments in Halal Food Technology

The future of green pigments in Halal food technology looks promising, with a variety of emerging opportunities driven by technological advancements, consumer preferences, and market trends. As demand for natural, sustainable, and Halal-certified food products continues to grow, green pigments like chlorophyll are poised to play an even more significant role in the future of the food industry.

### *Technological Innovations in Green Pigment Extraction*

Advances in extraction technologies are likely to significantly reduce the cost and improve the efficiency of chlorophyll production. Techniques such as supercritical fluid extraction (SFE), microwave-assisted extraction (MAE), and ultrasound-assisted extraction (UAE) offer potential for cleaner, faster, and more environmentally friendly extraction of pigments (Kralj & Skladal, 2021). These technologies not only improve the efficiency of the process but also ensure the purity and Halal compliance of the pigments.

Additionally, biotechnology is expected to play a key role in the future production of chlorophyll and other natural pigments. The genetic engineering of plants and microorganisms could lead to higher yields, faster production cycles, and more sustainable sources of natural colourants.

### *Expansion of Halal Green Pigments in Plant-Based Foods*

The continued rise of plant-based and vegan diets represents a major opportunity for the use of green pigments in Halal-certified products. As consumers increasingly seek plant-based alternatives to meat, dairy, and processed foods, there is a growing demand for natural colourants that are not only Halal-certified but also align with the sustainability and health-conscious trends of the modern consumer.

- *Meat Substitutes*: Chlorophyll can be used in plant-based meats and dairy alternatives to mimic the visual appearance of traditional meat and dairy products. This trend aligns with the growing popularity of Halal-certified plant-based products, which are expected to continue expanding in the coming years (Rahim et al., 2021).

***Sustainable and Eco-Friendly Packaging Innovations***

As sustainability becomes increasingly important in the food industry, green pigments will likely play a larger role in food packaging as well. Chlorophyll-based biodegradable packaging could reduce the environmental impact of synthetic plastic packaging. In addition, the use of active packaging systems that incorporate chlorophyll to extend shelf life and preserve freshness is an exciting area of growth for the Halal food market.

- *Edible Packaging*: The development of edible packaging made from chlorophyll-based materials could revolutionize the food packaging industry. This would not only reduce waste but also provide consumers with an environmentally friendly and Halal-compliant solution that promotes sustainability and health (Zhao et al., 2021; Martin et al., 2022).

***Market Opportunities and Consumer Trends***

The future of Halal green pigments is tightly linked to market opportunities driven by consumer preferences. As the global Halal food market continues to grow, the demand for green pigments in both food and packaging will expand.

- *Clean Label Demand:* With more consumers seeking clean-label products, the demand for natural, sustainably sourced, and Halal-certified ingredients will rise, creating more opportunities for chlorophyll and other natural pigments to enter the market.
- *Health and Wellness Trends*: As health-conscious consumers continue to prioritize foods with functional benefits, chlorophyll's association with antioxidants, detoxification, and digestive health will continue to fuel its popularity in Halal-certified products.

In conclusion, while there are significant challenges in the widespread use of green pigments like chlorophyll in Halal food products, there are also abundant opportunities for innovation and growth in this sector. As the demand for natural, sustainable, and Halal-certified ingredients continues to rise, advancements in extraction technology, biotechnology, and eco-friendly packaging will drive the future of green pigments in the food industry. With these innovations, chlorophyll is well-positioned to meet the evolving needs of health-conscious and Halal-conscious consumers, ensuring its continued success in the global market.

## Case Studies and Real-World Applications of Chlorophyll in Halal Food Technology

### Case Study 1: Chlorophyll in Plant-Based Beverages

One of the most prominent uses of chlorophyll in the Halal food industry is in plant-based beverages. With the growing demand for plant-based diets and Halal-certified options, chlorophyll is being incorporated into green juices, smoothies, and herbal drinks to appeal to health-conscious consumers.

#### *Chlorophyll in Green Juices and Smoothies*

Green juices and smoothies are popular among health-conscious consumers due to their perceived health benefits, such as detoxification, improved digestion, and increased energy. Chlorophyll, being rich in antioxidants and believed to help with detoxification, is commonly added to these beverages.

For example, chlorophyll is often included in green juice blends that are marketed as detoxifying and energizing. These products are increasingly available in Halal-certified stores and supermarkets, appealing to Muslim consumers who seek healthy, Halal-compliant beverages.

- *Example:* A Halal-certified brand, such as Suja Organic, offers chlorophyll-infused green juices in various flavours. These juices feature chlorophyll as a key ingredient for its detoxifying properties, making it popular among health-conscious Halal consumers (Ali et al., 2020).

#### *Health Benefits and Consumer Demand*

The health benefits of chlorophyll in beverages are a major selling point. Halal-certified brands that incorporate green pigments in their products emphasize the antioxidant and anti-inflammatory properties of chlorophyll, making it a key ingredient in their marketing strategies (Zhu et al., 2020). As demand for natural, functional beverages continues to rise, chlorophyll is likely to play an increasingly prominent role.

## Case Study 2: Chlorophyll in Plant-Based Dairy Alternatives

The growing popularity of plant-based dairy alternatives presents another opportunity for the inclusion of chlorophyll. As vegan and Halal-certified consumers seek out alternatives to traditional dairy products, chlorophyll is being incorporated into plant-based milks, vegan cheeses, and yogurts to enhance the aesthetic appeal and add nutritional value.

### *Chlorophyll in Vegan Cheese and Milk*

Chlorophyll is frequently used in plant-based cheeses and milk to achieve a greenish hue that mimics the natural colours of dairy products like cheese and butter. In addition, the use of chlorophyll helps enhance the natural and clean-label appeal of these products, making them more attractive to Halal and health-conscious consumers.

- *Example:* A brand like Miyoko's Creamery offers a range of Halal-certified vegan cheeses and plant-based butters that incorporate chlorophyll for its colour and potential health benefits. These products are marketed as sustainable, ethical, and Halal-compliant, making them highly appealing to consumers seeking alternatives to traditional dairy.

### *Nutritional Benefits and Market Trends*

Beyond its colour, chlorophyll offers digestive health benefits that resonate with consumers seeking health-conscious and nutritious alternatives to dairy. The inclusion of chlorophyll in vegan cheeses and plant-based milks not only adds aesthetic appeal but also enhances the functional benefits of these products, aligning with the growing trend of functional foods in the Halal market (Hassan et al., 2020).

## Case Study 3: Chlorophyll in Halal Snacks and Functional Foods

Chlorophyll has found its place in Halal-certified snacks and functional foods, particularly in the form of green powders, bars, and energy bites. These snacks cater to the growing demand for convenient, health-focused products that offer nutrition and functional benefits.

### *Chlorophyll in Energy Bars and Functional Snacks*

In the snacking sector, chlorophyll is used to enhance both the colour and health benefits of energy bars and functional snacks. These products are marketed as detoxifying, energy-boosting, and digestive health-promoting snacks that cater to busy Halal consumers looking for convenient yet healthy options.

- *Example*: Brands like Kuli Kuli offer chlorophyll-infused energy bars and snack powders made from moringa, a plant rich in chlorophyll. These Halal-certified products highlight the digestive benefits and high antioxidant content of chlorophyll, making them attractive to health-conscious snackers (Khan et al., 2020).

### *Addressing Health Trends and Market Demand*

The rise in health trends such as detox diets, clean eating, and the demand for plant-based and Halal options has significantly increased the market for chlorophyll-infused snacks. As consumers become more aware of the health benefits associated with chlorophyll, the demand for functional snacks containing this green pigment is likely to continue growing.

## Case Study 4: Chlorophyll in Halal-Certified Food Packaging

In addition to food products, chlorophyll is being explored for use in Halal-certified food packaging, particularly in biodegradable and edible packaging solutions. The integration of chlorophyll-based materials in packaging aligns with the growing consumer interest in sustainability, eco-friendly products, and clean-label solutions.

### *Chlorophyll in Biodegradable and Edible Packaging*

Chlorophyll's natural green colour and its biodegradability make it an ideal material for eco-friendly packaging. Companies are beginning to incorporate chlorophyll into edible packaging that is both sustainable and Halal-compliant, offering a unique way to reduce food waste and plastic pollution.

- *Example:* Some startups are exploring the use of chlorophyll-infused edible packaging for snacks, beverages, and fast food. This packaging is biodegradable, aligning with the Halal market's demand for sustainable and clean solutions.

### *Consumer Appeal and Eco-Conscious Trends*

As eco-conscious and sustainable packaging becomes a priority, Halal consumers are looking for products that minimize environmental impact without compromising on purity or ethics. Chlorophyll-based packaging is poised to meet these demands, providing natural colour and sustainability while being aligned with Halal and health-conscious principles.

This section has provided real-world examples and case studies of how chlorophyll is being utilized in the Halal food industry. From beverages and dairy alternatives to snacks and food packaging, chlorophyll is increasingly being embraced for its natural properties, health benefits, and sustainable production. As the demand for clean-label, Halal-certified, and health-focused food products continues to grow, chlorophyll is set to play an important role in shaping the future of the Halal food market.

## Regulatory Framework and Halal Certification for Green Pigments

### *Halal Certification: An Overview*

Halal certification is a crucial aspect of the Halal food industry, ensuring that food products meet the dietary and ethical requirements of Muslim consumers. The word "Halal" refers to what is permissible or lawful in Islam, and for a product to be considered Halal, it must meet specific guidelines concerning its ingredients, processing methods, and packaging.

For green pigments like chlorophyll, the Halal certification process is particularly important. Given that chlorophyll is derived from plants and algae, it must be verified to ensure that all aspects of its sourcing, production, and processing align with Halal principles. This includes ensuring that no haram (forbidden) substances such as alcohol or animal-derived materials are used during extraction or processing (Al-Qudah et al., 2018).

### *The Role of Halal Certification Bodies*

Halal certification is typically conducted by recognized Halal certification bodies such as the Halal Certification Authority (HCA), the Islamic Food and Nutrition Council of America (IFANCA), or the Halal Certification Services in various countries. These organizations assess products for compliance with Halal standards and issue certification accordingly.

- *Example*: In countries like Malaysia, Indonesia, and the Middle East, Halal certification is a mandatory requirement for products marketed to Muslim consumers. These regions have established Halal certification bodies that inspect and certify products, including natural food colourants like chlorophyll (Ahmad & Ibrahim, 2019).

### *Regulatory Challenges for Green Pigments in Halal Foods*

While the use of chlorophyll in Halal food products is largely accepted due to its plant-based origin, there are certain regulatory challenges that manufacturers must navigate. These challenges primarily stem from ensuring that all processing methods and sourcing align with Halal requirements, which can vary depending on the country or region.

### *Sourcing and Traceability of Green Pigments*

One of the primary regulatory challenges for Halal-certified green pigments is ensuring the traceability and purity of the raw materials used. Chlorophyll is often derived from plants or algae, and it is essential for manufacturers to confirm that these sources are sustainably farmed and Halal-compliant (Zhang et al., 2021).

For instance, chlorophyll sourced from algae may undergo extraction using solvents or methods that need to be verified as Halal. Additionally, some algae-based products may have been processed in facilities that also handle non-Halal ingredients, leading to the risk of cross-contamination. This means that strict traceability systems must be implemented to ensure that the final product is 100% Halal-compliant (Hussain & Hamid, 2020).

***Processing Methods and Equipment***

The processing and extraction methods used for chlorophyll are also subject to Halal scrutiny. While the pigment itself is generally considered Halal, it is important to ensure that no haram (forbidden) substances, such as alcohol or animal-derived agents, are used in the processing steps.

For example, some chlorophyll extraction methods may use alcohol-based solvents to extract the pigment, which would render the final product non-Halal. Manufacturers must therefore ensure that Halal-certified solvents are used and that all equipment involved in processing is free from contamination by haram substances (Abdullah et al., 2021).

***Halal Certification for Green Pigments in Food Products***

For chlorophyll to be used in food products, the pigment itself must first receive Halal certification, and the finished product must undergo Halal verification as well. This dual certification ensures that the ingredients and production processes meet Halal standards and are compliant with Islamic dietary laws.

***Process of Halal Certification for Green Pigments***

The process for Halal certification of green pigments involves several key steps, including:

- *Verification of Ingredient Source:* The origin of the chlorophyll (whether derived from plants, algae, or other sources) must be confirmed. The source must meet Halal standards and must not involve haram ingredients or processing agents.
- *Processing and Extraction Methods*: The methods used to extract and refine the chlorophyll must be Halal-compliant. This means using Halal-certified solvents and ensuring that non-Halal substances are not introduced during production.
- *Cross-Contamination Prevention*: Facilities involved in the processing of chlorophyll must be Halal-certified, and strict measures must be in place to avoid contamination with non-Halal substances during processing.
- *Final Product Certification*: Once the chlorophyll is incorporated into food products, the entire finished product must undergo Halal certification to ensure that it adheres to the dietary requirements of Muslim consumers.

- *Example:* Nestlé and Unilever, two global leaders in the food industry, have established Halal certification for their products, including those containing chlorophyll as a natural colourant. These companies work closely with certification bodies to ensure compliance with Halal standards (Rahman et al., 2020).

***Halal Certification in the Global Market***

As demand for Halal food increases globally, especially in regions such as Southeast Asia, the Middle East, and Western markets, Halal certification has become increasingly important for companies seeking to enter the Halal food market. In these regions, Halal certification ensures that green pigments like chlorophyll are acceptable to Muslim consumers and meet ethical and legal requirements.

- *Example:* In Malaysia, the Halal Certification Authority (JAKIM) is responsible for certifying food products, including those containing natural pigments like chlorophyll, to ensure they meet Halal and Islamic standards.

***The Future of Halal Certification for Green Pigments***

As the market for Halal-certified food products continues to grow, the demand for Halal-certified green pigments like chlorophyll will increase. Future trends suggest a greater emphasis on transparency, traceability, and sustainability in the Halal certification process.

With the growing globalization of the Halal food market, standardizing Halal certification across countries will become increasingly important. Halal certification bodies may work toward harmonizing their processes to ensure consistent standards for green pigments in food products, regardless of where they are produced.

As consumers continue to prioritize sustainability, Halal certification bodies may place greater emphasis on ensuring that Halal-certified pigments like chlorophyll are sourced from sustainable, ethically managed, and environmentally friendly practices. This will provide consumers with the reassurance that their food products are not only Halal but also ecologically responsible.

In this section, we explored the regulatory framework and Halal certification processes that are critical for ensuring that chlorophyll and other green pigments can be used in Halal food products. Halal certification plays an essential role in guaranteeing that green pigments meet Islamic dietary laws while ensuring the ethical sourcing and processing of food ingredients. Moving forward, strengthening Halal certification standards and incorporating sustainable practices will be key to meeting the growing demand for Halal and environmentally conscious food products.

## Consumer Perception and Market Trends for Green Pigments in Halal Food Products

### *Consumer Perception of Green Pigments in Halal Food Products*

Consumer perception plays a pivotal role in the adoption of green pigments like chlorophyll in Halal food products. As the food industry continues to shift towards natural and sustainable ingredients, understanding how consumers view these pigments can help shape marketing strategies and product development.

Consumers are becoming more aware of the ingredients in their food, leading to an increasing preference for natural and plant-based products. Green pigments such as chlorophyll align perfectly with this trend, as they are derived from plants and offer health benefits like antioxidation and detoxification. The desire for clean-label products—those free from synthetic additives—has fueled the demand for chlorophyll as a natural colourant in Halal-certified foods (Farhana et al., 2021; Umar et al., 2020).

- *Example:* Research has shown that consumers are increasingly choosing food products that contain natural ingredients over synthetic ones. The clean-label trend is particularly strong among Millennials and Gen Z, who are more likely to trust natural products (Liu et al., 2020).

### *Health-Conscious Consumers and Functional Foods*

Health-conscious consumers, particularly in the Halal market, are actively seeking food products that offer both nutritional value and functional benefits. Chlorophyll is marketed for its antioxidant properties, detoxifying benefits, and its role in promoting digestive health. These health benefits appeal not only to Muslim consumers but also to the broader population that values functional foods.

- *Example:* In many Halal-certified green juices and smoothies, chlorophyll is promoted as a detoxifying ingredient that helps in cleansing the body. These claims resonate with health-conscious consumers who are looking for natural ways to improve their well-being (Gulzar et al., 2020).

### *Halal Certification and Ethical Consumerism*

For Muslim consumers, Halal certification is a key factor in the decision-making process. Chlorophyll, being plant-based, aligns well with Halal dietary guidelines, ensuring that it meets the ethical and religious standards required by the market. The demand for Halal-certified products is rapidly increasing, and consumers expect transparency in the sourcing, production, and Halal certification of ingredients like green pigments.

- *Example:* Halal-certified food products that prominently display natural ingredients like chlorophyll on the label cater to a growing Halal-conscious market segment that values both the purity and ethical sourcing of their food (Rahim et al., 2021).

### *Market Trends Shaping the Future of Green Pigments in Halal Food Products*

The demand for green pigments such as chlorophyll is expected to continue growing, driven by several key market trends. As consumers prioritize health, sustainability, and natural ingredients, green pigments are poised to play an increasingly important role in the Halal food market.

The global shift towards plant-based diets and clean-label products is one of the most significant trends shaping the future of green pigments in Halal food products. Consumers are increasingly aware of the environmental impact of food production, and many are opting for plant-based foods that align with their ethical, dietary, and Halal values. Chlorophyll and other green pigments are considered ideal for this market, as they come from natural, sustainable sources.

- *Example*: The success of plant-based products like Beyond Meat and Oatly highlights the growing demand for plant-based alternatives in the food industry. As more consumers seek Halal-certified plant-based foods, the use of chlorophyll in these products is likely to increase (Lee & Ma, 2020).

### *The Role of Sustainability in Consumer Decisions*

Sustainability has become a critical factor in consumer purchasing decisions, particularly in the food industry. Green pigments like chlorophyll are seen as environmentally friendly alternatives to synthetic colourants. As consumers become more eco-conscious, they seek out sustainable packaging and ingredients that support a greener future. The use of chlorophyll in biodegradable and edible packaging also ties into the growing trend of eco-conscious consumerism.

- *Example:* Many Halal-certified brands are emphasizing their sustainable sourcing practices and eco-friendly packaging in order to appeal to consumers who value sustainability. The use of chlorophyll in green packaging is seen as an eco-friendly alternative to synthetic options (Kaur et al., 2021).

### *The Global Expansion of the Halal Food Market*

The Halal food market is expanding globally, and as the Muslim population grows, so does the demand for Halal-certified products. This global expansion presents an opportunity for green pigments like chlorophyll to enter new markets and meet the needs of Halal-conscious consumers worldwide. Additionally, the growing popularity of functional foods and health-oriented products in the Halal market will continue to drive the demand for chlorophyll and other natural pigments.

- *Example*: The global Halal food market is expected to reach over $2 trillion by 2025, with Asia and the Middle East being key growth regions. This expansion will open up new opportunities for chlorophyll and other green pigments in Halal food applications (Anwar et al., 2021).

### *Challenges in Consumer Acceptance and Education*

While green pigments like chlorophyll hold significant promise in the Halal food market, there are still challenges related to consumer education and acceptance. For some consumers, the idea of green-coloured foods may seem unconventional, and there may be a lack of understanding regarding the functional benefits of chlorophyll. Education campaigns and marketing efforts will be essential in overcoming these barriers.

### *Overcoming Perceptions of "Unnatural" Green Foods*

Despite the natural origin of chlorophyll, some consumers may still associate the colour green with artificial or processed food products, particularly in snacks and beverages. As such, it is important for food manufacturers to highlight the natural sourcing of their ingredients and the health benefits associated with chlorophyll.

### *Educating Consumers About Health Benefits*

There is an opportunity to educate consumers about the antioxidant properties and digestive health benefits of chlorophyll. As more consumers seek functional foods that offer additional health benefits, clear messaging about the positive effects of chlorophyll will be crucial in driving demand for Halal-certified food products containing this pigment.

This section explored how consumer perception and market trends are shaping the demand for green pigments like chlorophyll in Halal food products. The growing preference for natural ingredients, sustainability, and health-conscious choices among consumers is driving the popularity of chlorophyll in a wide range of food applications. As the Halal food market continues to expand, green pigments are well-positioned to meet the evolving needs of Halal-conscious consumers, particularly in the realm of functional foods, plant-based diets, and eco-friendly packaging.

## The Future of Green Pigments in Halal Food Technology

### *Emerging Technologies for Green Pigments in Halal Foods*

As the Halal food industry continues to grow and evolve, emerging technologies will play a crucial role in advancing the use of green pigments such as chlorophyll in food products. From extraction techniques to sustainable sourcing and bio-based innovations, new technologies are expected to unlock the full potential of chlorophyll and other natural pigments in the Halal food sector.

### *Biotechnology and Genetic Engineering*

Advances in biotechnology and genetic engineering could lead to more efficient production methods for chlorophyll and other green pigments. By utilizing genetically modified organisms (GMOs) or optimizing microbial fermentation processes, manufacturers may be able to produce chlorophyll on a larger scale with fewer environmental impacts.

For example, algae and other microorganisms could be genetically engineered to increase chlorophyll production, offering a more sustainable and cost-effective way to meet growing demand. This process could help Halal food producers meet consumer expectations for both natural colourants and sustainability without relying on synthetic chemicals.

- *Example*: In recent years, several companies have successfully used microbial fermentation to produce chlorophyll and chlorophyll derivatives on a commercial scale, reducing reliance on traditional plant-based extraction methods (Nguyen et al., 2020).

***Advances in Green Pigment Extraction***

Traditional methods of chlorophyll extraction from plants and algae can be costly and inefficient. Emerging technologies in extraction processes, such as supercritical fluid extraction (SFE) and enzymatic methods, hold the potential to make chlorophyll extraction more sustainable and cost-effective.

Supercritical CO2 extraction, for example, is an eco-friendly technique that can extract chlorophyll and other pigments from plant materials without the need for harmful solvents. This method, combined with enzymatic treatment, can improve extraction yields and reduce waste, making it an attractive option for Halal food manufacturers.

- *Example:* The use of supercritical CO2 extraction to extract chlorophyll from algae and plants has been gaining traction in the food and cosmetics industries due to its environmentally friendly nature and high efficiency (Lim et al., 2021).

***The Role of Green Pigments in the Future of Halal Food Products***

As consumer demand for health-conscious, sustainable, and natural food products increases, green pigments like chlorophyll are set to play a central role in the future of Halal food technology. From food to beverages and packaging, the potential applications for green pigments in the Halal food sector are vast and expanding (Ma et al., 2020).

### *Growth of Functional Foods and Beverages*

The growing demand for functional foods—foods that offer health benefits beyond basic nutrition—presents a significant opportunity for chlorophyll in Halal food products. Chlorophyll is already popular in green juices, smoothies, and detox drinks, and as the market for functional beverages expands, chlorophyll-infused drinks will become more prevalent. As consumers become more health-conscious, they seek products that promote detoxification, improved digestion, and general wellness. Chlorophyll is an ideal ingredient for functional foods because of its reputation for antioxidant properties, anti-inflammatory effects, and potential role in digestive health (Zhang et al., 2020; Zhang and Sun, 2021; Chowdhury et al., 2020).

The global market for functional beverages is projected to grow at a significant rate, and chlorophyll is well-positioned to play a role in this growth, particularly in Halal-certified functional drinks.

### *Sustainable Food Packaging and Eco-Friendly Solutions*

The future of Halal food packaging is increasingly focused on sustainability, with chlorophyll offering an innovative solution. Chlorophyll-based biodegradable and edible packaging will continue to gain popularity as the food industry seeks eco-friendly alternatives to traditional plastic packaging.

Green pigments like chlorophyll can be used in the development of active packaging systems, where the pigment serves a dual purpose of colouring and preserving food. By incorporating chlorophyll into packaging, manufacturers can provide Halal-conscious consumers with an environmentally responsible and functional packaging solution.

- *Example:* Chlorophyll-based biodegradable packaging has already been explored for products like fresh produce, snacks, and ready-to-eat meals, with promising results in extending shelf life and reducing waste (Santos et al., 2021).

### *Plant-Based and Vegan Trends in the Halal Food Market*

The rise of plant-based and vegan diets is one of the most significant trends shaping the future of the Halal food market. Chlorophyll is often used in plant-based and vegan food products, such as vegan cheeses, meat alternatives, and dairy-free products, to provide a natural green colour and functional benefits.

As more consumers shift towards plant-based and Halal-compliant foods, chlorophyll will play an increasing role in ensuring that these products are both visually appealing and nutritionally beneficial.

- *Example:* The plant-based food market is expected to grow exponentially, and chlorophyll will continue to be an essential ingredient in Halal-certified plant-based foods, helping to meet both consumer preferences and Halal dietary needs (Alkhuder et al., 2020).

### *Regulatory Advances and Global Standardization*

As the Halal food market continues to expand globally, regulatory advancements and the standardization of Halal certification for natural ingredients like chlorophyll will be essential in ensuring product consistency and quality.

### *Harmonization of Halal Standards*

The lack of consistent Halal standards across different countries has been a challenge for food manufacturers, particularly those seeking to export Halal-certified products. Chlorophyll and other green pigments will be subject to standardized certification processes to ensure that products meet Halal and food safety requirements globally.

- *Example:* Efforts to harmonize Halal certification standards are already underway, with organizations like the World Halal Council and International Halal Integrity Alliance working towards creating global certification standards for Halal ingredients like chlorophyll (Zaidi et al., 2020).

### *The Future of Green Pigments in the Halal Food Industry*

The future of green pigments like chlorophyll in Halal food products looks promising, with continued growth expected in the functional food, plant-based, and eco-friendly packaging sectors. As sustainability and health remain top priorities for consumers, chlorophyll will continue to be a key ingredient in Halal food products that meet the needs of the modern consumer.

### *Continuous Innovation in Green Pigments*

As the demand for natural, Halal, and functional food products continues to rise, manufacturers will focus on innovating and optimizing their use of green pigments. From new extraction technologies to the development of active food packaging, the future of chlorophyll in Halal food technology will be marked by sustainability, efficiency, and consumer demand.

- *Example:* Chlorophyll-infused food packaging that extends shelf life, coupled with plant-based food innovations, will continue to evolve as part of a broader trend toward sustainable food solutions (Bourne et al., 2021).

The future of green pigments like chlorophyll in the Halal food industry is bright, with advancements in sustainable sourcing, biotechnology, and food packaging driving innovation. As the Halal food market continues to grow, chlorophyll will play a central role in meeting the demand for natural, healthy, and environmentally conscious products. The evolution of Halal certification, coupled with new technological developments, will ensure that chlorophyll remains a key ingredient in the future of Halal food technology.

## Conclusion

In this chapter, we have journeyed through the multifaceted role of green pigments, particularly chlorophyll, in the Halal food industry. As consumer demands evolve and the food industry shifts toward more natural, sustainable, and health-conscious products, chlorophyll has emerged as a frontrunner among natural colourants and functional ingredients. As green pigments like chlorophyll gain greater acceptance in Halal food technologies, the future holds exciting possibilities. We can expect more innovative uses for chlorophyll, ranging from biodegradable packaging to functional foods with enhanced health benefits. Additionally, the demand for sustainably sourced and ethically produced pigments will drive the growth of green pigments in the Halal food industry, further integrating sustainability into food production.

The continual development of Halal certification standards and globalization of the Halal food market will create a dynamic environment for green pigments to thrive. Future research and advancements will undoubtedly explore new sources of green pigments, more efficient extraction processes, and the potential of chlorophyll in diverse food applications. As we continue to move towards a more conscious and health-focused world, chlorophyll and other green pigments will play a crucial role in shaping the future of Halal food technologies.

## References


Abdullah, M., Hassan, S., Ibrahim, N., & Rashid, M. (2021). Halal certification for natural food colorants: Challenges and prospects. *Journal of Halal Studies, 3*(1), 45-50.

Ahmad, M., & Ibrahim, M. (2019). Regulatory perspectives on Halal certification of food ingredients. *Food Safety and Quality Journal, 10*(4), 122-130.

Ahmed, N., Khan, S., Ali, M., & Rehman, S. (2019). Plant-based food colorants in the Halal industry. *Food Technology and Science, 28*(5), 60-68.

Alkhuder, M., Ibrahim, A., Al-Shamiri, H., & Hassan, M. (2020). The rise of plant-based foods in the Halal market: Opportunities for natural ingredients. *Halal Food Trends, 17*(4), 220-225.

Al-Qudah, M., Al-Dalahmeh, M., Al-Shawabkeh, A., & Al-Rawashdeh, M. (2018). Challenges in Halal food certification for natural pigments. *International Journal of Food Science and Technology, 53*(12), 2981-2990.

Anwar, S., Malik, A., Hussain, F., & Tariq, R. (2021). The global expansion of the Halal food market and its impact on natural ingredients. *Journal of Halal Food Research, 4*(2), 45-52.

Atta-Delgado, M. X., et al. (2024). Assessing consumer perceptions of sustainable packaging. *Journal of Cleaner Production, 254*, 120123. https://doi.org/10.1016/j.jclepro.2020.120123.

Ben, M., Khan, J., Patel, L., & Ahmed, Z. (2020). Sustainability and bio-based alternatives to artificial colorants in Halal foods. *Sustainability in Food Production, 5*(4), 124-130.

Bourne, J., Patel, R., Zhang, L., & Thompson, K. (2021). Sustainable food packaging solutions: The role of green pigments. *Journal of Sustainable Food Packaging, 9*(1), 32-38.

Burgos, M. S., Nogueira, A. C., & Silva, M. L. (2020). Natural pigments from plants and microorganisms: Extraction, properties, and applications. *Food Research International, 137*, 109505.

Caffin, N., Li, S., Zhang, T., & Wong, K. (2020). Natural food colorants in the Halal food industry. *Food Quality and Safety, 4*(3), 71-80.

Chowdhury, R., Islam, M., Karim, A., & Ahmed, Z. (2020). Natural green pigments in functional foods and beverages. *Food Science and Nutrition, 8*(2), 234-245.

Ding, L., Zhang, J., Li, H., & Liu, W. (2021). Chlorophyll and its applications in food products. *Food and Bioprocess Technology, 14*(1), 1-12.

European Food Safety Authority. (2015). Re-evaluation of chlorophylls (E 140(i)) as food additives. *EFSA Journal, 13*(5), 4089. https://doi.org/10.2903/j.efsa.2015.4089.

Farhana, K., Hassan, R., Ali, M., & Rahman, S. (2021). Green pigments for natural food products. *Natural Food Colorants Review, 8*(2), 115-120.

Farouk, M., & Yousaf, F. (2021). Halal certification: Challenges, opportunities, and future directions. *Journal of Halal Research, 5*(3), 170-181.

Gulzar, M., et al. (2020). Health consciousness and the adoption of green pigments in Halal beverages. *Journal of Health and Food Sciences, 15*(3), 98-104.

Hussain, M., & Hamid, A. (2020). Traceability in Halal certification: Ensuring purity in green pigments. *Food Quality Assurance Journal, 21*(6), 56-63.

Ibrahim, N., Al-Mansoori, A., Hassan, S., & Karim, M. (2021). Chlorophyll and other green pigments: Application in health foods. *Food Chemistry and Technology, 4*(6), 312-321.

Jiang, X., Wang, J., Li, Y., & Liu, H. (2021). Biodegradable films for food packaging applications: A comprehensive review. *Food Research International, 140*, 109838. https://doi.org/10.1016/j.foodres.2020.109838.

Kaur, R., Singh, A., Sharma, P., & Rathi, S. (2021). Sustainability and eco-friendly packaging in the Halal food industry. *Sustainability Journal, 12*(10), 3215-3222.

Khandeparkar, A. S. (2024). Eco-friendly innovations in food packaging: A sustainable approach. *Science Progress, 107*(2), 003685042110041. https://doi.org/10.1177/003685042110041.

Khanna, P., Sharma, R., Gupta, S., & Mehta, V. (2021). Bioactive properties of chlorophyll and its impact on health. *Journal of Plant-Based Products, 12*(3), 90-97.

Kouadio, J., Li, Z., Zhang, Y., & Wang, Q. (2021). Influence of antioxidants on the stability of chlorophylls in food products. *Food Research International, 138*, 109805. https://doi.org/10.1016/j.foodres.2020.109805.

Kralj, S., & Skladal, P. (2021). Technological innovations in green pigment extraction. *Journal of Food Science and Technology, 58*(5), 1632–1645. https://doi.org/10.1007/s11483-021-01630-2.

Lee, S., & Ma, Y. (2020). The rise of plant-based products and the role of Halal certification in the global market. *Food Industry Trends, 21*(7), 44-48.

Lim, S., Tan, Y., Lee, J., & Wong, H. (2020). Chlorophyll-based packaging solutions in the Halal industry. *Journal of Eco-Friendly Packaging, 19*(1), 10-15.

Lim, S., Tan, Y., Lee, J., & Wong, H. (2021). Advances in green pigment extraction: Supercritical CO2 and enzymatic methods. *Food Science and Technology, 45*(6), 134-139.

Liu, H., et al. (2020). Consumer trends towards clean-label products: Implications for the food industry. *International Journal of Food Science and Technology, 55*(9), 2719-2726.

Ma, Y., Zhang, T., Liu, X., & Wang, H. (2020). The role of green pigments in Halal food certification. *Journal of Halal Food Certification, 22*(3), 35-40.

Mandal, B. K., Sharma, R., Gupta, S., & Mehta, V. (2023). Analysis of chlorophylls/chlorophyllins in food products. *Journal of Food Science and Technology, 60*(7), 2201–2211. https://doi.org/10.1007/s11483-023-08447-7.

Martin, A., Liu, Z., Zhang, J., & Wang, Y. (2022). Green innovations in food packaging: Antioxidant-rich materials for sustainable solutions. *Food Chemistry, 366*, 130518. https://www.sciencedirect.com/science/article/pii/S0022354922004322.

Mohd Hatta, F. A., Mat Ali, Q. A., Mohd Kashim, M. I. A., Othman, R., Abd Mutalib, S., & Mohd Nor, N. H. (2023). Recent advances in Halal bioactive materials for intelligent food packaging indicator. *Foods, 12*(12), 2387. https://doi.org/10.3390/foods12122387.

Nguyen, T., Tran, H., & Hoang, P. (2020). Microbial fermentation for the production of green pigments in the food industry. *Biotechnology Advances, 38*(3), 142-148.

Patel, R., Kumar, S., Sharma, A., & Singh, P. (2020). The rise of biodegradable packaging materials: A review of environmental impact and food safety. *Sustainability, 12*(2), 392-407.

Pereira, A., Silva, M., Lima, J., & Santos, R. (2021). Encapsulation techniques for the preservation of bioactive compounds in food systems. *Food Chemistry, 342*, 128392. https://doi.org/10.1016/j.foodchem.2020.128392.

Purnama, R., Wijaya, H., & Sulaiman, M. (2020). Plant-based natural colorants: Sustainable and Halal compliant. *Sustainable Food Innovation, 4*(3), 112-118.

Rahim, M., Ahmad, F., Zainuddin, A., & Noor, S. (2021). The future of Halal food products: Trends and consumer preferences. *Halal Consumer Review, 19*(4), 15-20.

Rahman, M., & Malik, R. (2019). Challenges in natural food colorant application in Halal food processing. *Food Engineering and Technology, 5*(7), 40-45.

Rahman, M., et al. (2020). Halal food certification practices and challenges in the global market. *Food Technology and Biotechnology, 58*(4), 456-464.

Safi, M., Ahmad, N., Ismail, M., & Ali, R. (2020). Halal food certification and its market: A review of the current status in Malaysia and Indonesia. *Trends in Food Science & Technology, 99*, 287-295.

Santos, M., Oliveira, P., Ferreira, D., & Costa, E. (2021). Green pigments in active food packaging: Role of chlorophyll. *Food Packaging and Shelf Life, 28*(3), 100-110.

Simons, A., Lee, J., Patel, R., & Wong, L. (2020). Halal certification and consumer behavior in the food industry. *International Journal of Food Marketing, 17*(8), 167-175.

Siti, M., & Muhammad, I. (2021). Green pigments in Halal foods: A sustainable perspective. *Sustainable Food Science Journal, 6*(2), 75-80.

Tan, K., Lee, H., Lim, S., & Wong, Y. (2020). Extraction methods for natural green pigments in Halal food products. *Journal of Food Extraction, 9*(3), 145-151.

U.S. Food and Drug Administration. (2021). Listing of color additives exempt from certification. 21 CFR 73.125. Retrieved from https://www.ecfr.gov/current/title-21/chapter-I/subchapter-A/part-73/subpart-B/section-73.125.

Umar, S., Khan, A., Ahmad, N., & Ali, M. (2020). Consumer perception of natural pigments in Halal foods. *Food Consumer Research, 11*(3), 67-75.

Zaidi, F., Khan, S., Ahmad, R., & Tariq, M. (2020). Global harmonization of Halal certification: Implications for the food industry. *International Journal of Halal Food Certification, 6*(2), 55-62.

Zhang, H., Li, Y., Wang, J., & Zhang, L. (2020). Functional foods and beverages: The role of chlorophyll in health and wellness. *Functional Foods Journal, 18*(7), 212-219.

Zhang, L., & Li, Q. (2021). Biodegradable packaging with natural green pigments. *Food Packaging Innovation, 10*(2), 89-94.

Zhang, X., Liu, Z., Wang, S., & Chen, Y. (2020). The impact of clean-label trends on Halal food certification. *International Journal of Food and Consumer Sciences, 14*(9), 321-330.

Zhang, Y., & Sun, Y. (2021). The role of green pigments in plant-based foods and beverages. *Food and Beverage Journal, 11*(2), 88-95.

Zhao, Y., Liu, X., Wang, J., & Zhang, L. (2022). Sustainable packaging materials: A review of the use of plant-derived bioplastics and chlorophyll-based films. *Science of the Total Environment, 801*, 149567. https://doi.org/10.1016/j.scitotenv.2021.149567.

Zhao, Z., Li, H., Wang, M., & Chen, Y. (2021). Sustainable practices in Halal food packaging and the role of natural pigments. *Journal of Sustainable Food and Packaging Technologies, 6*(1), 15-21.

Zheng, Y., Wang, L., Li, X., & Zhang, T. (2021). Halal certification and green pigments: The intersection of religion, sustainability, and food technology. *Halal Food Science and Technology Journal, 3*(1), 112-118.

Zhu, L., Li, Z., Chen, S., & Wang, J. (2020). Health benefits of chlorophyll and its role in Halal food applications. *Journal of Halal Functional Foods, 7*(4), 145-153.

Zubaidi, M. A. (2023). Halal food standards and guidelines. *Journal of Halal Quality and Certification*. Retrieved from https://hrcak.srce.hr/file/478556.

# About the Editors

**Ts. Dr Rashidi Othman** is a distinguished professor, certified professional technologist, and expert horticulturist at the Department of Landscape Architecture, International Islamic University Malaysia (IIUM). With a Doctorate in Philosophy from Lincoln University, New Zealand, Dr. Rashidi brings a wealth of knowledge and innovative thinking to his field. His multidisciplinary expertise spans plant biotechnology, environmental sciences, landscape ecology, and ethnobotany, where he seamlessly integrates ecological sustainability with cutting-edge scientific research. As a leader in his field, Dr. Rashidi's work explores the profound connections between humans and the natural world, focusing on how plant-based solutions can foster environmental resilience and biodiversity conservation. His contributions not only advance academic knowledge but also drive practical, sustainable strategies for managing landscapes and ecosystems. With a passion for bridging the gap between traditional ecological wisdom and modern scientific advancements, Dr. Rashidi remains committed to inspiring the next generation of environmental leaders while advancing the holistic understanding of our planet's ecosystems.

**Dr. Qurratu Aini Mat Ali** is currently a Research and Development Executive at Tropica Beauty Sdn. Bhd. in Selangor, Malaysia. She holds a Doctor of Philosophy from the Institute of Islam Hadhari, The National University of Malaysia (2025), and a Master of Science in Built Environment from the International Islamic University Malaysia (2016). Dr. Qurratu Aini specializes in halal colorants and natural products, ethnobotany, environmental science, green technology, and landscape ecology. Her work aims to integrate sustainable and halal practices into various industries, with a focus on developing eco-friendly solutions that align with Islamic principles.

**Dr. Razanah Ramya** Dr. Razanah Ramya is a research fellow at the Institute of the Malay World and Civilization (ATMA), The National University of Malaysia. She holds a Doctor of Philosophy in Halal Industry (Science) from the International Institute for Halal Research and Training, International Islamic University Malaysia (IIUM), as well as a Master of Science in Built Environment and a Bachelor of Landscape Architecture, both from IIUM. Her research interests include landscape ecology, ethnobotany, and natural products. She is particularly focused on exploring the sustainable use of plant-based resources in traditional and modern contexts. Dr. Razanah is dedicated to bridging the gap between scientific research and the cultural heritage of the Malay world, contributing to both environmental and community well-being.

**Ts. Dr. Farah Ayuni Mohd Hatta** is a Research Fellow at the Institute Islam Hadhari, Universiti Kebangsaan Malaysia, and a certified Professional Technologist in Green Technology, accredited by the Malaysia Board of Technologists. She holds a PhD in Halal Industry (Science) from the International Islamic University Malaysia. With a multidisciplinary expertise in plant biotechnology, halal science, biopackaging, and environmental sustainability, her work bridges innovative solutions for both the halal industry and sustainable practices.

**Dr. Wan Syibrah Hanisah Wan Sulaiman** Dr. Wan Syibrah Hanisah Wan Sulaiman is an assistant professor at the International Institute for Halal Research and Training (INHART), International Islamic University Malaysia (IIUM). She earned her Doctor of Philosophy from IIUM in 2019 and holds a Master of Environmental Management and Sustainability from the University of South Australia (2013). Her research focuses on environmental sustainability, halal lifestyle, phytoremediation, and green technology in the built environment. Additionally, she investigates halal ingredients, including the development of halal colorants from natural pigments. Her interdisciplinary work encompasses halal tourism, inclusive design, and bioactive compounds, with the goal of aligning ecological innovation with Islamic values to promote sustainability in the halal industry.

**Dr. Nur Hanie Mohd Latiff** is an assistant professor at INHART, International Islamic University Malaysia (IIUM), where she has been teaching since 1 October 2019. She earned her PhD in Halal Industry from IIUM and holds both a Bachelor of Landscape Architecture (Hons) and a Master of Science in Built Environment, also from IIUM. A dedicated researcher in environmental science, Dr. Nur Hanie's areas of expertise include landscape ecology, soil quality, heavy metal analysis, phytoremediation, renewable energy, and wastewater treatment. Her current research interests extend to the halal industry, focusing on halal lifestyle, Muslim-friendly hospitality, halal tourism, and sustainable development.

# Index

## β



## A

## B

## C

## F

## G



## H

## I

## O



## P

## Q



## R



## S

## T



## U



## V